IEE CONTROL ENGINEERING SERIES 40

Series Editors: Professor P. J. Antsaklis
Professor D. P. Atherton
Professor K. Warwick

Deterministic control of uncertain systems

Other volumes in this series:

Volume 1 **Multivariable control theory** J. M. Layton
Volume 2 **Elevator traffic analysis, design and control** G. C. Barney and S. M. dos Santos
Volume 3 **Transducers in digital systems** G. A. Woolvet
Volume 4 **Supervisory remote control systems** R. E. Young
Volume 5 **Structure of interconnected systems** H. Nicholson
Volume 6 **Power system control** M. J. H. Sterling
Volume 7 **Feedback and multivariable systems** D. H. Owens
Volume 8 **A history of control engineering, 1800-1930** S. Bennett
Volume 9 **Modern approaches to control system design** N. Munro (Editor)
Volume 10 **Control of time delay systems** J. E. Marshall
Volume 11 **Biological systems, modelling and control** D. A. Linkens
Volume 12 **Modelling of dynamical systems—1** H. Nicholson (Editor)
Volume 13 **Modelling of dynamical systems—2** H. Nicholson (Editor)
Volume 14 **Optimal relay and saturating control system synthesis** E. P. Ryan
Volume 15 **Self-tuning and adaptive control: theory and application** C. J. Harris and S. A. Billings (Editors)
Volume 16 **Systems modelling and optimisation** P. Nash
Volume 17 **Control in hazardous environments** R. E. Young
Volume 18 **Applied control theory** J. R. Leigh
Volume 19 **Stepping motors: a guide to modern theory and practice** P. P. Acarnley
Volume 20 **Design of modern control systems** D. J. Bell, P. A. Cook and N. Munro (Editors)
Volume 21 **Computer control of industrial processes** S. Bennett and D. A. Linkens (Editors)
Volume 22 **Digital signal processing** N. B. Jones (Editor)
Volume 23 **Robotic technology** A. Pugh (Editor)
Volume 24 **Real-time computer control** S. Bennett and D. A. Linkens (Editors)
Volume 25 **Nonlinear system design** S. A. Billings, J. O. Gray and D. H. Owens (Editors)
Volume 26 **Measurement and instrumentation for control** M. G. Mylroi and G. Calvert (Editors)
Volume 27 **Process dynamics estimation and control** A. Johnson
Volume 28 **Robots and automated manufacture** J. Billingsley (Editor)
Volume 29 **Industrial digital control systems** K. Warwick and D. Rees (Editors)
Volume 30 **Electromagnetic suspension—dynamics and control** P. K. Sinha
Volume 31 **Modelling and control of fermentation processes** J. R. Leigh (Editor)
Volume 32 **Multivariable control for industrial applications** J. O'Reilly (Editor)
Volume 33 **Temperature measurement and control** J. R. Leigh
Volume 34 **Singular perturbation methodology in control systems** D. S. Naidu
Volume 35 **Implementation of self-tuning controllers** K. Warwick (Editor)
Volume 36 **Robot control** K. Warwick and A. Pugh (Editors)
Volume 37 **Industrial digital control systems (revised edition)** K. Warwick and D. Rees (Editors)
Volume 38 **Parallel processing in control** P. J. Fleming (Editor)
Volume 39 **Continuous time controller design** R. Balasubramanian
Volume 40 **Deterministic control of uncertain systems** A. S. I. Zinober (Editor)
Volume 41 **Computer control of real-time processes** S. Bennett and G. S. Virk (Editors)
Volume 42 **Digital signal processing: principles, devices and applications** N. B. Jones and J. D. McK. Watson (Editors)
Volume 43 **Trends in information technology** D. A. Linkens and R. I. Nicolson (Editors)
Volume 44 **Knowledge-based systems for industrial control** J. McGhee, M. J. Grimble, A. Mowforth (Editors)

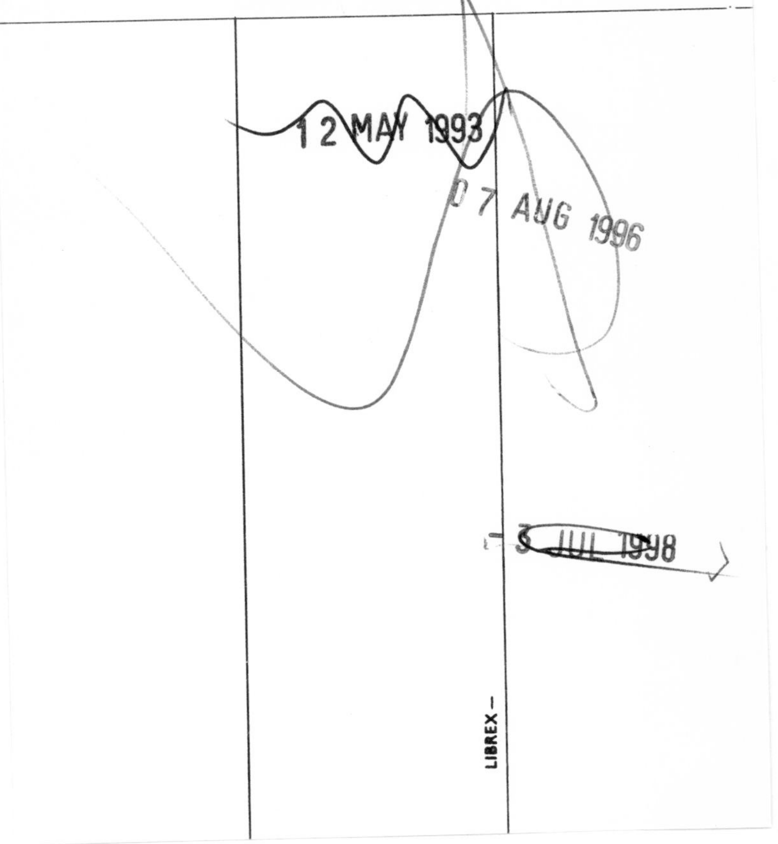

Deterministic control of uncertain systems

Edited by
A.S.I. Zinober

Peter Peregrinus Ltd. on behalf of the Institution of Electrical Engineers

Published by: Peter Peregrinus Ltd., London, United Kingdom

Peter Peregrinus Ltd.,
Michael Faraday House,
Six Hills Way, Stevenage,
Herts. SG1 2AY, United Kingdom

British Library Cataloguing in Publication Data
Deterministic control of uncertain systems
1. Variable-structure systems
I. Zinober, Alan
629.8′312

ISBN 0 86341 170 3

Printed in England by Short Run Press Ltd., Exeter

Contents

Preface

1	**An introduction to variable structure control.** A. S. I. Zinober	1
	1.1 Introduction	1
	1.2 Sliding motion	2
	1.3 A second-order scalar control problem	3
	1.4 Phase canonic system with scalar control	9
	1.5 Multivariable variable structure control systems	12
	1.5.1 Sliding motion	13
	1.6 Properties in the sliding mode	15
	1.7 Conclusion	19
	1.8 Appendix	20
	1.9 References	21
2	**Sliding mode control with switching command devices.** H. Bühler	27
	2.1 Introduction	27
	2.2 Sliding mode control of scalar control systems with state variable feedback	28
	2.2.1 Basic structure and switching strategy	28
	2.2.2 Sliding mode conditions	30
	2.2.3 State equation in sliding mode	30
	2.2.4 Pole assignment	31
	2.2.5 Sliding mode domain	32
	2.2.6 Switching frequency	34
	2.3 Application to the position control of a DC drive	36
	2.3.1 State equations of the controller system	36
	2.3.2 Control structure	37
	2.3.3 Pole assignment	37
	2.3.4 Practical results	39
	2.4 Sliding mode control of multivariable systems with state variable feedback	40
	2.4.1 Basic structure and switching strategy	40
	2.4.2 Sliding mode conditions	41
	2.4.3 State equation in sliding mode	42
	2.4.4 Pole assignment	43
	2.4.5 Sliding mode domain	46

2.5 Example of a multivariable system sliding mode control 46
2.5.1 Basic relations 46
2.5.2 State feedback design (pole assignment) 47
2.5.3 Feedforward matrix design 48
2.5.4 Switching planes 48
2.5.5 Trajectory in the sliding mode 49
2.5.6 Sliding mode domain 50
2.6 References 50

3 **Hyperplane design and CAD of variable structure control systems.** C. M. Dorling and A. S. I. Zinober 52

3.1 Introduction 52
3.2. Variable structure control systems 53
3.3 The system 53
3.4. A canonical form for VSCS design 55
3.5 Methods for hyperplane design 56
3.6 Hyperplane design by quadratic minimisation 57
3.7 Hyperplane design by eigenstructure assignment 58
3.8 Hyperplane design by robust eigenvector assignment 61
3.8.1 Redundancy in the sliding mode system 62
3.8.2 Hyperplane design by robust eigenvalue assignment 62
3.9 Control scheme designs 65
3.9.1 Unit vector control 66
3.9.2 Practical implementation of VSC 67
3.10 Examples 68
3.10.1 Example 1 68
3.10.2 Example 2 71
3.11 Conclusion 77
3.12 References 78

4 **Subspace attractivity and invariance: Ultimate attainment of prescribed dynamic behaviour.** E. P. Ryan 80

4.1 Introduction 80
4.2 Background concepts 81
4.3 The system 82
4.4 Stabilisation by generalised feedback 83
4.4.1 Problem formulation 83
4.4.2 Outline of approach 84
4.4.3 The subspace of $\mathscr{S}$ 84
4.4.4 Generalised feedback $\mathscr{F}$ 85
4.4.5 Global uniform asymptotic stability of the zero state of the feedback system 86
4.4.6 Positive invariance and finite-time invariance of $\mathscr{S}$ 87
4.4.7 Linear motion in $\mathscr{S}$ 88
4.4.8 Summary of results 89
4.5 Tracking in the presence of uncertainty 89
4.5.1 The tracking error system 90
4.5.2. Asymptotic tracking of x^* 91
4.6 Model following in the presence of uncertainty 91
4.6.1 The model-following error system 93
4.6.2 Asymptotic model-following 93

4.7 Discussion 93
4.8 References 94

5 **Model-following control of time-varying and nonlinear avionics systems.** S. K. Spurgeon, M. K. Yew, A. S. I. Zinober and R. J. Patton 96

5.1 Introduction 96
5.2 Model-following control systems 97
5.3 Variable structure model-following control systems 97
5.4 Example 1: Lateral motion stability augmentation system design 101
5.5 Example 2: Pitchpointing/vertical translation control system design 105
5.6 Conclusion 111
5.7 References 113

6 **Canonical formulation and general principles of variable structure controllers**. L Guzzella and H. P. Geering 115

6.1 Introduction 115
6.2 Plant description and transformations 116
6.3 Regulator synthesis 118
6.4 Stability conditions and equivalent system 120
6.5 Some remarks 120
6.6 Reference model tracking control 122
6.7 A numerical example 123
6.8 Conclusion 126
6.9 References 127
6.10 Appendix 127

7 **Variable structure controllers for robots**. L. Guzzella and H. P. Geering 132

7.1 Introduction 132
7.2 System description 133
7.3 Control system structure 135
7.4 Sliding system 136
7.5 Stability conditions 137
7.6 Example 139
7.7 References 143

8 **Applications of output feedback in variable structure control.** B. A. White 144

8.1 Introduction 144
8.2 State feedback 144
8.2.1 Range space dynamic equations 145
8.2.2 Range space dynamic stability 146
8.3 Output feedback 150
8.3.1 Range space dynamics 150
8.3.2 Range space stability 152
8.4 Lateral motion autopilot for a remotely piloted vehicle 156
8.4.1 Aircraft dynamic model 156

8.4.2 Output VSCS control 158
8.5 Flexible structure control 160
8.5.1 Beam model 161
8.5.2 Variable structure control system design 162
8.5.3 Improving the VSCS design 164
8.6 References 166
8.7 Appendices 167
8.7.1 Variation in the A matrix with roll and pitch angle 167
8.7.2 A, B and C matrices for the flexible beam model 168

9 **The hyperstability approach to VSCS design**. A. Balestrino and M. Innocenti 170

9.1 Introduction 170
9.2 Preliminaries: Hyperstability 171
9.3 Design objectives 173
9.4 A SISO example 174
9.5 A MIMO example: Nonlinear observers 176
9.6 HVSCS design 180
9.7 Design procedure in the limit case $K_R = 0$ 186
9.8 Applications in robotics 188
9.9 A design example in aircraft control 192
9.10 Conclusion 198
9.11 References 199

10 **Nonlinear continuous feedback control for robust tracking.** F. Garofalo and L. Glielmo 201

10.1 Introduction 201
10.2 Statement of the problem 202
10.3 Stability analysis 203
10.4 Design of the controller 212
10.5 Design consideration 214
10.6 Conclusions 218
10.7 References 218

11 **Deterministic control of uncertain systems. A Lyapunov theory approach.** M. Corless and G. Leitmann 220

11.1 Introduction 220
11.2 Basic notions 221
11.2.1 Existence and extension of solutions 221
11.2.2 Boundedness and stability 222
11.2.3 Lyapunov functions and a sufficient condition for GUAS 223
11.2.4 Systems with control 224
11.3 Uncertain systems 225
11.4 Initial problem statement: Stabilisation 226
11.4.1 Problem statement 226
11.4.2 A useful theorem for the synthesis of stabilising controllers 227
11.5 L – G controllers 227
11.5.1 Original class of uncertain systems 228

11.5.2 L – G controllers 230
11.5.3 Extension of original system class 231
11.5.4 Properties of systems with L – G controllers 231
11.6 Relaxed problem statement. Practical stabilisation 232
11.7 Modified L – G controllers 234
11.7.1 Systems under consideration 234
11.7.2 Modified L – G controllers 235
11.8 Tracking controllers for uncertain mechanical systems 237
11.8.1 Systems under consideration and problem statement 237
11.8.2 Proposed controllers 238
11.8.3 Application to a Manutec r^3 robot 240
11.9 Appendix 246
11.9.1 Caratheodory functions 246
11.9.2 *K* and *KR* functions 247
11.10 References 247

12 **Control of uncertain systems with neglected dynamics.** M. Corless, G. Leitmann and E. P. Ryan 252

12.1 Introduction 252
12.2 The full-order system 253
12.3 The reduced-order system 254
12.4 Problem formulation 256
12.5 Nonlinear output feedback 257
12.6 A compact attractor for the output feedback controlled reduced-order system 258
12.7 A compact attractor for the output feedback controlled full-order system 260
12.8 Example: Uncertain system with actuator and sensor dynamics 265
12.9 References 268

13 **Nonlinear composite control of a class of nominally linear singularly perturbed uncertain systems.** F. Garofolo and G. Leitmann 269
13.1 Introduction 269
13.2 The singularly perturbed uncertain dynamical system 272
13.3 Design objectives and mathematical preliminaries 272
13.4 The 'slow controller' for practical stability 273
13.5 The 'fast controller' for practical stabilisation 277
13.6 Composite control and practical stability of the complete system 280
13.7 Example 286
13.8 Conclusions 287
13.9 Acknowledgment 287
13.10 References 287

14 **Some extensions of variable structure control theory for the control of nonlinear systems.** G. Bartolini and T. Zolezzi 289

14.1 Introduction 289
14.2 VSC of nonlinear ordinary differential equations 290
14.3 Output feedback with observers 294
14.4 Asymptotic linearisation by VSC for uncertain systems 297
14.5 VSC for model reference control of non-minimum phase linear systems 299

14.6 VSC of semilinear distributed systems 305
14.7 Conclusions 307
14.8 References 307

15 **Continuous self-adaptive control using variable structure design techniques.** J. A. Burton and A. S. I. Zinober 309

15.1 Introduction 309
15.2 The variable structure control system 310
15.3 The smoothed variable structure control system 311
15.3.1 Analysis of the system response in a SmVSC system 311
15.4 Continuous self-adaptive control of scalar control systems 313
15.4.1 Self-adaptive control 313
15.4.2 CSAC of second-order scalar systems 314
15.4.3 CSAC of higher order scalar systems 318
15.5 Continuous self-adaptive control of multivariable control systems 320
15.5.1 Justification of the adaptive control scheme 321
15.5.2 Method of adaptation 322
15.5.3 Stability of the adaptive control scheme 326
15.6 Application of the CSAC 326
15.7 Conclusion 332
15.8 References 332

16 **State observation of nonlinear control systems via the method of Lyapunov.** S. H. Żak and B. L. Walcott 333

16.1. Introduction 333
16.2 System description and notation 335
16.3 The Thau observer 335
16.4 Observation of dynamical systems in the presence of bounded uncertainties 337
16.5 Sliding mode properties of min–max observers 339
16.6 Duality between min–max controllers and observers 341
16.7 Sufficiency conditions for the semidefiniteness of the matrix $\hat{Q}$ 343
16.8 VSC performance in the presence of small unmodelled inertias 345
16.9 Sliding modes in uncertain systems with asymptotic observers 347
16.10 Improved performance through the use of a state observer 347
16.10.1 Dynamical sensor and asymptotic observer 347
16.10.2 Dynamical actuator and asymptotic observer 348
16.11 References 349

17 **Control of infinite-dimensional plants.** V. I. Utkin 351

17.1 Introduction 351
17.2 Examples 351
17.3 Control of oscillation process 354
17.4 Control of interconnected heat processes 355
17.5 Sliding mode equations 358
17.6 References 361
Index

Preface

The system analyst models the salient features of a system in the real world by means of a mathematical model. This model contains uncertainties due to lack of knowledge of parameter values, which may be constant or time-varying and imprecisely known, and uncertainties arising from imperfect knowledge of system inputs and inaccuracies in the mathematical modelling. One of the main avenues of study in the control of dynamical systems has been that of the effective control of practical time-varying systems with uncertain parameters and external disturbances. In stochastic adaptive control systems, such as self-tuning controllers, the parameter values and disturbances are constantly monitored using online identification algorithms and appropriate controllers are implemented.

In direct contrast to these adaptive controllers the deterministic control of uncertain time-varying systems control is achieved using fixed nonlinear feedback control functions, which operate effectively over a specified magnitude range of a class of system parameter variations. There is no requirement for on-line identification of the values of the system parameters. Furthermore no statistical information of the system variations is required to yield the desired robust dynamic behaviour. This deterministic approach contrasts sharply with many other adaptive control schemes, which need to have global parameter convergence properties. Furthermore, if the parameter variations satisfy certain matching conditions, complete insensitivity to system variations can be achieved.

The two main approaches are variable structure control (VSC) and Lyapunov control. Variable structure control with a sliding mode was first described by Russian authors, notably Emel'yanov and Utkin. Draženović established some early results on the invariance of VSC in the sliding mode to a class of disturbances and parameter variations. In recent years the subject has attracted the attention of numerous research workers because of these excellent invariance properties. In the 1970s VSC design was extended to multivariable control systems and in the 1980s CAD packages have become available. VSC is particularly well suited to the deterministic control of uncertain control systems. Some of the major interests have been the use of VSC and allied techniques in model-following and model reference adaptive control, tracking control and observer systems.

The essential property of VSC is that the nonlinear feedback control has a discontinuity on one or more manifolds in the state space. Thus the structure of the feedback system is altered or switched as its state crosses each discontinuity surface. Sliding motion occurs when the system state repeatedly crosses and immediately re-crosses a switching surface, because all motion in the neighbourhood of the manifold is directed inwards towards the manifold. More recently the discontinuous controller has been replaced in many practical applications with continuous nonlinear control which yields a dynamic response arbitrarily close to the discontinuous controller, but without undesirable chatter motion.

The first stage of the design of a VSC system entails the choice of the sliding subspace. Stable sliding motion is assured by the suitable selection of sliding hyperplanes so that desired closed-loop dynamic response is achieved. This may be specified by a quadratic performance criterion or by employing eigenvalue (and eigenvector) assignment. The second stage of the design procedure involves the selection of the control which will ensure that the chosen sliding mode is attained and maintained for the expected range of system uncertainty.

Lyapunov control follows the approach of early research workers such as Leitmann, Corless, Gutman, Palmor and Ryan. Using a Lyapunov function and specified magnitude bounds on the uncertainties, a nonlinear control law is developed to ensure uniform ultimate boundedness of the closed-loop feedback trajectory to achieve sufficient accuracy. The resulting controller is a discontinuous control function, with generally continuous control in a boundary layer in the neighbourhood of the switching surface. The boundary layer control prevents the excitation of high-frequency unmodelled parasitic dynamics. Controllers have been devised for numerous types of system with many different classes of uncertainty. The control is designed using a Lyapunov design approach and allows for a range of expected system variation.

The contents of this book reflect the research output of many authors in the area of deterministic control. The chapters include material of an introductory nature as well as some of the latest research results. Attention has also been focussed upon some of the main areas of application of deterministic control of uncertain systems, which include electric motor drives, robotics and flight control systems. The book should prove useful to control designers, theoreticians and graduate students.

I would like to thank all the contributors for their hard work and cooperation in meeting deadlines. In particular I wish to thank Gene Ryan for his scholarly advice. I thank my family for their encouragement during the long gestation period of this book. Finally I thank Madeleine Floy for her secretarial and administrative assistance in the preparation of the manuscript.

University of Sheffield
February 1990

Alan Zinober

List of Contributors

Chapters 1, 3, 5 and 15
A.S.I. Zinober
Dept. of Applied and Computational Maths.
University of Sheffield
Sheffield
United Kingdom

Chapter 2
H. Bühler
Swiss Federal Institute of Technology
Industrial Electronics Laboratory
Lansanne
Switzerland

Chapter 3
C.M. Dorling
Topexpress Ltd.
Poseidon House
Castle Park
Cambridge
United Kingdom

Chapters 4 and 12
E.P. Ryan
School of Mathematical Sciences
University of Bath
Claverton Down
Bath
United Kingdom

Chapter 5
S.K. Spurgeon
Department of Mathematical Sciences
Loughborough University of Technology
Loughborough
Leicestershire
United Kingdom

M.K. Yew
Dept. of Applied and Computational Maths.
University of Sheffield
Sheffield
United Kingdom

R.J. Patton
Dept. of Electronics
University of York
Heslington
York
United Kingdom

Chapters 6 and 7
L. Guzzella and H.P. Geering
Measurement and Control Laboratory
Swiss Federal Institute of Technology (ETH)
Zürich
Switzerland

Chapter 8
B.A. White
Control and Guidance Group
School of Electrical Engineering and Science
Royal Military College of Science
Shrivenham
Wiltshire
United Kingdom

Chapter 9
A. Balestrino and M. Innocenti
University of Pisa
Pisa
Italy

Chapters 10 and 13
F. Garofalo
Dipartimento di Informatica e Sistemistica
University of Naples
Italy

Chapter 10
L. Glielmo
Dipartimento di Informatica e Sistemistica
University of Naples
Italy

Chaper 11 and 12
M. Corless
School of Aeronautics and Astronautics
Purdue University
West Lafayette
Indiana
USA

Chapters 11, 12 and 13
G. Leitmann
Department of Mechanical Engineering
University of California
Berkeley
California
USA

Chapter 14
G. Bartolini and T. Zolezzi
University of Genoa
Genoa
Italy

Chapter 15
J.A. Burton
Simulation Modelling and Control Ltd.
7 Ty-Segur
Neath
United Kingdom

Chapter 16
S.H. Żak
School of Electrical Engineering
Purdue University
West Lafayette
Indiana 47907
USA

B.L. Walcott
Department of Electrical Engineering
University of Kentucky
Lexington
KY 40506
USA

Chapter 17
V.I. Utkin
Institute of Control Sciences
Moscow
USSR

To
Brenda
Madeleine and Rebecca

Chapter 1

An introduction to variable structure control

A. S. I. Zinober

University of Sheffield

1.1 Introduction

Variable structure control (VSC) with a sliding mode was first described by Soviet authors including Emel'yanov et al., (1964, 1967), Utkin (1971, 1972, 1978) and Itkis (1976). A survey paper by Utkin (1977) references many of the early contributions available in translation. Recent survey and tutorial papers with numerous references have been written by Utkin (1983, 1987) and DeCarlo *et al.* (1988), and a book in French has been published by Bühler (1986).

Draženović (1969) established early results on the invariance of VSC systems to a class of disturbances and parameter variations. More recently the subject has attracted great interest because of these excellent invariance properties. Consequently, VSC is particularly suited to the deterministic control of uncertain and nonlinear control systems. Numerous theoretical advances and practical applications have been reported in the literature (see the numerous papers cited in the Section 1.9). There have been major studies in the use of VSC and allied techniques in, for instance, model-following and model reference adaptive control (Devaud and Caron, 1975; Young, 1977, 1978; Zinober, El-Ghezawi and Billings, 1982; Zinober, 1984) and the development of variable structure tracking control (Young and Kwatny, 1981, 1982; Calise and Raman, 1981, 1982). Some other areas of concern have been the relationships between VSC, singularly perturbed and linear high-gain feedback systems (Young, Kokotović and Utkin, 1977; Saksena, O'Reilly and Kokotović, 1984); the geometrical interpretation of the sliding mode in VSC (El-Ghezawi, Billings and Zinober, 1983; El-Ghezawi, Zinober and Billings, 1983; El-Ghezawi et al., 1982); and the control of uncertain dynamical systems using Lyapunov control (Ryan, 1983, 1988; Ryan and Corless, 1984). Many of these topics and numerous applications are discussed in other Chapters of the book, and further references are listed in Section 1.9 regarding research in related topics and applications in areas such as robotics, avionics and electrical drives.

The essential feature of a variable structure control system is the non-linear feedback control which has a discontinuity on one or more manifolds in the state space. Thus the structure of the feedback system is altered or switched as its state crosses each discontinuity surface; in consequence of which, the closed-loop system is described as a variable structure control system. A VSC system may be regarded as a combination of subsystems, each with a fixed structure and each operating in a specified region of the state space.

In this introductory Chapter attention will be focused upon the regulator, where the aim of the design is to regulate the system state to zero. Much of the theory may be applied with suitable modifications to model-following and tracking control systems.

1.2 Sliding motion

The central feature of VSC is sliding motion. This occurs when the system state repeatedly crosses and immediately re-crosses a switching manifold, because all motion in the neighbourhood of the manifold is directed inwards (i.e. towards the manifold). Depending on the form of the control law selected, sliding motion may occur on individual switching surfaces in the state space, on a selection of surfaces, or on all the switching manifolds together. When the last of these cases occurs, the system is said to be in the sliding mode.

The term 'sliding mode' is often also used for the sliding subspace. The dynamic motion of the system is then effectively constrained to lie within a certain subspace of the full state space. The system is thus formally equivalent to an unforced system of lower order, termed the equivalent system. The motion of this equivalent system is different from that of each of the constituent subsystems.

The design of a VSC system entails the choice of the switching surfaces, the specification of the discontinuous control functions and the determination of the switching logic associated with the discontinuity surfaces. The switching surfaces are usually fixed hyperplanes in the state space passing through the state space origin, the intersection of which forms the sliding subspace. The objective of the design is to drive the state of the system from an arbitrary initial condition to the intersection of the switching surfaces. Once the state starts sliding, the action of the control is required only to maintain the state on (or in the neighbourhood of) the intersection manifold.

The equivalent system must be asymptotically stable to ensure that the state approaches the state space origin within the sliding mode. Stable sliding motion is assured by the suitable selection of the switching hyperplanes. This forms the first stage of the VSC design process. The problem of determining a set of hyperplanes providing suitable behaviour in the sliding mode is described as the existence problem. It is concerned solely with fixing a set of hyperplanes such that the sliding mode on their intersection gives a specified dynamic performance to the equivalent lower order system. The solution of the existence problem may be completed without any assumptions on the form of the control functions to be employed by the system.

Once the existence problem has been solved, the second stage of the design procedure involves the selection of the control which will ensure that the chosen sliding mode is attained. For this reason, the problem of determining a control structure and associated gains, which ensure the reaching or hitting of the sliding mode, is called the reachability problem. In theory there is an infinite number of forms which the control functions may take, but only a limited number of combinations have been tried in practice. The solution of the reachability problem is dependent on the switching hyperplanes since the control functions are required to be discontinuous on the switching surfaces, and so cannot be achieved until the existence problem has been solved.

The transient motion of a VSC system therefore consists of two independent stages: a (preferably rapid) motion bringing the state of the system to the manifold in which sliding occurs; and a slower sliding motion during which the state slides towards the state space origin while remaining in the sliding subspace. This two-stage behaviour can help to resolve the conflict between the opposing requirements of static and dynamic accuracy which are encountered when designing a linear control system, because a VSC system may be designed to give: a rapid response with no loss of stability, asymptotic state regulation, insensitivity to a class of parameter variations and invariance to certain external disturbances.

During the sliding mode the discontinuous control chatters about the switching surface at high frequency. This phenomenon is usually undesirable for most practical applications and a smoothed continuous nonlinear control can be substituted with little alteration in the dynamic behaviour of the system (see for instance Burton and Zinober, 1986). In addition the excitation of unmodelled parasitic high frequency dynamics can be avoided.

1.3 A second-order scalar control problem

To illustrate the basic concepts of VSC let us consider the second-order system in phase canonic form

$$\begin{aligned} \dot{x}_1 &= x_2 \\ \dot{x}_2 &= -a_1 x_1 - a_2 x_2 + bu \qquad (b > 0) \end{aligned} \tag{1.1}$$

where x_1 and x_2 are the state variables and u the scalar control. Let the state vector be $\boldsymbol{x} = (x_1 x_2)^T$. The a_i and b are constants or time-varying parameters and their precise values may be unknown.

Consider the discontinuous control

$$u = \begin{cases} u^+ & cx_1 + x_2 > 0 \\ u^- & cx_1 + x_2 < 0 \end{cases} \tag{1.2}$$

with $c > 0$ and $u^+ \neq u^-$.

The switching function is

$$s = cx_1 + x_2 \tag{1.3}$$

and the line

$$s = 0 \tag{1.4}$$

is the surface on which the control u has a discontinuity. Suppose at time $t = 0$ that $s > 0$. It can be readily shown that for suitable u the state x reaches the switching line $s = 0$ in a finite time τ as depicted in Fig. 1.1.

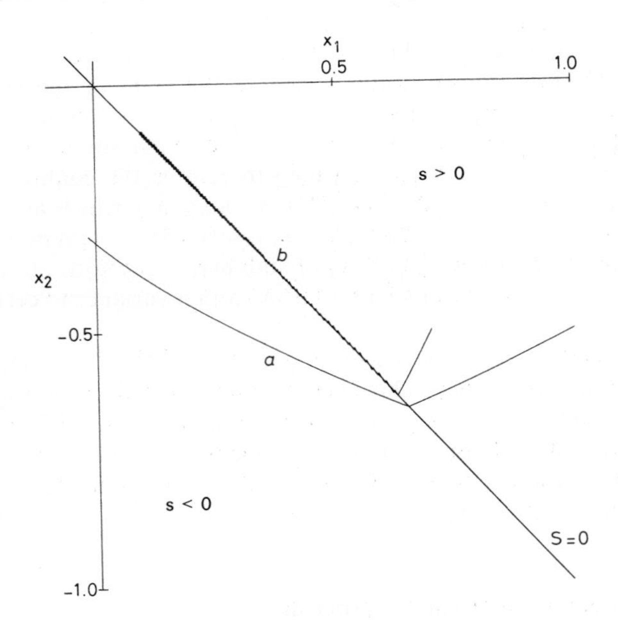

Fig. 1.1 *Double-integrator with relay control*
a Bang-bang *b* Sliding

The state $\boldsymbol{x}$ crosses the switching line and enters the region $s < 0$, resulting in the value of u being altered from u^+ to u^-. Depending upon the values of the system parameters and c the state trajectory may continue in the region $s < 0$, yielding bang-bang control (see Fig. 1.1).

Alternatively the state trajectory may immediately re-cross the switching line and enter the region $s > 0$. This yields sliding (or chatter) motion (Flügge-Lotz,

1953; André and Seibert, 1956; Weissenberger, 1966; Fuller, 1967). Assuming that the switching logic works infinitely fast, the state $\boldsymbol{x}$ is constrained to remain on the switching line $s = 0$ by the control which oscillates between the values u^+ and u^-.

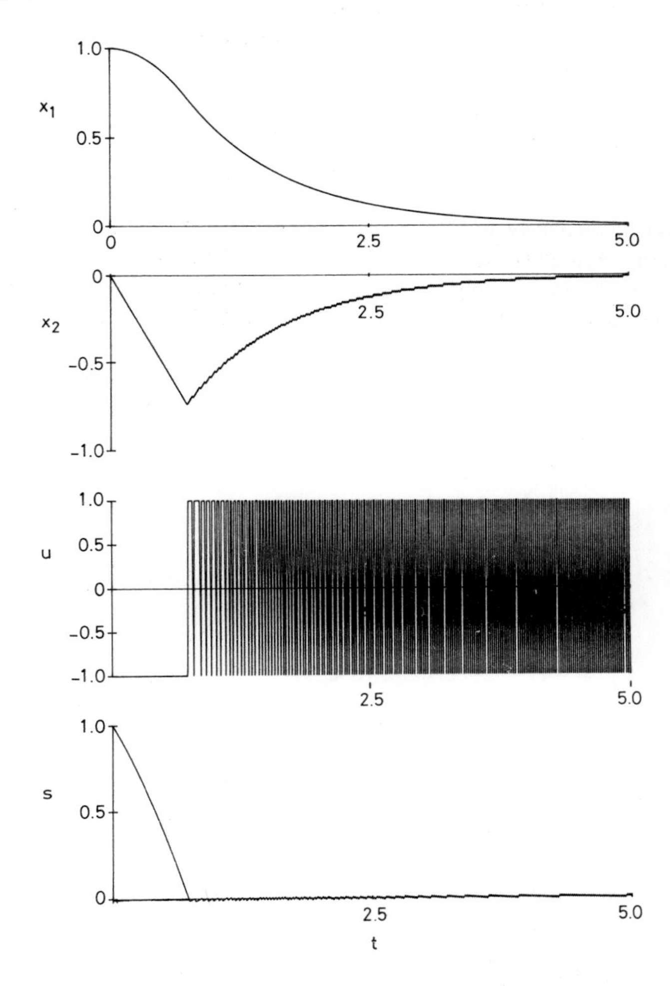

Fig. 1.2 *Double-integrator with relay control (d = 0)*

For sliding motion to occur we need

$$\lim_{s \to 0^+} \dot{s} < 0 \quad \text{and} \quad \lim_{s \to 0^-} \dot{s} > 0 \tag{1.5}$$

on opposite sides of the switching line. This ensures that the motion of the state $\boldsymbol{x}$ on either side of the switching line $s = 0$ in the neighbourhood of the switching line is towards the switching line. The two conditions may be combined to give

$$s\dot{s} \leqslant 0 \tag{1.6}$$

in the neighbourhood of the switching line.

During sliding motion s remains zero, so we have the differential equation

$$s = cx_1 + \dot{x}_1 = 0 \tag{1.7}$$

governing the system dynamics, with solution

$$x_1(t) = x_1(\tau)e^{-c(t-\tau)} \tag{1.8}$$

Thus the second-order system behaves like an asymptotically stable first-order system with eigenvalue (pole) at $-c$ during the sliding mode. The dynamics are independent of the system parameter b over a specified magnitude range.

For the symmetric relay controller, $u^+ = -1$ and $u^- = 1$, with $c = 1$, the state trajectory is depicted in Figs. 1.2 and 1.3. The chatter motion of the control

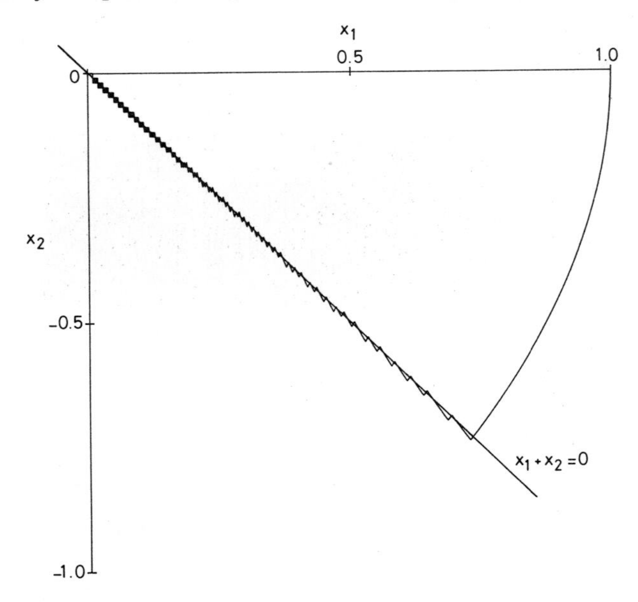

Fig. 1.3 *Double-integrator with relay control (d = 1)*

is clear displayed in Fig. 1.2. The state $\boldsymbol{x}$ approaches the switching (sliding) line, $s = x_1 + x_2 = 0$, within 0·75 s and thereafter (effectively) remains on the sliding line until the origin is attained.

To eliminate the chatter motion we may use the smoothed control (Burton and Zinober, 1986):

$$u = -s/(|s| + d) \tag{1.9}$$

where the smoothing term d is positive and d is small. (In the limit $d = 0$ and we have the original relay control, $u = -\operatorname{sgn} s = -s/|s|$). The resulting trajectories with the control (1.9) are shown in Figs. 1.4 and 1.5. The phase plane protrait (Fig. 1.5) indicates that the deviation from the ideal relay control is very small. The state lies within a small neighbourhood of the sliding subspace, $cx_1 + x_2 = 0$, once the sliding subspace has been attained.

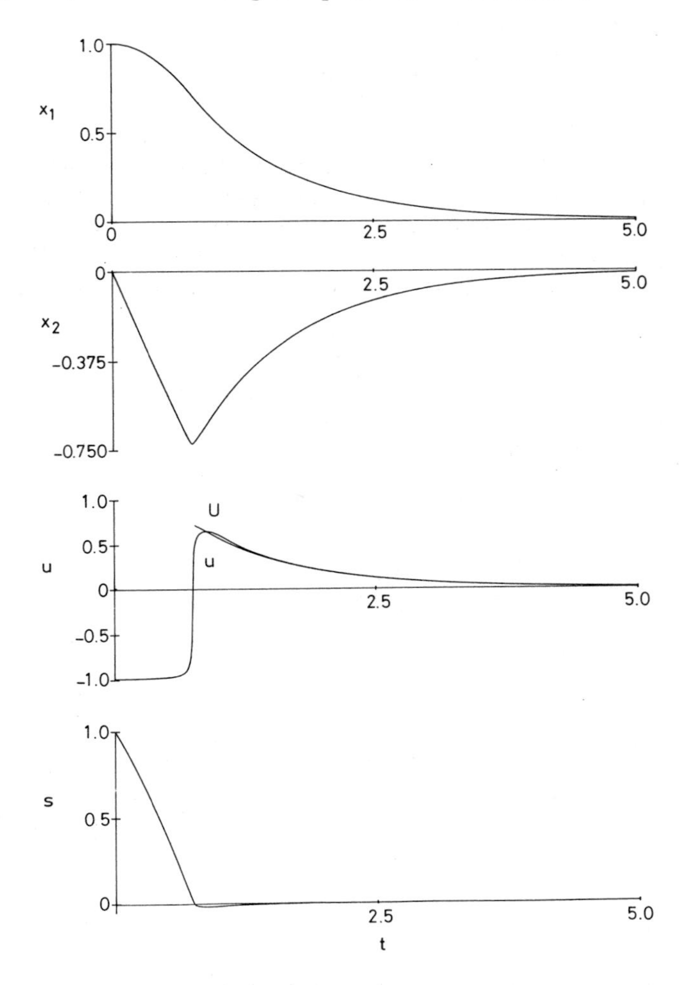

Fig. 1.4 *Double-integrator plant with smoothed control ($d = 0{\cdot}01$)*

Let us next consider the equivalent control approach (Utkin, 1971) used in the analysis of this VSC system. During sliding

$$s = 0 \quad \text{and} \quad \dot{s} = 0 \tag{1.10}$$

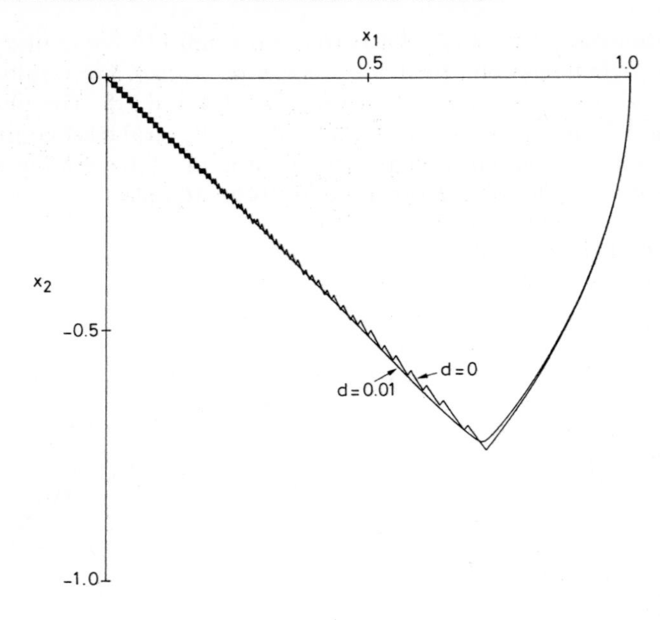

Fig. 1.5 *Double-integrator*

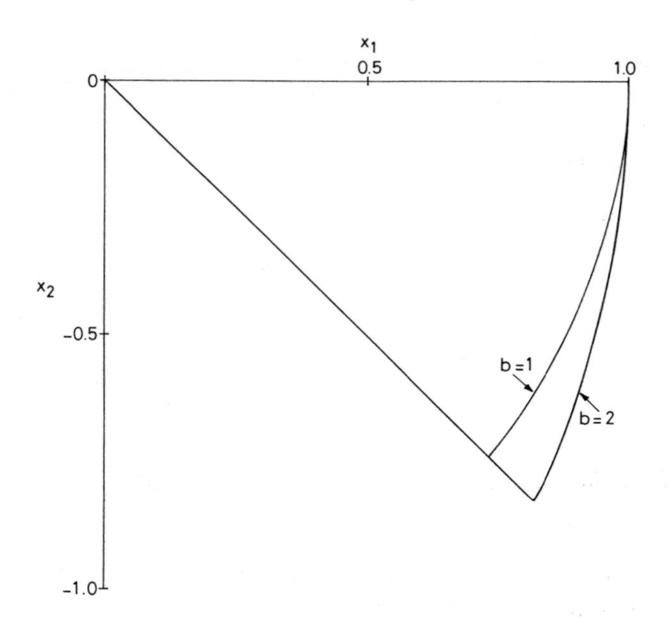

Fig. 1.6 *Double-integrator (d = 0·01)*

The equivalent linear control U which would maintain the state on the sliding line is from, (1.1) and (1.10), readily shown to be

$$U = -x_2/b \tag{1.11}$$

The smoothed control (1.9), which can be considered to be close to the averaged value of the discontinuous control, is practically identical to the theoretically derived value of the equivalent control U (see Fig. 1.4).

In Appendix 1.8 (see Fig. 1.9) the Filippov approach is outlined for determining the effective velocity vector $\dot{\boldsymbol{x}}$ of this system of differential equations with discontinuous right hand side in the sliding mode. A new approach uses the theory of differential inclusions (Ryan, 1988).

The control maintains the state on the sliding line for a magnitude range of parameter variations (and disturbances) (Draženović, 1969). Note, however, that a linear control of the form (1.11) would not maintain the state on (or in the neighbourhood of) the sliding line in the presence of parameter variations or disturbances. An example of this invariance of VSC is shown in Fig. 1.6 where simulations of the cases $b = 1$ and $b = 2$ are given for the same control with $d = 0{\cdot}01$. Similar results can be obtained for a range of gain parameter values b, and for time-varying $b(t)$.

1.4 Phase canonic system with scalar control

We next consider the nth order system in phase canonic form

$$\begin{aligned} \dot{x}_i &= x_{i+1} \qquad (i = 1, 2, \ldots, n-1) \\ \dot{x}_n &= -\sum_{j=1}^{n} a_j x_j + bu \qquad (b > 0) \end{aligned} \tag{1.12}$$

where a_j and b may be time-varying and uncertain.

Relay control may be used or other nonlinear control functions such as

$$u = -\sum_{i=1}^{k} \psi_i x_i \qquad (1 \leqslant k \leqslant n) \tag{1.13}$$

where

$$\psi_i = \begin{cases} \alpha_i & x_i s > 0 \\ \beta_i & x_i s < 0 \end{cases} \tag{1.14}$$

$s = \boldsymbol{c}^T\boldsymbol{x}$ where $\boldsymbol{c} = (c_1 c_2 \ldots c_n)^T$ and $\boldsymbol{x} = (x_1 x_2 \ldots x_n)^T$. The parameters c_i are design constants chosen to yield suitable dynamic response in the sliding mode. The parameters c_i should be chosen to yield asymptotically stable motion in the sliding mode (see Utkin, 1977; Itkis, 1976 for further details).

For sliding on the hyperplanes $s = 0$ the required inequality conditions on the values of the α_i and β_i can be readily obtained using the necessary sliding condition $s\dot{s} < 0$ and the expected magnitude range of the plant parameters (see Utkin, 1977; Itkis, 1976). Sliding motion can be attained for suitably chosen

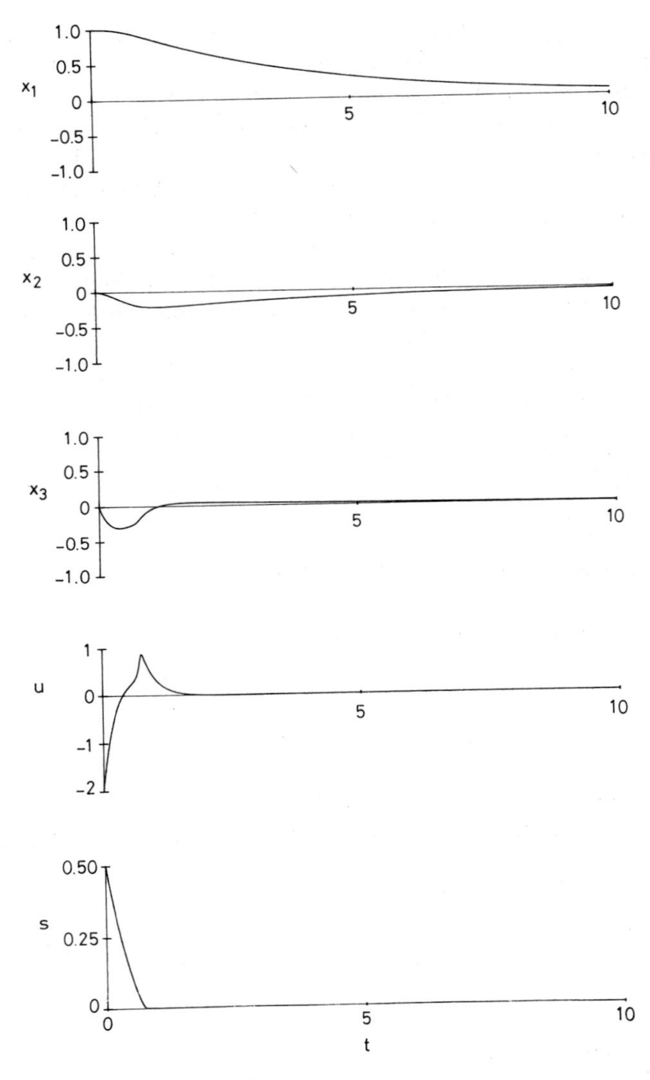

Fig. 1.7 *Triple-integrator (d = 0·01)*
$s + x_1 + 3x_2 + x_3$

values of the α_i and β_i provided the range of the expected parameter values a_j and b are known. It is essential to ensure that the state trajectory $\boldsymbol{x}(t)$ reaches the sliding hyperplane $s = 0$ from all initial conditions in the state space.

In the sliding mode

$$s = c_1x_1 + c_2x_2 + \ldots + c_{n-1}x_{n-1} + c_n\dot{x}_{n-1} = 0. \tag{1.15}$$

So the $(n - 1)$th order differential equation

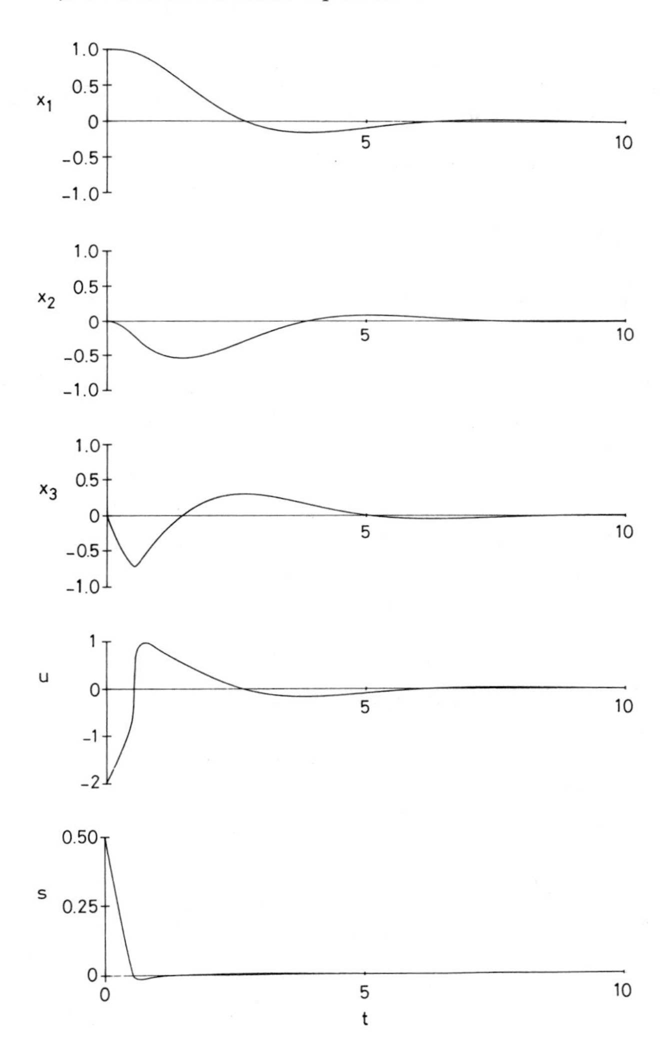

Fig. 1.8 *Triple-integrator ($d = 0{\cdot}01$)*
$s = x_1 + x_2 + x_3$

$$\dot{x}_{n-1} = -(c_1x_1 + c_2x_2 + \ldots + c_{n-1}x_{n-1}) \tag{1.16}$$

describes the dynamics of the system. Thus the system dynamics are independent of the system parameters a_j in the sliding mode. The parameters c_i specify the $(n - 1)$ eigenvalues of the system in the sliding mode. As an example consider the system

$$\dot{x}_i = x_{i+1} \qquad i = 1, 2$$

$$\dot{x}_3 = u$$

with

$$u = -s/(|s| + 0{\cdot}01).$$

Simulation results are presented in Fig. 1.7 for the switching function

$$s = x_1 + 3x_2 + x_3$$

$$= x_1 + 3\dot{x}_1 + \ddot{x}_1$$

So on the switching plane in the sliding mode

$$\ddot{x}_1 + 3\dot{x}_1 + x_1 = 0$$

with eigenvalues (poles) corresponding to an overdamped second-order linear system with damping ratio 1·5 and undamped natural frequency 1. In Fig. 1.8 the switching function

$$s = x_1 + x_2 + x_2$$

is shown to yield dynamic response in the sliding mode corresponding to an underdamped second-order system with damping ratio 0·5.

1.5 Multivariable variable structure control systems

The multivariable system to be considered satisfies in its most general form the differential equations

$$\dot{x}(t) = f(x, t) + Bu(t) + Dh(x, t) \qquad (1.17)$$

where $x \in R^n$, $h \in R^l$ represents disturbances and the effects of time-varying parameters, and $u \in R^m$ is the vector control. The discontinuous control has the form

$$u_i = \begin{cases} u_i^+(x) & s_i(x) > 0 \\ u_i^-(x) & s_i(x) < 0 \end{cases} \qquad (1.18)$$

where u_i is the ith component of $\boldsymbol{u}$ and $u_i^+ \neq u_i^-$. $s_i(x) = 0$ is the ith of the m switching hyperplanes

$$s(x) = Cx = 0 \qquad (1.19)$$

1.5.1 Sliding motion

For sliding on the ith hyperplane we need

$$s_i \dot{s}_i < 0 \tag{1.20}$$

in the neighbourhood of $s_i(x) = 0$. In the sliding mode the system satisfies the equations

$$s_i(x) = 0 \quad \text{and} \quad \dot{s}_i(x) = 0 \qquad \mathrm{i} = 1, 2, \ldots m \tag{1.21}$$

and the system has invariance properties (Draženović, 1969) yielding motion which is independent of certain system parameters and disturbances. Thus variable structure systems are usefully employed in systems with uncertain and time-varying parameters.

Before studying the nature of the control u enforcing sliding for the case of constant C, i.e. fixed switching hyperplanes, the detailed behaviour of the sliding dynamics needs to be considered. Suppose the sliding mode exists on all the hyperplanes. Then, during sliding

$$s = Cx = 0 \quad \text{and} \quad \dot{s} = C\dot{x} = 0 \tag{1.22}$$

The system equations (1.17) and (1.18) have discontinuous right hand sides and do not satisfy the classical theorems on the existence and uniqueness of the solutions. In the idealised equations non-ideal properties such as switching time delays and hysteresis are not considered. These properties determine the system dynamics in the neighbourhood of the discontinuity surfaces. As the non-ideal properties tend to zero, the motion tends to the ideal sliding mode on the intersection of the switching hyperplanes (Utkin, 1977). Utkin developed the equivalent control technique to study the sliding mode (see also Appendix 1.8). One sets

$$\dot{s} = 0 \tag{1.23}$$

and solves the resulting equations for the control vector termed the equivalent control U which is the control function yielding the sliding mode equations for the system. The value of U is effectively the average value of u which maintains the state on the discontinuity surfaces $s = 0$. The actual control u consists of a low frequency (average) and a high frequency component. The equivalent control is the control without the high frequency component. Further discussion of this topic may be found in Utkin (1971). Substituting the equivalent control U for the original control u gives the sliding mode equations

$$\dot{x} = f + BU + Dh \tag{1.24}$$

Solving for U from $\dot{s} = C\dot{x} = 0$ yields

$$U = -(CB)^{-1}C(f + Dh) \tag{1.25}$$

assuming $(CB)^{-1} \neq 0$. Substitution into the original equation yields the nth order equation

$$\dot{x} = [I - B(CB)^{-1}C](f + Dh) \tag{1.26}$$

[Utkin (1978) has also considered the singular case $(CB)^{-1} = 0$.]

Consider next the linear disturbance-free case $f = Ax$ and $h = 0$. Suppose (A, B) is a controllable pair with B having full rank. Then

$$U = -(CB)^{-1}CAx \tag{1.27}$$

and

$$\dot{x} = [I - B(CB)^{-1}C]Ax = A_{eq}x \tag{1.28}$$

The state vector x remains on the $(n - m)$-dimensional manifold of the intersection of the m discontinuity surfaces. Therefore we can obtain an $(n - m)$th order set of system equation describing the sliding mode, by eliminating m state variables using the equations $s = 0$. Young *et al.* (1977) have studied high-gain feedback systems and obtained results linking sliding motion and high-gain systems. The motion of the equivalent system is identical to that of the 'slow' and 'fast' subsystems of a high-gain system. Using the transformation

$$x^* = \begin{bmatrix} x_1^* \\ x_2^* \end{bmatrix} = Mx \qquad x_1^* \in R^{n-m}, \qquad x_2^* \in R^m \tag{1.29}$$

where

$$\begin{aligned} M &= \begin{bmatrix} M_1 \\ M_2 \end{bmatrix} \\ M_1 B &= 0 \\ MAM^{-1} &= \begin{bmatrix} A_{11} & A_{12} \\ A_{21} & A_{22} \end{bmatrix} \\ MB &= \begin{bmatrix} 0_1 \\ B_2 \end{bmatrix} \\ CM^{-1} &= (C_1 \, C_2) \end{aligned} \tag{1.30}$$

an expression for the equivalent control system is obtained. In terms of x_1^* and x_2^*

$$U = -(C_2B_2)^{-1}\{(C_1A_{11} + C_2A_{21})x_1^* + (C_1A_{12} + C_2A_{22})x_2^*\} \tag{1.31}$$

and

$$s = C_1x^* + C_2x_2^* = 0 \tag{1.32}$$

Thus

$$\dot{x}_1^* = A_{11}x_1^* + A_{12}x_2^*, \qquad s = 0 \text{ and } \dot{s} = 0, \tag{1.33}$$

yield the $(n - m)$th order system

$$\dot{x}_1^* = (A_{11} - A_{12}C_2^{-1}C_1)x_1^* \tag{1.34}$$

corresponding to the 'slow' subsystem. The eigenvalues of this system are the transmission zeros of the triple (C, A, B) considering C as the output matrix (Young *et al.*, 1977; El-Ghezawi *et al.*, 1983). The eigenvalues are determined by the choice of switching hyperplanes. Assuming (A, B) a controllable pair, then so is (A_{11}, A_{12}) (Young *et al.*, 1977) and the eigenvalues of (1.34) can be placed arbitrarily in the complex plane by suitable choice of C. The 'fast' subsystem is given by

$$\dot{x}_2^* = A_{21}x_1^* + A_{22}x_2^* + B_2u \tag{1.35}$$

The matrix C specifies the location of the hyperplanes and the associated eigenvalues and eigenvectors of the sliding system. Design techniques are discussed by Dorling and Zinober (1986, 1988). Alternatively Utkin and Young (1978) have shown that the hyperplanes may be chosen to minimise the (infinite-time) optimal control functionals

$$I_1 = \int_{t_s}^{\infty} x^T Qx \, dt \tag{1.36}$$

or

$$I_2 = \int_{t_s}^{\infty} (x^T Qx + U^T RU) \, dt \tag{1.37}$$

where Q and R are positive semi-definite symmetric matrices and t_s is the starting time of the sliding mode.

1.6 Properties in the sliding mode

We shall next sutdy some of the properties in the sliding mode using the concept of projectors. This simplifies the proof and understanding of these properties. We shall consider here the linear system with $f = Ax$.

Definition: Given a decomposition of the space S into subspaces S_1 and S_2 so that for any $x \in S$

$$(x = x_1 + x_2; \qquad x_1 \in S_1, \quad x_2 \in S_2$$

the linear operator P that maps x into x_1 is called a projector on S_1 along S_2, i.e.

$$Px = x_1, \qquad Px_2 = 0$$

Properties of projectors: Some useful projectors are now listed:

(*a*) A linear operator P is a projector if and only if it is idempotent, i.e. if

$$P^2 = P$$

(*b*) If P is the projector on S_1 along S_2, then $(I - P)$ is the projector on S_2 along S_1.

(*c*) If P is the projector on $R(P)$ (range of P) along $N(P)$ (null space of P), then $(I - P)$ is the projector on $N(P)$ along $R(P)$.

(*d*) For any $x \in R(P)$

$$Px = x$$

and

$$(I - P)x = 0$$

$$\text{rank}\ (P) = \text{trace}\ (P)$$

and

$$\text{rank}\ (I - P) = n - \text{rank}\ (P)$$

$$R(P) = N(I - P)$$

and

$$N(P) = R(I - P).$$

Certain matrix operators encountered in VSC are projectors:

(i) $[B(CB)^{-1}C]$ is a projector.

Proof: Since

$$[B(CB)^{-1}C]^2 = B(CB)^{-1}CB(CB)^{-1}C = B(CB)^{-1}C$$

$B(CB)^{-1}$ is idempotent and consequently a projector. $B(CB)^{-1}$ projects R^n on $R(B)$ along $N(C)$ because

$$R[B(CB)^{-1}C] = R(B)$$

since $R(BK) = R(B)$ if B and K are full rank. In this case, $K = (CB)^{-1}C$ which is full rank since B and CB are full rank. Similarly

$$N[B(CB)^{-1}C] = N(C)$$

since nullity (HC) = nullity (C), if H and C are full rank, where $H = B(CB)^{-1}$.

(ii) $[I - B(CB)^{-1}C]$ is a projector (see El-Ghezawi *et al.*, 1983).

Both of the above projectors are invaluable when exploring the basic features of variable structure systems.

(iii) If A is an $m \times n$ matrix and A^g is a generalised inverse of A, then

$$AA^g,\ A^gA,\ I_m - AA^g,\ I_n - A^gA$$

are all idempotent and therefore projectors. A^g is the $\{1\}$-inverse and satisfies $AA^gA = A$. The proof follows immediately from the definition of the $\{1\}$-inverse of a matrix, i.e.

$$(AA^g)^2 = AA^gAA^g = AA^g$$

AA^g and A^gA are projectors on $R(A)$ and $R(A^g)$, respectively and $(I_m - AA^g)$, $(I_m - A^gA)$ are projectors on $N(A^g)$ and $N(A)$, respectively.
(iv) The matrix $B(CB)^{-1}$ qualifies as a right inverse of C (see El-Ghezawi *et al.*, 1983).

We shall now apply the above to obtain some important properties of the sliding mode of VSC.

Order reduction: In the sliding mode the equation describing the system is given by

$$\dot{x} = [I - B(CB)^{-1}C]Ax = A_{eq}x$$

and $[I - B(CB)^{-1}C]$ is a projector which maps all the columns of A on $N(C)$. The order of the system has therefore been reduced because the state vector is now constrained to lie in $N(C)$ which is an $(n - m)$th dimensional subspace.

The invariance principle: The invariance principle formulated by Draženović (1969) states that for the system given by

$$\dot{x} = Ax + Bu + Dh \tag{1.38}$$

$$s = Cx \tag{1.39}$$

to be invariant to disturbances $h \in R^l$ in the sliding mode, the columns of matrix D should belong to the range space of B, i.e. $\text{col}(D) \in R(B)$. This principle will now be re-examined and a more general version derived. This generalisation extends the theory to the case where CB is singular (assuming sliding exists).

Theorem: The system given by (1.38) and (1.39) is invariant with respect to the disturbance h in the sliding mode if

$$\text{col}\,(D) \in R[B(CB)^{-1}C] \tag{1.40}$$

or

$$\text{col}\,(D) \in R[B(CB)^gC] \quad \text{if } CB \text{ is singular} \tag{1.41}$$

where col $(\cdot)$ stands for columns of $(\cdot)$.

Proof: The system in the sliding mode satisfies

$$\dot{x} = [I - B(CB)^{-1}C]Ax + [I - B(CB)^{-1}C]Dh \tag{1.42}$$

For the system to be invariant to h, $[I - B(CB)^{-1}C]D$ should be zero. Suppose $|CB| \neq 0$. If

$$\text{col}(D) \in R[B(CB)^{-1}C] \tag{1.43}$$

then

$$[I - B(CB)^{-1}C]D = 0 \tag{1.44}$$

as required. Conversely, if

$$[I - B(CB)^{-1}C]D = 0 \tag{1.45}$$

then

$$\text{col}(D) \in N[I - B(CB)^{-1}C] \tag{1.46}$$

For $|CB| = 0$ we can replace $(CB)^{-1}$ by $(CB)^g$ in the above proof.

Condition (1.43) is identical to that given by Draženović for the case where $|CB| \neq 0$ because

$$R[B(CB)^{-1}C] = R(B) \tag{1.47}$$

and condition (1.43) is equivalent to rank (B, D) = rank B (Draženović, 1969).

When CB is singular the invariancy is weakened since

$$R[B(CB)^g C] \subset R(B) \tag{1.48}$$

In this case, therefore, there will be no rejection in the sliding mode to any disturbance that belongs to $R(B)$ but not to $R[B(CB)^g C]$.

It is well known that the scalar system

$$\dot{x}_i = x_{i+1}, \qquad i = 1, \ldots, n-1$$

$$x_n = -\sum_{j=1}^{n} a_j x_j + bu$$

is invariant to parameter variations when it is in the sliding mode. This is because all variations in the a_i and b belong to $R(B)$ where $B = [0, 0, \ldots, b]^T$.

Physical insight into the invariance principle is readily achieved using projector theory. Let $P = B(CB)^{-1}C$, and from the previous definitions the projector P decomposes the state space X into the direct sum

$$X = R(P) \oplus N(P) \tag{1.50}$$

$$X = R(P) \oplus R(I - P) \tag{1.51}$$

Alternatively

$$R(P) \cap R(I - P) = \{\phi\} \tag{1.52}$$

Since $x \in R(I - P)$ during sliding, for x not to be affected by any disturbance h, the disturbance should lie in the complementary subspace of $(I - P)$, i.e. $h \in R(P)$, which is the condition of invariance.

Effect of sliding on the system zeros: Young (1977) has shown for scalar variable structure systems that the system zeros are unaffected by the sliding mode. This is to be expected since the sliding mode results from state feedback, and it is well known that state feedback cannot affect the system zeros. However, it is instructive to demonstrate that this is also the case here.

Given the system $S(A, B, C)$, we wish to show that the system zeros are not affected by the organisation of a sliding mode on the intersection of the hyperplane

$$s = Gx = 0 \tag{1.53}$$

Sliding resulting from the application of state feedback

$$U = -(GB)^{-1}CAx \tag{1.54}$$

yields the closed-loop system

$$\dot{x} = [I - B(GB)^{-1}G]Ax = A_G x \tag{1.55}$$

We shall show that the zeros of $S(A_G, B, C)$ are identical to the zeros $S(A, B, C)$. The zeros of $S(A_G, B, C)$ are given (see El-Ghezawi *et al.*, 1982, 1983) by the n − m eigenvalues of

$$\begin{aligned} M^g[I - B(CB)^{-1}C]A_G M &= M^g[I - B(CB)^{-1}C][I - B(GB)^{-1}G]AM \\ &= M^g[I - B(GB)^{-1}G - B(CB)^{-1}C \\ &\quad + B(CB)^{-1}CB(GB)^{-1}G]AM \\ &= M^g[I - B(CB)^{-1}C]AM \end{aligned} \tag{1.56}$$

where $M = N(C)$. But the eigenvalues of (1.56) are the zeros of the system $S(A, B, C)$ and $S(A_G, B, C)$. Therefore, sliding does not alter the system zeros.

Projector theory provides a neat method of studying many properties of VSC in the sliding mode. It also exposes the relationships between recurring themes associated with VSC in the sliding mode. These themes involve the closed-loop eigenvector matrix $W = (w_1, w_2, \ldots, w_{n-m})$ of A_{eq}, the input matrix B and the projector matrix $P = B(CB)^{-1}C$, together with the generalised inverses of W and B. The relationships are invaluable when formulating methods for constructing the switching hyperplane matrix C (see El-Ghezawi *et al.*, 1983 for further details).

1.7 Conclusion

In this Chapter some of the basic features and properties of variable structure control have been discussed. Numerous applications to a wide variety of practical problems have appeared in the literature and many are listed in Section 1.9.

The deterministic control of uncertain time-varying systems control is ac-

achieved using nonlinear feedback control functions, which operate effectively over a specified magnitude range of a class of system parameter variations, without the need for on-line identification of the values of the parameters. Statistical information of the nature of the system variations is not required. If the parameter variations satisfy certain matching conditions, total invariance can be achieved. This control philosophy contrasts sharply with stochastic adaptive control systems in which the control law is altered whilst system parameter values are calculated using online identification algorithms.

The two main approaches are VSC and Lyapunov control. Lyapunov control follows the approach of researchers such as Gutman and Palmor (1982), Corless and Leitmann (1981), Ryan (1988) and many other researchers. Using a Lyapunov function and specified magnitude bounds on the uncertainties, a nonlinear control law is developed to ensure uniform ultimate boundedness of the closed-loop feedback trajectory to achieve sufficient accuracy. The resulting controller is a discontinuous control function, with generally a continuous control in a boundary layer in the neighbourhood of the switching surface. The boundary layer control prevents the excitation of high-frequency unmodelled parasitic dynamics. Lack of space prevents a detailed analysis here of the extensive theory of this important field. The reader is invited to study further details in other chapters in this book and a comparative study presented by DeCarlo *et al.* (1988).

1.8 Appendix

Filippov (1960) has defined a solution for the general autonomous system, with discontinuous feedback control $u = u(x)$,

$$\frac{dx}{dt} = f(x, u(x)) = g(x) \tag{1.57}$$

where $g(x)$ is a function discontinuous on switching surfaces $G_i(x) = \{x : s_i(x) = 0\}$ in the state space, $x \in R^n$.

A vector function $x(t)$ defined on the interval $[t_1, t_2]$ is defined as a solution of (1.57) in the sense of Filippov, if it is absolutely continuous, and for almost all $t \in [t_1, t_2]$

$$\frac{dx}{dt} \in H(g(t, x)) \tag{1.58}$$

where

$$H(g(t, x)) = \bigcap_{\delta>0} \bigcap_{\mu N=0} \overline{\text{conv}}\, g(t, B(x, \delta) - N)$$

with $\overline{\text{conv}}$ denoting the closed convex hull, $B(x, \delta)$ the ball of radius centred at x and μ the Lebesgue measure. The notation $\bigcap_{\mu N=0}$ indicates the intersection over all sets of zero measure. The differential equation (1.57) has been replaced by the differential inclusion (1.58).

Consider for notational convenience a single switching surface (Fig. 1.9). Let G be a smooth surface (manifold) separating the space into regions G^+ and G^-. Suppose that G is regular, so that it can be divided by a smooth real-valued function $s(x)$ (i.e. $G = \{x : s(x) = 0\}$), such that $g(t, x)$ is bounded, and for any fixed t its limiting values $f^+(t, x)$ and $f^-(t, x)$ exist when G is approached from

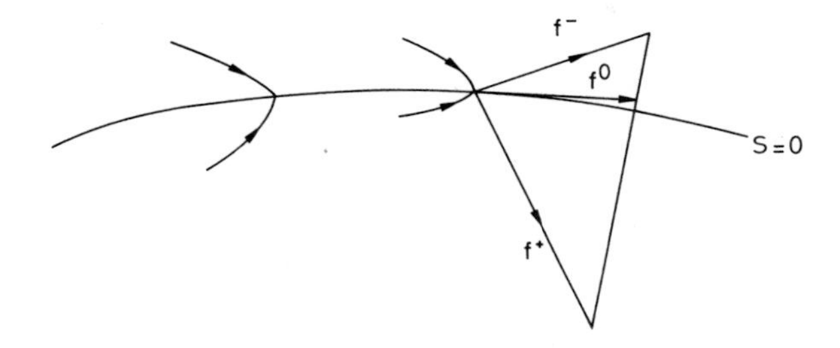

Fig. 1.9 *Filippov method*

G^+ and G^-, respectively. Let $f_0^+(f_0^-)$ be the projection of $f^+(f^-)$ on the normal to the surface G directed towards $G^-(G^+)$. Then for an absolutely continuous $x(t) \in G$ satisfying $f_0^-(t, x(t)) \geqslant 0$, $f_0^+(t, x(t)) \leqslant 0$ and $f_0^- - f_0^+ > 0$, the trajectories pointing towards G, are a solution of (1.57) according to the previous definition if and only if

$$\frac{dx(t)}{dt} = \alpha(t) f_0^+(t, x(t)) + (1 - \alpha(t)) f_0^-(t, x(t)) \tag{1.59}$$

where

$$\alpha(t) = \frac{f_0^-(t, x(t))}{f_0^-(t, x(t)) - f_0^+(t, x(t))}$$

where $f_0^+(t, x) = \langle \nabla s, f^+(t, x)\rangle$ and $f_0^-(t, x) = \langle \nabla s, f^-(t, x)\rangle$. The right-hand side of (1.59) is orthogonal to ∇s, so the solution remains on the surface. The values taken by $f(t, x)$ in the neighbourhood of G yield solutions constrained to slide on the surface G.

Further details can be obtained from André and Seibert (1956), and Ryan (1982); also see Ryan (1988).

1.9 References

AIZERMAN, M. A., and PYATNISKII, E. S. (1974): 'Foundations of a theory of discontinuous systems I', *Automat. Remote Contr.*, **35**, pp. 1066–1080

AMBROSINO, G., CELENTANO, G., and GAROFALO, F. (1984): 'Variable structure model reference adaptive control systems', *Int. J. Control*, **39**, pp. 1339–1349

AMBROSINO, G., CELENTANO, G., and GAROFALO, F. (1985): 'Decentralized PD controllers for tracking control of uncertain multivariable systems'. Proc. 7th IFAC/IFORS Symp. on Identification & System Parameter Estimation, York, pp. 1907–1912

AMBROSINO, G., CELENTANO, G., and GAROFALO, F. (1985): Robust model tracking control for a class of non-linear plants', *IEEE Trans.* **AC-30**, pp. 275–279

AMBROSINO, G., CELENTANO, G., and GAROFALO, F. (1986): 'Ultimate boundedness of unknown SISO plants with polynomial disturbances'. Proc. 25th IEEE Conf. on Decision and Control, Athens, pp. 317–321

ANDRÉ, J., and SEIBERT, P. (1956): 'Über stuckweise linear Differentialgleichungen, die bei Regelungsproblemen auftreten; I and II', *Arch. Math.*, **7,** pp. 148–165

ANDRÉ, J., and SIEBERT, P. (1960): 'After end-point motions of general discontinuous control systems and their stability problems'. Proc. 1st IFAC Congress, Moscow, **2**, pp. 919–922

ASLIN, P. P., PATTON, R. J., and DORLING, C. M. (1985): 'The design of a sliding mode controller for Dutch roll damping in a non-linear aircraft system'. Proc. IEE Conf. Control '85, Cambridge, pp 435–39

BAILEY, E., and ARAPOSTATHIS, A. (1987): 'Simple sliding mode control scheme applied to robot manipulators', *Int. J. Contr.*, **45,** pp. 1197–1209

BALESTRINO, A., DE MARIA, G., and SCIAVICCO, L. (1982): 'Hyperstable adaptive model following control of nonlinear plants', *Syst. and Control Lett.*, **1,** pp. 232–236

BALESTRINO, A., DE MARIA, G., and SCIAVICCO, L. (1983): 'Adaptive control design in servosystems'. 3rd IFAC Symposium on Control in Power Electronics & Electrical Drives, Lausanne, pp. 125–131

BALESTRINO, A., DE MARIA, G., and SCIAVICCO, L. (1983): 'An adaptive model following control for robot manipulators', *Trans. ASME G,* **105,** pp. 143–151

BALESTRINO, A., DE MARIA, G., and SICILIANO, B. (1985): 'Hyperstable variable structure control for a class of uncertain systems'. Proc. 7th IFAC/IFORS Symp. on Identification and System Parameter Estimation, York, pp. 1913–1920

BALESTRINO, A., DE MARIA, G., and ZINOBER, A. S. I. (1984): 'Nonlinear adaptive model-following control', *Automatica,* **20,** pp. 559–568

BARMISH, B. R., CORLESS, M., and LEITMANN, G. (1983): 'A new class of stabilizing controllers for uncertain dynamical systems', *SIAM J. Contr. Optimization,* **21,** 246–255

BARTOLINI, G., and ZOLEZZI, T. (1985): 'Variable structure nonlinear in the control law', *IEEE Trans.,* **AC-30,** pp. 681–684

BARTOLINI, G., and ZOLEZZI, T. (1986): 'Some development of variable structure control for nonlinear systems'. Proc. 25th IEEE Conf. on Decision and Control, Athens, pp. 328–333

BREGER, A. M., BUTKOVSKII, A. G., KUBYSHKIN, V. A., and UTKIN, V. I. (1980): 'Sliding modes for control of distributed parameter entities subjected to a mobile multiple signal', *Automation & Remote Control,* **41,** 346–355

BÜHLER, H. (1986): 'Reglage par mode de glissement', (Presses Polytechniques Romandes, Lausanne)

BURTON, J. A., and ZINOBER, A. S. I. (1986): 'Continuous approximation of variable structure control', *Int. J. Systems Science*, **17**, pp. 876–885

BURTON, J. A., and ZINOBER, A. S. I. (1988): 'Continuous self-adaptive control using a smoothed variable structure controller', *Int. J. Systems Science,* **19,** pp. 1515–28

CALISE, A. J. (1980): 'Hyperstability in variable structure systems'. Proc. 19th IEEE Conf. on Decision & Control, Albuquerque, New Mexico, pp. 1107–1109

CALISE, A. J., and KRAMER, F. (1982): 'A variable structure approach to robust control of VTOL aircraft'. Proc. American Control Conf., pp. 1046–1052

CALISE, A. J., and RAMAN, K. V. (1981): 'A servo compensator design approach for variable structure systems'. Proc. 19th Allerton Conf., Univ. of Illinois, pp. 452–460

CALISE, A. J., and RAMAN, K. V. (1982): 'A servo compensator design approach for variable structure systems'. Proc. 21st IEEE Conf. on Decision & Control, Orlando, pp. 1014–1019

CHOUINARD, J. P., DAUER, J. P., and LEITMANN, G. (1985): 'Properties of matrices used in uncertain linear control systems', *SIAM J. Contr. Optimization,* **23,** pp. 381–389

CORLESS, M., GOODALL, D. P., LEITMANN, G., and RYAN, E. P. (1985): 'Model-following controls for a class of uncertain dynamical systems'. Proc. 7th IFAC/IFORS Symp. on Identification & System Parameter Estimation, York, pp. 1895–1900

CORLESS, M. J., and LEITMANN, G. (1981): 'Continuous state feedback guaranteeing uniform ultimate boundedness for uncertain dynamical systems', *IEEE Trans.,* **AC-26,** pp. 1139–1144

DAVAUD, F. M., and CARON, J. Y. (1975): 'Asymptotic stability of model reference systems with bang-bang control', *IEEE Trans.,* **AC-20,** pp. 694–696

DECARLO, R. A., ŻAK, S. H., and MATTHEWS, G. P. (1988): 'Variable structure control of nonlinear multivariable systems: a tutorial', *Proc. IEEE,* **76,** pp. 212–232

DORLING, C. M., and ZINOBER, A. S. I. (1983): 'A comparative study of the sensitivity of observers', Proc. IASTED Symp. on Applied Control & Estimation, Copenhagen, pp. 632–637

DORLING, C. M. and ZINOBER, A. S. I. (1984): 'Computer aided design of variable structure systems'. Proc. IMC Workshop on Computer Aided Control System Design, Brighton, pp. 67–71

DORLING, C. M., and ZINOBER, A. S. I. (1985): 'Computer aided design of robust multivariable variable structure control systems'. Proc. 3rd IFAC Symp. on CAD in Control & Engineering Systems, Copenhagen, pp. 402–407

DORLING, C. M., and ZINOBER, A. S. I. (1985): 'Hyperplane design in model-following variable structure control systems'. Proc. 7th IFAC/IFORS International Symp. on Identification & System Parameter Estimation, York, pp. 1901–1906

DORLING, C. M., and ZINOBER, A. S. I. (1986): 'Two approaches to hyperplane design of multivariable variable structure control systems', *Int. J. Control,* **44,** pp. 65–82

DORLING, C. M., and ZINOBER, A. S. I. (1988): 'Robust hyperplane design in multivariable variable structure control systems', *Int. J. Control,* **48,** pp. 2043–2054

DRAŽENOVIĆ, B. (1969): 'The invariance conditions in variable structure systems', *Automatica,* **5,** pp. 287–295

EL-GHEZAWI, O. M. E., BILLINGS, S. A., and ZINOBER, A. S. I. (1983): 'Variable-structure systems and system zeros', *Proc. IEE,* **130D,** pp. 1–5

EL-GHEZAWI, O. M. E., ZINOBER, A. S. I., and BILLINGS, S. A. (1983): 'Analysis and design of variable structure systems using a geometric approach', *Int. J. Con.,* pp. 657–671

EL-GHEZAWI, O. M. E., ZINOBER, A. S. I., OWENS, D. H., and BILLINGS, S. A. (1982): 'Computation of the zeros and zero directions of linear multivariable systems', *Int. J. Control.,* **36,** pp. 833–843

EMEL'YANOV, S. V. (1964): 'Design of variable structure control systems with discontinuous switching functions', *Eng. Cybern.,* **1,** 156–60.

EMEL'YANOV, S. V. et al. (1967): 'Variable structure control systems (in Russian)' (Nauka, Moscow) also (Oldenburg Verlag, Munchen–Wien) (in German).

EMEL'YANOV, S. V., and UTKIN, V. I. (1967) 'Invariant solutions of differential equations with discontinuous coefficients. Design principles of variable structure control systems' *in* 'Mathematical theory of control' (Academic Press, NY)

ERSCHLER, J., ROUBELLAT, F., and VERNHES, J. P. (1974): 'Automation of a hydroelectric power station using variable structure systems', *Automatica,* **10,** pp. 27–36

ESPANA, M. D., ORTEGA, R. S., and ESPINO, J. J. (1984): 'Variable structure systems with chattering reduction: a microprocessor based approach', *Automat.*, **20**, pp. 133–134

FILIPPOV, A. G. (1960): 'Application of the theory of differential equations with discontinuous right-hand sides to non-linear problems in automatic control', Proc. 1st IFAC Congress, Moscow, **2**, pp. 923–927

FLÜGGE-LOTZ, I. (1953): 'Discontinuous automatic control' (Princeton University Press)

FURUTA, K., and MORISADA, M. (1988): 'Implementation of sliding mode control by a digital computer'. Proc. IECON 88, pp. 453–458

FULLER, A. T. (1967): 'Linear control of nonlinear systems', *Int. J. Contr.,* **5,** pp. 197–243

GOLEMBO, B. Z., EMEL'YANOV, S. V., UTKIN, V. I., and SHUBLADZE, A. M. (1976): 'Application of piecewise-continuous dynamic systems to filtering problems', *Autom. Remote*

Control, **37,** pp. 369–377

GOODALL, D. P., and RYAN, E. P. (1988): 'Feedback controlled differential inclusions and stabilization of uncertain dynamical system', *SIAM J. Control & Optimisation,* **26,** pp. 1431–1441

GOUGH, N. E., and Ismail, Z. M. (1982): 'Computer-aided design of variable structure control systems', *Control and Computers,* **10,** pp. 71–75

GOUGH, N. E., ISMAIL, Z. M., and KING, R. E. (1983): 'Analysis of variable structure systems with sliding modes', *Int. J. Sys. Sci.,* **15,** pp. 403–409

GUTMAN, S. (1976): 'Uncertain dynamical systems – a differential game approach'. NASA Tech. Memo. TMX 73, pp. 135

GUTMAN, S. (1979): 'Uncertain dynamical systems – a Lyapunov min-max approach', *IEEE Trans.,* **AC-24,** pp. 437–443

GUTMAN, S., and PALMOR, Z. (1982): 'Properties of min-max controllers in uncertain dynamical systems', *SIAM J. Control Optimization,* **20,** pp. 850–861

GUZZELLA, L., and GEERING, H. P. (1986): 'Model following variable structure control for a class of uncertain mechanical systems. Proc. 25th Conf. on Decision and Control, Athens, pp. 312–316

HA, I. (1989): 'New matching conditions for output regulation of a class of uncertain nonlinear systems', *IEEE Trans,* **AC-34,** pp. 116–119

HAJEK, O. (1979): 'Discontinuous differential equations: I', *J. Diff. Equations,* **32,** pp. 149–170

HAJEK, O. (1979): 'Discontinuous differential equations: II', *J. Diff. Equations,* **32,** pp. 171–185

HSU, Y. Y., and CHAN, W. C. (1984): 'Optimal variable-structure controller for DC motor speed control', *Proc. IEE,* **131D,** pp. 233–237

ITKIS, U. (1976): 'Control Systems of Variable Structure' (Wiley, NY)

KAUTSKY, J., NICHOLS, N. K., and VAN DOOREN, P. (1985): 'Robust pole assignment in linear state feedback', *Int. J. Control,* **41,** 1129–1155

KERMIN ZHOU and PRAMAD KERGONEKAR, P. K. (1988): 'On the stabilization of uncertain linear systems via bound invariant Lynapunov functions', *SIAM J. Control & Optimisation,* **26,** pp. 1265–1273

KRAVARIS, C., and PALANKI, S. (1988): 'A Lyapunov approach to robust nonlinear state feedback synthesis', *IEEE Trans.,* **AC-33,** pp. 1188–1191

LEFEBVRE, S., RICHTER, S., and DECARLO, R. (1983): 'Decentralized variable structure control design for a two-pendulum system', *IEEE Trans.,* **AC-28,** pp. 112–114

LEITMANN, G. (1979): 'Guaranteed asymptotic stability of some linear systems with bounded uncertainties', *Trans ASME G,* **101,** pp. 212–216

LEITMANN, G. (1981): 'On the efficiency of nonlinear control in uncertain linear systems', *Trans. ASME G,* **103,** pp. 95–102

LEITMANN, G., RYAN, E. P., and STEINBERG, A. (1986): 'Feedback control of uncertain systems: robustness with respect to neglected actuator and sensor dynamics', *Int. J. Control,* **43,** pp. 1243–1256

LOZGACHEV, G. I. (1972): 'The construction of a Lyapunov function for variable structure systems', *Automat. Remote Contr.,* **33,** pp. 1391–1393

MARINO, R. (1985): 'High-gain feedback in nonlinear control systems', *Int. J. Contr.,* **42,** pp. 1369–1385

MATTHEWS, G., DECARLO, R., HAWLEY, P. and LEFEBVRE, S. (1986): 'Toward a feasible variable structure control design for a synchronous machine connected to an infinite bus', *IEEE Trans.* **AC-31,** pp. 1159–1163

MATTHEWS, G. and DECARLO, R. (1988): 'Decentralized tracking for a class of interconnected nonlinear systems using variable structure control', *Automatica,* pp. 187–193

MILOSAVLJEVIC, C. (1985): 'General conditions for the existence of a quasisliding mode on the switching hyperplane in discrete variable structure systems', *Automat. Remote Contr.,* **46,** pp. 307–314

MORGAN, R. G., and OZGUNNER, U. (1985): 'A decentralized variable structure control algorithm for robotic manipulators', *IEEE J. Robotics Automat.,* **RA-1,** pp. 57–65

MUDGE, S. K., and PATTON, R. J. (1988): 'Analysis of the technique of robust eigenstructure

assignment with application to aircraft control', *Proc. IEE,* **135D,** pp. 275–281
ORLOV, Y. V. (1983): 'Application of Lyapunov method in distributed systems', *Automat. Remote Contr.,* **44,** pp. 426–431
ORLOV, Y. V., and UTKIN, V. I. (1982): 'Use of sliding modes in distributed system control problems', *Automat. Remote Contr.,* **43,** pp. 1127–1135
ORLOV, Y. V., and UTKIN, V. I. (1987): 'Sliding mode control in infinite dimensional systems', *Automatica,* **23,** pp. 753–757
PADEN, B. E., and SASTRY, S. S. (1987): 'A calculus for computing Filippov's differential inclusion with application to the variable structure control of robot manipulators', *IEEE Trans.* **CAS-34,** pp. 73–82
PENG LI, J., and AHMED, N. U. (1988): 'On exponential stability of infinite dimensional linear systems with bounded or unbounded perturbations', *Applicable Analysis,* **30,** pp. 175–187
PETERSEN, I. R. (1985): 'Structural stabilization of uncertain systems: Necessity of the matching condition', *SIAM J. Contr. Optimization,* **23,** pp. 286–296
PETERSEN, I. R. (1987): 'A procedure for simultaneously stabilizing a collection of single input linear systems using nonlinear state feedback control, *Automatica,* **23,** pp. 33–40
PETERSEN, I. R. (1987): 'Notions of stabilizability and controllability for a class of uncertain linear systems', *Int. J. Control,* **46,** pp. 409–422
PETERSEN, I. R. (1988): 'Stabilization of an uncertain linear system in which uncertain parameters enter into the input matrix', *SIAM J. Control Optimisation,* **26,** pp. 1257–1264
RYAN, E. P. (1982): 'Optimal relay and saturating control system analysis' (Peter Peregrinus)
RYAN, E. P. (1983): 'A variable structure approach to feedback regulation of uncertain dynamical systems', *Int. J. Control,* **38,** 1121–1134
RYAN, E. P. (1988): 'Adaptive stabilization of a class of uncertain nonlinear systems: a differential inclusion approach', *Systems and Control Lett.,* **10,** pp. 95–101
RYAN, E. P., and CORLESS, M. (1984): 'Ultimate boundedness and asymptotic stability of a class of uncertain dynamical systems via continuous and discontinuous feedback control', *IMA J. Math. Control Information,* **1,** pp. 223–242
SAKSENA, V. R., O'REILLY, J., and KOKOTOVIĆ, P. V. (1984): 'Singular perturbations and time-scale methods in control theory: survey 1976–1983', *Automatica,* **20,** pp. 273–293
SCHMITENDORF, W. E., and BARMISH, B. R. (1986): 'Robust asymptotic tracking for linear systems with unknown parameters:, *Automatica,* **22,** pp. 335–360
SIRA-RAMIREZ, H. (1987): 'Sliding motion in bilinear switched networks', *IEEE Trans,* **CAS-34,** pp. 919–933
SIRA-RAMIREZ, H. (1988): 'Periodic sliding motions', *IEEE Trans.,* **AC-33,** pp. 1191–1194
SIRA-RAMIREZ, H., and DWYER, T. A. (1987): 'Variable structure controller design for spacecraft mutation damping', *IEEE Trans.,* **AC-32,** pp. 435–438
SIRA-RAMIREZ, H. (1987): 'Harmonic response of variable-structure-controlled Van der Pol oscillators', *IEEE Trans.,* **CAS-34,** pp. 103–106
SIVARAMAKRISHNAN, A. Y., HARIHARAN, M. V., and SRISAILAM, M. C. (1984): 'Design of variable-structure load-frequency controller using pole assignment technique', *Int. J. Control,* **40,** pp. 487–498
SLOTINE, J. J., and SASTRY, S. S. (1983): 'Tracking control of non-linear systems using sliding surfaces, with application to robot manipulators', *Int. J. Control,* pp. 465–492
SLOTINE, J. J. (1984): 'Sliding controller design for nonlinear systems', *Int. J. Contr.,* **40,** pp. 421–434
STEINBERG, A. (1988): 'A sufficient condition for output feedback stabilization of uncertain dynamical systems', *IEEE Trans.,* **AC-33,** pp. 676–677
UTKIN, V. I. (1971): 'Equations of the slipping regime in discontinuous systems: I', *Autom. Remote Control,* **32,** pp. 1897–1907
UTKIN, V. I. (1972): 'Equations of the slipping regime in discontinuous systems: II', *Autom. Remote Control,* **33,** pp. 211–219
UTKIN, V. I. (1977): 'Variable structure systems with sliding modes', *IEEE Trans.,* **AC-22,** pp. 212–222

UTKIN, V. I. (1978): 'Sliding modes and their applications in variable structure systems', (MIR, Moscow)

UTKIN, V. I. (1981): 'Principles of identification using sliding regimes', *Sov. Phys. Dokl.*, **26**, pp. 271–272

UTKIN, V. I. (1983): 'Variable structure systems present and future', *Automat. Remote Contr.*, **44**, pp. 1105–1120

UTKIN, V. I. (1987): 'Variable structure control state of the art', *Proc. IFAC Conf.* (Munich).

UTKIN, V. I., and YANG, K. (1978): 'Methods for constructing discontinuity planes in multi-dimensional variable structure systems', *Automat. Remote Contr.*, **26**, pp. 1466–1470

WALCOTT, B. L., and ŻAK, S. H. (1987): 'State observation of nonlinear uncertain dynamical systems', *IEEE Trans.*, **AC-32**, pp. 166–170

WEISSENBERGER, S. (1966): 'Stability-boundary approximation for relay-control systems with a steepest ascent construction of Lyapunov functions', *J. Bas. Engng*, **88**, pp. 419–428

WILLEMS, J. C. (1981):'Almost invariant subspaces: an approach to high gain feedback design. Part I: Almost controlled invariant subspaces', *IEEE Trans.*, **AC-26**, pp. 235–252

WHITE, B. A. (1983): 'Reduced order switching functions in variable structure control theory', *Proc. IEEE*, **130D**, pp. 33–40

WHITE, B. A. (1985): 'Some problems in outback feedback in variable structure control systems'. Proc. 7th IFAC/IFORS International Symposium on Identification and System Parameter Estimation, York, pp. 1921–1927

WHITE, B. A., PATTON, R. J., MUDGE, S. K., and ASLIN, P. R. (1986): Reduced order variable structure control of the lateral motion of an aircraft'. Proc. 25th IEEE Conf. on Decision and Control, Athens, pp. 322–327

WONHAM, W. M., and JOHNSON, C. D. (1963): 'Optimal bang-bang control with quadratic performance index'. Proc. 4th Joint Automatic Control Conf., Minneapolis, pp. 101–112

YEW, M. K., ZINOBER, A. S. I., and PATTON, R. J. (1988): 'An assessment of the application of sliding mode control to aircraft flight systems'. SERC Research Report GR/D/55436, pp. 128

YONG, J. (1988): Feedback stabilisation of nonlinear uncertain dynamic systems', *J. Math. Anal. Appl.*, **129**, pp. 153–165

YOUNG, K.-K. D. (1977): 'Asymptotic stability of model reference systems with variable structure control', *IEEE Trans.*, **AC-22**, pp. 279–281

YOUNG, K.-K. D. (1978): 'Design of variable structure model-following control systems', *IEEE Trans.*, **AC-23**, pp. 1079–1085

YOUNG, K.-K. D. (1978): 'Controller design for a manipulator using the theory of variable structure systems', *IEEE Trans. Syst. Man. Cybern.*, **8**, pp. 101–109

YOUNG, K.-K. D., and KWATNY, H. G. (1981): 'Formulation and dynamic behaviour of a variable structure servomechanism'. Proc. 22nd JACC, Charlottesville, Va, Paper WP-3F, pp. 315–321

YOUNG, K.-K. D., and KWATNY, H. G. (1982): 'Variable structure servomechanism design and applications to overspeed protection control', *Automatica*, **18**, pp. 385–400

YOUNG, K.-K. D., KOKOTOVIĆ, P. V., and UTKIN, V. I. (1977): 'A singular perturbation analysis of high-gain feedback systems', *IEEE Trans.*, **AC-22**, pp. 931–937

ZINOBER, A. S. I.: 'Controller design using the theory of variable structure systems' *in* BILLINGS, S. A., and HARRIS, C. (Eds.) (1982): 'Self tuning and adaptive control' (Peter Peregrinus) Chap. 9

ZINOBER, A. S. I., EL-GHEZAWI, O. M. E., and BILLINGS, S. A., (1982): 'Multivariable variable structure adaptive model-following control systems', *Proc. IEE*, **129D**, pp. 6–12.

ZINOBER, A. S. I. (1984): 'Properties of adaptive discontinuous model-following control systems,' Proc. 4th IMA International Conference on Control Theory (Cambridge), pp. 204–29.

ZINOBER, A. S. I. (1988): 'Variable structure model-following control of a robot manipulator,' Proc. 4th IFAC Symp. on Computer Aided Design in Control Systems (CADCS '88), Beijing, pp. 175–180.

ZINOBER, A. S. I. (1986): 'Self-adaptive control of the time-varying nonlinear crane problem with disturbances and state constraints', *Control and Computers* (IASTED), **14**, pp. 103–109

Chapter 2

Sliding mode control with switching command devices

H. Bühler

Swiss Federal Institute of Technology

2.1 Introduction

The sliding mode is a particular mode of operation in variable structure control systems (Utkin, 1977 and 1987). Two switching principles based on state vari-

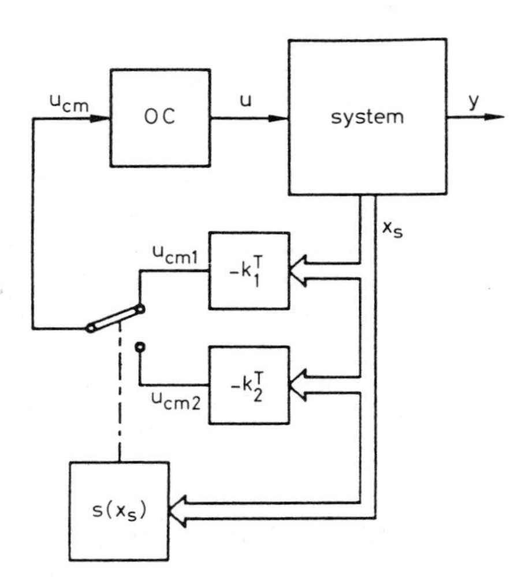

Fig. 2.1 *Control with a variable structure depending on switching of state variable feedbacks*

able feedback will be described. In the first case the variable structure is realised by switching between two different state variable feedback laws according to the switching strategy $s(\boldsymbol{x})$ depicted in Fig. 2.1

This method has the disadvantage of overstressing the command device (a continuous actuator). This kind of actuator is not designed for the high variations at high frequency, which appear in sliding mode operation.

Developments in switching power devices, for example DC choppers or pulsation inverters, allow the designer advantageously to use their individual switching properties to define the variable structure as in Fig. 2.2. A switching strategy $s(\boldsymbol{x})$ based on state variable feedback determines the state ($u = u_{max}$ or u_{min}) of the command device (Bühler, 1986)

In this context a pulsation voltage convertor (DC chopper) can be used to feed a DC motor (Sabanovic, 1983; Feller and Benz, 1987) and a 3-phase pulsation inverter to feed an induction motor (Bilalovic, 1983; Utkin, 1987).

In this Chapter this second type of variable structure control system is described, beginning with scalar control systems. The basic structure, the sliding

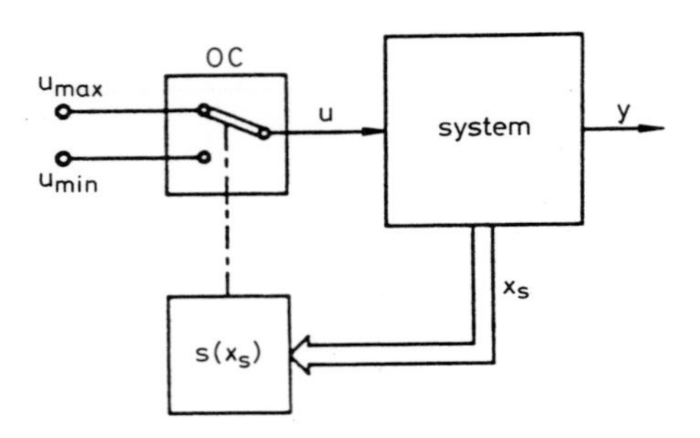

Fig. 2.2 *Control with a variable structure depending on switching of a switching command device*

mode conditions and the state equations in the sliding mode will be considered. The pole assignment method will be used to design the switching strategy. Afterwards the sliding mode domain and the switching frequency will be studied. These theoretical considerations will be illustrated with a practical example, the position control of a DC drive.

Next, for multivariable systems, the sliding mode conditions, the design of the multivariable switching strategy with a pole assignment method and robustness, i.e. invariance properties with regard to parameter variations, will be described.

2.2 Sliding mode control of scalar control systems with state variable feedback

2.2.1 Basic structure and switching strategy

Fig. 2.3 shows the basic structure with a switching command device and a switching strategy based on state variable feedback with an integrator regulator.

Similarly to linear state feedback control, the switching strategy is given by

$$s(\boldsymbol{x}) = -\boldsymbol{k}^T\boldsymbol{x} + k_w w \tag{2.1}$$

where

$$\boldsymbol{x} = \begin{bmatrix} \boldsymbol{x}_S \\ x_R \end{bmatrix} \tag{2.2}$$

is the state n-vector of the complete system with $n = n_s + 1$, where n_s is the order of the controlled system and w is the reference input. $\boldsymbol{x}_s$ is the state vector of the controlled system S described by the state equations

$$\dot{\boldsymbol{x}}_s = \boldsymbol{A}_s\boldsymbol{x}_s + \boldsymbol{b}_s u + \boldsymbol{b}_{sv} v \tag{2.3}$$

$$y = \boldsymbol{c}_s^T \boldsymbol{x}_s \tag{2.4}$$

where v is a disturbance variable acting on the system. x_R is a state variable describing the integrator regulator R by

$$\frac{dx_R}{dt} = \frac{1}{T_i}(w - y) \tag{2.5}$$

where T_i is the integrator regulator time constant and y the output variable of the controlled system.

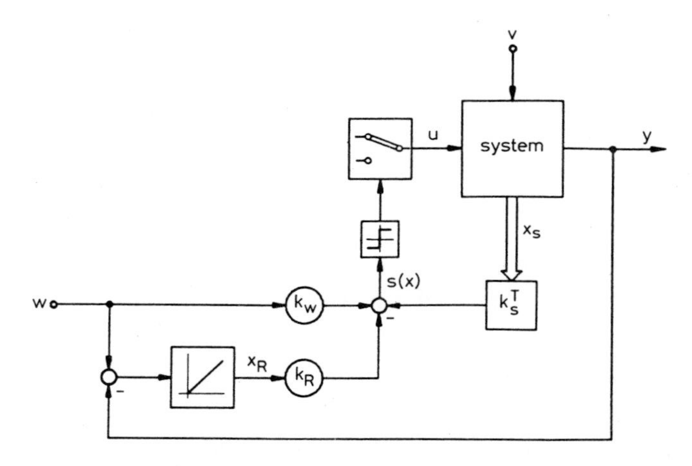

Fig. 2.3 *Basic structure for a scalar control system with state variable feedback and integrator regulator*

The row vector $\boldsymbol{k}^T$ in (2.1) contains the state fedback coefficients comprising the row vector $\boldsymbol{k}_s^T$ and the coefficient $-k_R$ for the integrator regulator feedback. k_w is the feedforward coefficient of the reference variable w.

The complete system can be described by its state equations

$$\dot{\boldsymbol{x}} = \boldsymbol{A}\boldsymbol{x} + \boldsymbol{b}u + \boldsymbol{b}_v v + \boldsymbol{b}_w w \tag{2.6}$$

$$y = \boldsymbol{c}^T \boldsymbol{x} \tag{2.7}$$

with the combined matrix and vectors

$$A = \begin{bmatrix} A_s & O \\ -\frac{1}{T_i} c_s^T & O \end{bmatrix}; \quad b = \begin{bmatrix} b_s \\ O \end{bmatrix}; \quad b_v = \begin{bmatrix} b_{sv} \\ O \end{bmatrix}; \quad b_w = \begin{bmatrix} O \\ \frac{1}{T_i} \end{bmatrix}$$

$$c^T = [c_s^T \quad O] \tag{2.8}$$

The control variable u switches between u_{max} and u_{min} according to the switching strategy $s(x)$.

$$u = \begin{cases} u_{max} & \text{for} \quad s(x) > O \\ u_{min} & \text{for} \quad s(x) < O \end{cases} \tag{2.9}$$

2.2.2 Sliding mode conditions

For a theoretical description we assume that in the sliding mode the control variable u switches between u_{max} and u_{min} at an infinite frequency. Then the state point describes a trajectory on the switching hyperplane given by $s(x) = 0$. Therefore, $\dot{s}(x) = 0$. From (2.1) and (2.6), we obtain

$$\dot{s}(x) = -k^T(Ax + bu + b_v v + b_w w) + k_w \dot{w} = 0 \tag{2.10}$$

At infinite switching frequency the control variable u can be replaced by its mean value u_{eq}, the equivalent control variable, which is a continuous variable. By (2.10) we can write (Bühler, 1986)

$$u_{eq} = -\frac{1}{k^T b}[k^T(Ax + b_v v + b_w w) - k_w \dot{w}] \tag{2.11}$$

with $k^T b \neq 0$ a necessary condition for the sliding mode existence. Another condition for the existence of the sliding mode is

$$u_{min} \leqslant u_{eq} \leqslant u_{max} \tag{2.12}$$

2.2.3 State equation in sliding mode

Introducing (2.11) in the state equation (2.6), we obtain the state equation in the sliding mode

$$\dot{x} = A^*x + b_v^* v + b_w^* w + b_{\dot{w}}^* \dot{w} \tag{2.13}$$

with

$$\begin{aligned} A^* &= \left(I - \frac{1}{k^T b} bk^T\right) A \\ b_v^* &= \left(I - \frac{1}{k^T b} bk^T\right) b_v \\ b_w^* &= \left(I - \frac{1}{k^T b} bk^T\right) b_w \end{aligned} \tag{2.14}$$

$$b_w^* = \frac{k_w}{k^T b} b$$

The matrix A^* is singular because of the linear dependence of the state variables introduced by $s(x) = 0$.

2.2.4 Pole assignment

For continuous or sampling state controller design the state feedback coefficients can be determined by a pole assignment method (Ackermann, 1977; Bühler, 1983). This method can also be applied to sliding mode controller design.

The dynamic behaviour in the sliding mode is determined by the characteristic equation

$$P(s) = \det(sI - A^*) = \\ = s^n + \alpha_{n-1}s^{n-1} + \ldots + \alpha_1 s + \alpha_0 = 0 \qquad (2.15)$$

The poles p_i are linked to the α_i coefficients by

$$P(s) = (s - p_1)(s - p_2) \ldots (s - p_n) \qquad (2.16)$$

with n the degree of the complete system.

Using a linear transformation $x_t = Tx$, the state equation (2.6) can be transformed to the canonical form. The transformation matrix T can be shown to be (Bühler, 1986)

$$T = \begin{bmatrix} e^T \\ e^T A \\ \vdots \\ e^T A^{n-1} \end{bmatrix} \qquad (2.17)$$

where e^T is an auxiliary row vector defined by

$$e^T Q_c = [0 \quad 0 \ldots 0 \quad 1] \qquad (2.18)$$

$Q_c = [b \quad Ab \ldots A^{n-1}b]$ is the controlability matrix for the complete system.

Using an approach similar to linear state feedback (Bühler, 1983 or Ackermann, 1985) the feedback coefficients k_i are linked with the α_i coefficients by

$$k^T = c[\alpha_1 \quad \alpha_2 \ldots \alpha_{n-1} \quad 1]\, T \qquad (2.19)$$

(Bühler, 1986). Because A^* is singular, one pole has to be assigned to zero and a coefficient, the factor c, can be chosen freely. The remaining $n - 1$ poles can also be chosen freely (real or in complex conjugate pairs).

2.2.5 Sliding mode domain

The sliding mode does not exist on the whole switching hyperplane $s(\boldsymbol{x}) = 0$. The sliding mode domain is limited by $u_{eq} = u_{lim}$ with $u_{lim} = u_{max}$ or u_{min}. From (2.10) the state vector $\hat{\boldsymbol{x}}$ on these limits is given by

$$\boldsymbol{k}^T(\boldsymbol{A}\hat{\boldsymbol{x}} + \boldsymbol{b}u_{lim} + \boldsymbol{b}_v v + \boldsymbol{b}_w w) - k_w \dot{w} = 0 \tag{2.20}$$

Of course, the switching strategy

$$s(\hat{\boldsymbol{x}}) = -\boldsymbol{k}^T\hat{\boldsymbol{x}} + k_w w \tag{2.21}$$

has also to be satisfied.

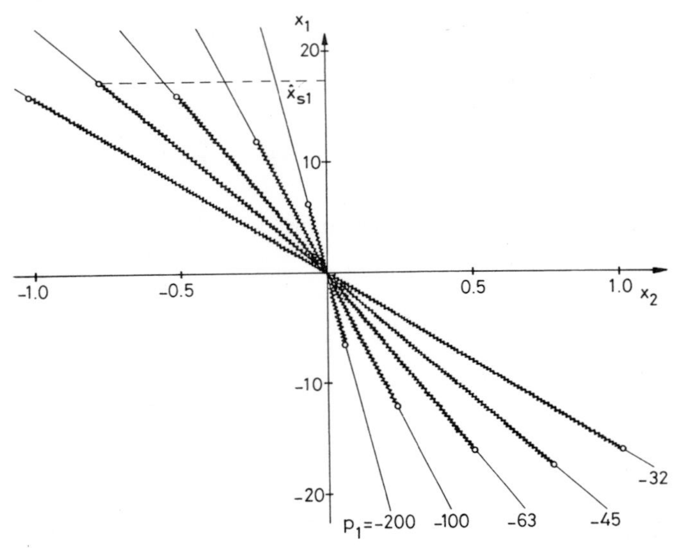

Fig. 2.4 *Influence of the chosen pole on the sliding mode domain for a second order system ($n = 2$)*

Looking at (2.20), note that the row vector $\boldsymbol{k}^T$ given by the pole values, the maximum values of the system variables ($\boldsymbol{x}$, w, $\dot{w}$, v) and the coefficient k_w determine these limits and therefore affect the size of the sliding mode domain in the state space.

For a second-order system ($n = 2$) the sliding mode domain is delimited by two points on the sliding straight line (Fig. 2.4). The inclination of this straight line and also the size of the sliding mode domain in the sliding straight line depend on the value of the freely chosen pole p_1.

In fact, for a second order system the switching strategy (2.1) becomes with $w = 0$

$$k_1 x_1 + k_2 x_2 = 0 \tag{2.22}$$

Setting $k_2 = 1$ arbitrarily we can see that

$$x_2 = -k_1 x_1 \tag{2.23}$$

If the second order system is given by

$$A = \begin{bmatrix} a_{11} & a_{12} \\ a_{21} & a_{22} \end{bmatrix}; \quad b = \begin{bmatrix} b_1 \\ 0 \end{bmatrix} \tag{2.24}$$

the matrix A^* in the sliding mode according to (2.14) will be

$$A^* = \begin{bmatrix} -a_{21}/k_1 & -a_{22}/k_1 \\ a_{21} & a_{22} \end{bmatrix} \tag{2.25}$$

This singular matrix has the poles

$$p_1 = -(a_{21}/k_1 + a_{22}); \quad p_2 = 0 \tag{2.26}$$

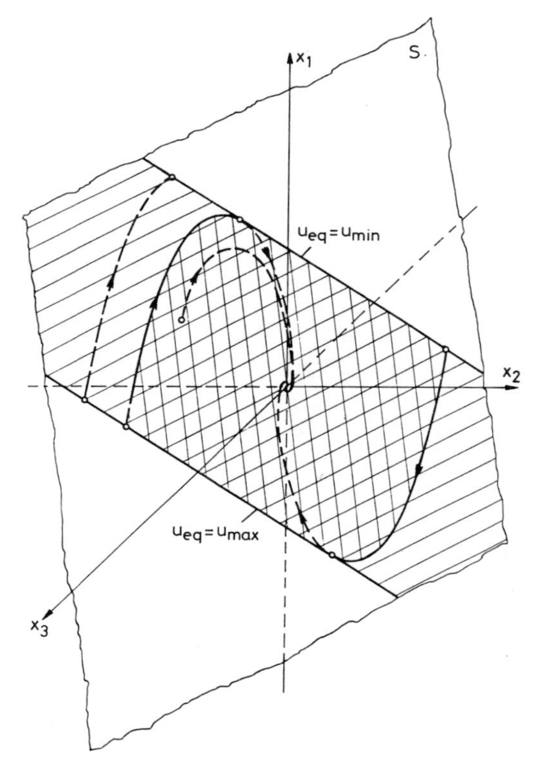

Fig. 2.5 *Sliding mode domain for a third order system (n = 3)*

The limits of the sliding mode domain are derived from (2.20). With $w = 0$ and $v = 0$ we obtain, taking into account (2.23)

$$\hat{x}_1 = -\frac{b_1}{a_{11} - k_1 a_{12} + a_{21}/k_1 - a_{22}} u_{lim} \tag{2.27}$$

If $k_1 << 1$ we can simplify this relation to

$$\hat{x}_1 \cong -\frac{b_1}{a_{11} - 2a_{22} - p_1} u_{lim} \cong \frac{b_1}{p_1} u_{lim} \tag{2.28}$$

Normally, if the (absolute) value of the pole p_1 increases, the size of the sliding domain in the state space decreases. If the sliding mode domain is too small, the pole has to be moved to the right and the transient phenomenon becomes slower. If on the other hand the sliding mode domain is too large, it is possible to move the pole to the left and the transient phenomenon becomes faster. Similar relations exist for higher order systems.

Fig. 2.5 shows the sliding mode domain with single hatching for a third order complete system ($n = 3$). In this case the sliding mode domain is bounded by two parallel straight lines in the switching plane S. It is possible that a trajectory leaves the sliding mode domain. We can also define a restricted sliding mode domain (represented with double hatching in Fig. 2.5). For a trajectory whose initial point is situated inside this restricted domain, the sliding mode is always maintained. For more details the reader should consult Bühler, (1986).

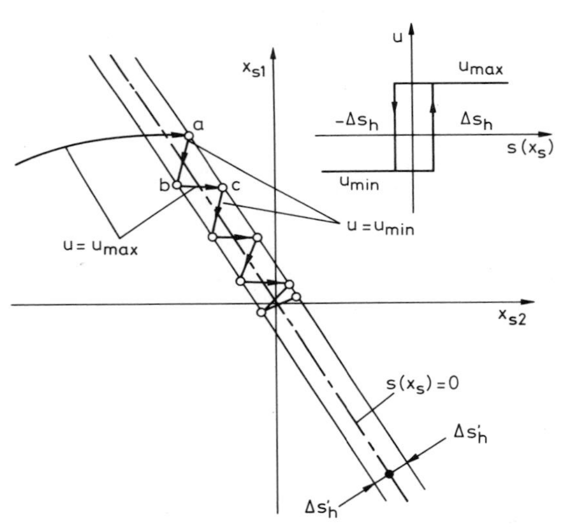

Fig. 2.6 *Sliding mode trajectory for a second order system*

2.2.6 Switching frequency

In the above theoretical study the switching frequency was assumed to be infinite. In a practical realisation the switching is realised by a hysteresis on–off device. Therefore, the switching frequency becomes finite. Fig. 2.6 shows the trajectory of a second order system entering in the sliding mode and Fig. 2.7 shows the relation between equivalent control variable u_{eq} and the pulsed real one u.

If the switching frequency is finite, the transient motion is not situated in the switching hyperplane $s(\boldsymbol{x}) = 0$ but in its vicinity. For a constant value $u_{lim} = u_{max}$ or u_{min} of the control variable the equation (2.10) becomes

$$\dot{s}(\boldsymbol{x}) \quad = \quad -\boldsymbol{k}^T(\boldsymbol{A}\boldsymbol{x} + \boldsymbol{b}u_{lim} + \boldsymbol{b}_v v + \boldsymbol{b}_w w) + k_w \dot{w} \quad \neq \quad 0 \tag{2.29}$$

In the ideal case of infinite switching frequency we have with $u = u_{eq}$

$$0 \quad = \quad -\boldsymbol{k}^T(\boldsymbol{A}\boldsymbol{x} + \boldsymbol{b}u_{eq} + \boldsymbol{b}_v v + \boldsymbol{b}_w w) + k_w \dot{w} \tag{2.30}$$

In the real case we can assume little variation around the switching hyperplane. The trajectory of the state vector $\boldsymbol{x}$ is also practically the same as that of the ideal case. With the same value for $\boldsymbol{x}$ (and also for w and v) in the equations (2.29) and (2.30) the difference between both equations give

$$\dot{s}(\boldsymbol{x}) \quad = \quad -\boldsymbol{k}^T\boldsymbol{b}(u_{lim} - u_{eq}) \tag{2.31}$$

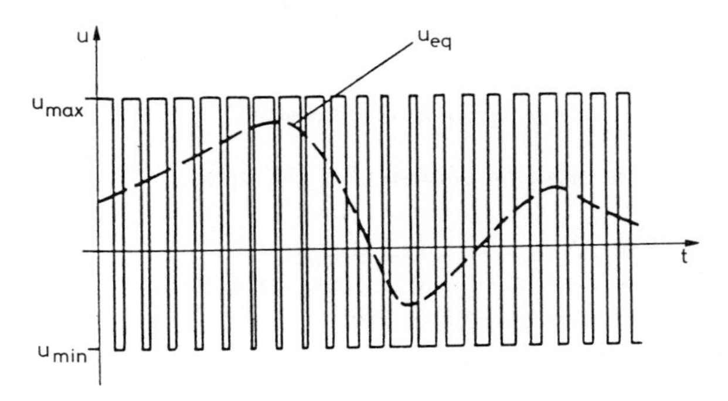

Fig. 2.7 *Equivalent and real control variables*

Fig. 2.8 shows the temporal variation of $s(\boldsymbol{x})$. As a consequence of the hysteresis, $s(\boldsymbol{x})$ is varying between $\pm\Delta s_h$, the threshold of the symmetrical on–off device. During the on-state u_{lim} is equal to u_{max} and for the switching on-time t_e we obtain

$$t_e \quad = \quad \frac{2\Delta s_h}{-\dot{s}(\boldsymbol{x})} \quad = \quad \frac{2\Delta s_h}{\boldsymbol{k}^T\boldsymbol{b}(u_{max} - u_{eq})} \tag{2.32}$$

u_{eq} is considered as constant in this short interval. During the off-state u_{lim} will be equal to u_{min} and the switching off-time t_d is given by

$$t_d \quad = \quad \frac{2\Delta s_h}{\dot{s}(\boldsymbol{x})} \quad = \quad \frac{2\Delta s_h}{\boldsymbol{k}^T\boldsymbol{b}(u_{eq} - u_{min})} \tag{2.33}$$

Therefore, the switching frequency f_c will be

$$f_c \quad = \quad \frac{1}{t_e + t_d} \tag{2.34}$$

or substituting t_e and t_d by (2.32) and (2.33) we obtain

$$f_c \quad = \quad \frac{\boldsymbol{k}^T\boldsymbol{b}}{2\Delta s_h} \frac{(u_{max} - u_{eq})(u_{eq} - u_{min})}{u_{max} - u_{min}} \tag{2.35}$$

The switching frequency can be chosen in a wide range, varying either the threshold value Δs_h or the value of the state feedback vector $\boldsymbol{k}^T$ through the choice of the factor c in the equation (2.19), without variation of the poles, which depend only on the relative values of the state feedback coefficients.

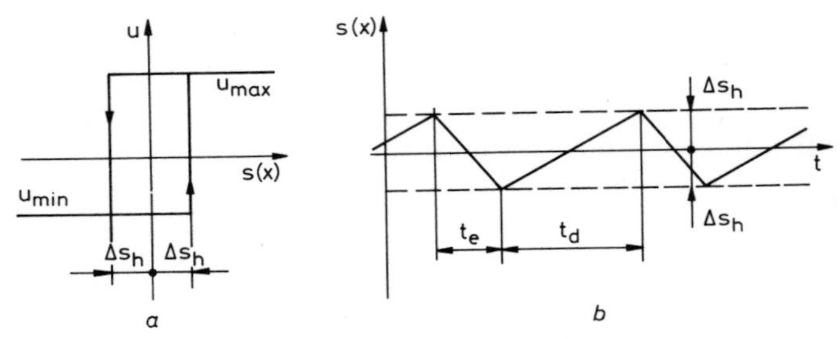

Fig. 2.8 *Switching law with a hysteresis on–off device*

2.3 Application to the position control of a DC drive

2.3.1 State equations of the controlled system

To illustrate the above theory, a practical example, the position control of a DC drive, is presented (Feller and Benz, 1987).

The DC motor behaviour can be described by a state equation in relative values (p.u.)

$$\dot{\boldsymbol{x}}_s = \boldsymbol{A}_s \boldsymbol{x}_s + \boldsymbol{b}_s u + \boldsymbol{b}_{sv} m_r \tag{2.36}$$

and an output equation

$$\theta = \boldsymbol{c}_s^T \boldsymbol{x}_s \tag{2.37}$$

The state vector $\boldsymbol{x}_x$ is given by $\boldsymbol{x}_s = [i \quad n \quad \theta]^T$ where i is the the armature current, n the rotating speed and θ the angular position of the shaft; m_r is the load torque (disturbance variable). The matrix $\boldsymbol{A}_s$, and the vectors $\boldsymbol{b}_s$, $\boldsymbol{b}_{sv}$ and $\boldsymbol{c}_s^T$ are defined by (Bühler, 1986)

$$\boldsymbol{A}_s = \begin{bmatrix} -\dfrac{1}{T_a} & -\dfrac{1}{r_a T_a} & 0 \\ \dfrac{1}{T_m} & 0 & 0 \\ 0 & \dfrac{1}{T_\theta} & 0 \end{bmatrix}; \quad \boldsymbol{b}_s = \begin{bmatrix} \dfrac{1}{r_a T_a} \\ 0 \\ 0 \end{bmatrix} \tag{2.38}$$

$$\boldsymbol{b}_{sv} = \begin{bmatrix} 0 \\ -\dfrac{1}{T_m} \\ 0 \end{bmatrix}; \quad \boldsymbol{c}_s^T = [0 \quad 0 \quad 1]$$

where r_a and T_a are, respectively, the armature resistance and time constant; T_m is the mechanical time constant and T_θ is a time constant introduced to define θ in relative units (p.u.). The motor is supposed to be excited at nominal flux.

2.3.2 Control structure

The sliding mode control structure is illustrated in Fig. 2.9, which shows the complete position control structure with integrator regulator to avoid static deviation in steady state and with two limiters L_i and L_n limiting, respectively, the armature current and the rotating speed. The limit values of these limiters have to be calculated according to the value of the feedback coefficients k_i and k_n. Furthermore, the state x_R of the integrator has to be corrected when one or both limiters are active to avoid overshooting (in other words the windup phenomenon).

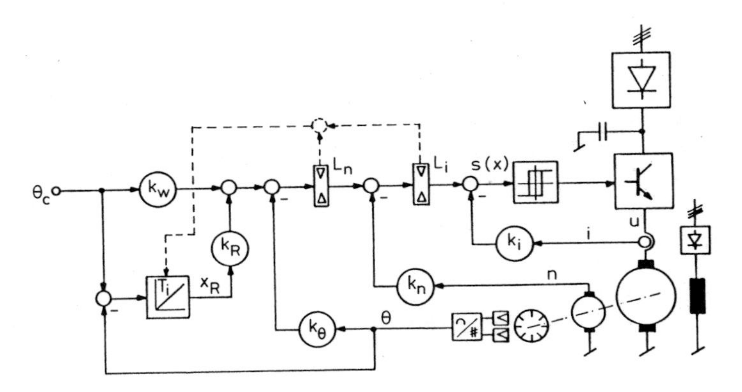

Fig. 2.9 *Sliding mode control structure of the angular position*

The state vector $\boldsymbol{x}$ describing the complete system becomes

$$\boldsymbol{x} = [\boldsymbol{x}_s^T \quad x_R]^T = [i \quad n \quad \theta \quad x_R]^T \tag{2.39}$$

The complete open-loop system is now described by the state equation

$$\dot{\boldsymbol{x}} = \boldsymbol{A}\boldsymbol{x} + \boldsymbol{b}u + \boldsymbol{b}_v m_r + \boldsymbol{b}_w \theta_c \tag{2.40}$$

where $\boldsymbol{A}$, $\boldsymbol{b}$, $\boldsymbol{b}_v$ and $\boldsymbol{b}_w$ are defined by (2.8), introducing in this case for $\boldsymbol{A}_s$, $\boldsymbol{b}_s$, $\boldsymbol{b}_{sv}$ and $\boldsymbol{c}_s^T$ the expressions given by eqn. (2.38).

2.3.3 Pole assignment

After computation of the transformation matrix $\boldsymbol{T}$ by (2.17) and (2.18), the state feedback vector $\boldsymbol{k}^T = [k_i \quad k_n \quad k_\theta \quad -k_R]$ is given by (2.19). For this fourth-order system, we obtain

$$\boldsymbol{k}^{\mathrm{T}} = c[\alpha_1 \quad \alpha_2 \quad \alpha_3 \quad 1]\,\boldsymbol{T} \tag{2.41}$$

and explicitly

$$\begin{aligned} k_i &= cr_a T_a \\ k_n &= cr_a T_a T_m \alpha_3 \\ k_\theta &= cr_a T_a T_m T_\theta \alpha_2 \\ k_R &= cr_a T_a T_m T_\theta T_i \alpha_1 \end{aligned} \tag{2.42}$$

The characteristic equation

$$(s - p_1)(s - p_2)(s - p_3)(s - p_4) = s^4 + \alpha_3 s^3 + \alpha_2 s^2 + \alpha_1 s + \alpha_0 = 0 \tag{2.43}$$

depends on the assigned poles.

The DC motor parameters are $r_a = 0{\cdot}02$, $T_a = 0{\cdot}05s$, $T_m = 0{\cdot}5s$, $T_\theta = 0{\cdot}2s$, $T_i = 0{\cdot}01s$ and the power device is characterised by $u_{lim} = \pm 1{\cdot}2$.

The desired working range is defined by $-2 \leqslant i \leqslant 2$, $-1 \leqslant n \leqslant 1$ and $-1 \leqslant m_r \leqslant 1$. For the following assigned poles $p_1 = 0$, $p_2 = p_3 = p_4 = -55{\cdot}5$, the sliding mode is guaranteed in the whole desired domain. For these purposes we can apply the relations developed in the Section 2.2.5.

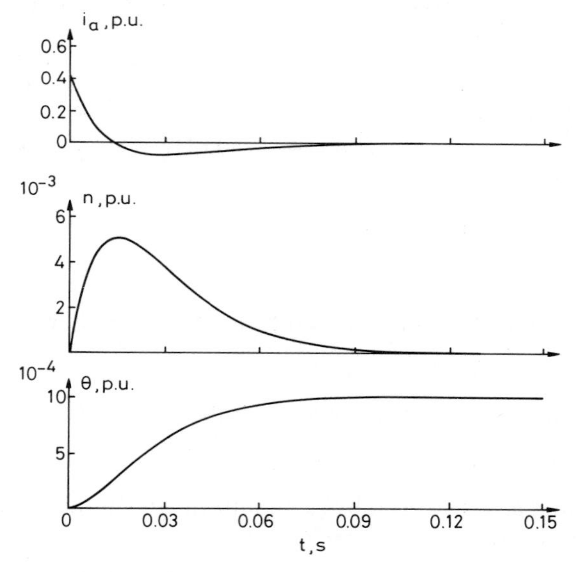

Fig. 2.10 *Responses for a small step variation of θ_c without actuating the limitations and load torque $m_r = 0$*

The factor $c = 100$ in (2.41) is arbitrarily chosen to obtain a realisable value Δs_h for the threshold of the hysteresis on–off device. This threshold is computed according to (2.35) to impose the maximum frequency $f_{cmax} = 4\,\text{kHz}$.

From (2.42) the state feedback coefficients can now be calculated to give

$$k_i = 0{\cdot}1, \quad k_n = 8{\cdot}33, \quad k_\theta = 92{\cdot}6, \quad k_R = 17{\cdot}1.$$

The feedforward coefficient k_w of the reference variable θ_c is experimentally chosen to be $k_w = 45$ with the help of a numerical simulation to obtain the fastest step response without overshooting. Responses of the current i, the speed n and the position θ for a step variation of the reference variable θ_c are illustrated in Figs. 2.10 and 2.11, respectively without and with limits of the internal variables i and n, from numerical simulation.

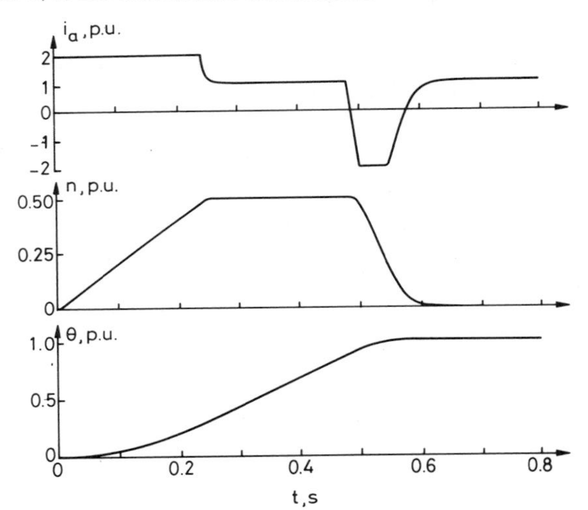

Fig. 2.11 *Responses for a large step variation of θ_c with actuating the limitations and load torque $m_r = 1$*

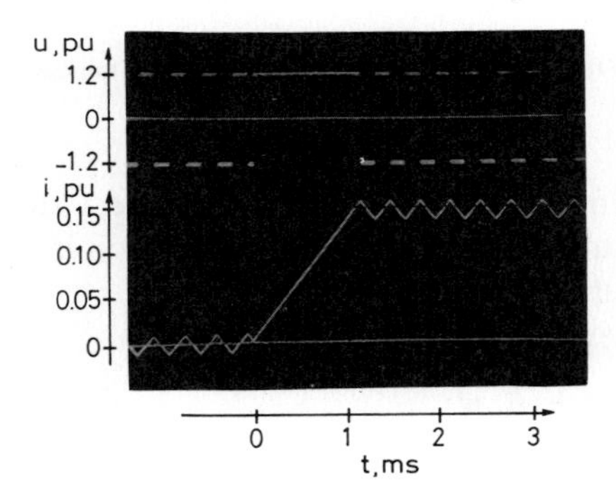

Fig. 2.12 *Transient phenomenon illustrating a transient motion leaving the sliding mode domain*

2.3.4 Practical results

Practical results measured on a fully real-time analogue simulator confirm simulation results illustrated in Figs. 2.10 and 2.11. Fig. 2.12 shows the control variable u and the current i during transient motion due to a step change of the reference variable θ_c. The system leaves the sliding mode domain and has $u = u_{max}$ for 1·1 ms before reaching the switching hyperplane where it begins to slide again.

2.4 Sliding mode control of multivariable systems with state variable feedback

2.4.1 Basic structure and switching strategy

The above sliding mode control can also be applied to multivariable systems. Fig. 2.13 shows the basic structure with a switching command device for each

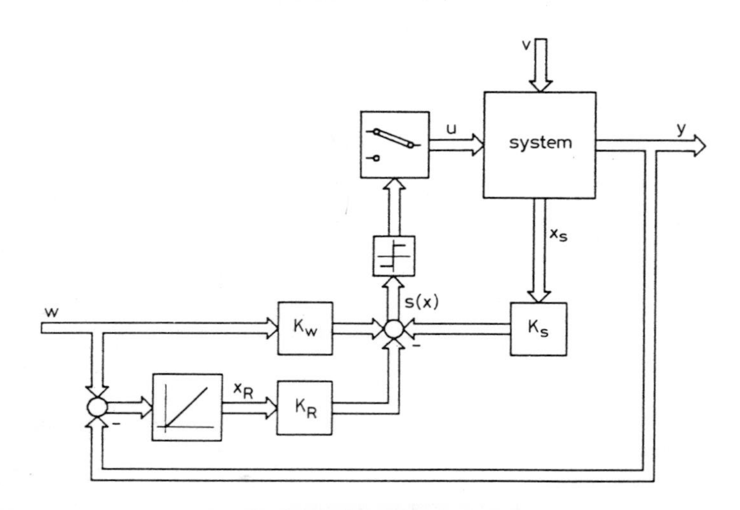

Fig. 2.13 *Basic structure for a multivariable system with state variable feedback and integrator regulator*

component of the command vector $\boldsymbol{u}$. Even in this case, a state variable feedback with a superposed multivariable integrator regulator is applied to form the switching strategy (Bühler, 1986).

The switching strategy (switching law) is

$$\boldsymbol{s}(\boldsymbol{x}) = -\boldsymbol{K}\boldsymbol{x} + \boldsymbol{K}_w\boldsymbol{w} \tag{2.44}$$

where

$$\boldsymbol{K} = [\boldsymbol{K}_s \; -\boldsymbol{K}_R] \tag{2.45}$$

is the state variable feedback matrix and

$$\boldsymbol{x} = \begin{bmatrix} \boldsymbol{x}_s \\ \boldsymbol{x}_R \end{bmatrix} \tag{2.46}$$

is the state vector of the complete system as an ($n = n_s + m$)-vector, with n_s the order of the controlled system S and m the number of control variables $\boldsymbol{u}$ equal to the number of controlled variables $\boldsymbol{y}$.

The controlled system is described by the state equations

$$\dot{\boldsymbol{x}}_s = \boldsymbol{A}_s \boldsymbol{x}_s + \boldsymbol{B}_s \boldsymbol{u} + \boldsymbol{B}_{sv} \boldsymbol{v} \tag{2.47}$$

$$\boldsymbol{y} = \boldsymbol{C}_s \boldsymbol{x}_s \tag{2.48}$$

and the multivariable integrator regulator by

$$\dot{\boldsymbol{x}}_R = \boldsymbol{B}_R(\boldsymbol{w} - \boldsymbol{y}) \tag{2.49}$$

$\boldsymbol{B}_R$ is a $m \times m$ diagonal matrix, where the elements $b_{Rkk} = 1/T_{i,k}$ are given by the inverse of the integrator time constants.

For the complete system, the state equation is given by

$$\dot{\boldsymbol{x}} = \boldsymbol{A}\boldsymbol{x} + \boldsymbol{B}\boldsymbol{u} + \boldsymbol{B}_v \boldsymbol{v} + \boldsymbol{B}_w \boldsymbol{w} \tag{2.50}$$

$$\boldsymbol{y} = \boldsymbol{C}\boldsymbol{x} \tag{2.51}$$

with

$$\begin{aligned} \boldsymbol{A} &= \begin{bmatrix} \boldsymbol{A}_s & \boldsymbol{0} \\ -\boldsymbol{B}_R \boldsymbol{C}_s & \boldsymbol{0} \end{bmatrix}; \quad \boldsymbol{B} = \begin{bmatrix} \boldsymbol{B}_s \\ \boldsymbol{0} \end{bmatrix}; \\ \boldsymbol{B}_v &= \begin{bmatrix} \boldsymbol{B}_{sv} \\ \boldsymbol{0} \end{bmatrix}; \quad \boldsymbol{B}_w = \begin{bmatrix} \boldsymbol{0} \\ \boldsymbol{B}_R \end{bmatrix} \\ \boldsymbol{C} &= [\boldsymbol{C}_s \quad \boldsymbol{0}] \end{aligned} \tag{2.52}$$

Each component u_k of the control vector $\boldsymbol{u}$ switches between u_{kmax} and u_{kmin} according to the switching strategy

$$u_k = \begin{cases} u_{kmax} & \text{for} \quad s_k(\boldsymbol{x}) > 0 \\ u_{kmin} & \text{for} \quad s_k(\boldsymbol{x}) < 0 \end{cases} \qquad k = 1 \ldots m \tag{2.53}$$

2.4.2 Sliding mode conditions

In the case of a multivariable system, we assume that in the sliding mode each component u_k of the command vector switches between u_{kmax} and u_{kmin} with an infinite switching frequency. The state point describes a trajectory on the switching hyperplane given by $\boldsymbol{s}(\boldsymbol{x}) = \boldsymbol{0}$, and also $\dot{\boldsymbol{s}}(\boldsymbol{x}) = \boldsymbol{0}$.

Considering (2.44) and (2.50), we obtain

$$\dot{\boldsymbol{s}}(\boldsymbol{x}) = -\boldsymbol{K}(\boldsymbol{A}\boldsymbol{x} + \boldsymbol{B}\boldsymbol{u} + \boldsymbol{B}_v \boldsymbol{v} + \boldsymbol{B}_w \boldsymbol{w}) + \boldsymbol{K}_w \dot{\boldsymbol{w}} = 0 \tag{2.54}$$

The control vector $\boldsymbol{u}$ can be replaced by its mean value $\boldsymbol{u}_{eq}$, the equivalent control vector, which is a continuous variable vector yielding

$$u_{eq} = -(KB)^{-1}K(Ax + B_v v + B_w w) + (KB)^{-1}K_w \dot{w} \tag{2.55}$$

A first condition for the existence of the sliding mode is the nonsingularity of the matrix KB. Another condition is

$$u_{kmin} \leqslant u_{keq} \leqslant u_{kmax} \tag{2.56}$$

for all the m control variables to be operating in the sliding mode.

However, in multivariable systems it is possible that a partial sliding mode exists. In this case only some of the m control variables u_k are switched. The other ones are fixed at u_{kmax} or u_{kmin}. Describing the different possibilities in a general way is very difficult. It is judicious to examine these different possibilities for each particular application.

2.4.3 State equation in sliding mode

As in the case of a scalar system we can substitute (2.55) in (2.50) to obtain the equivalent state equation in the sliding mode. However, for a multivariable system it is judicious to apply another procedure.

The state vector x of the complete system is decomposed into a partial $(n - m)$-vector x_a and a partial m-vector x_b, corresponding to the number of control variables. So the state equation (2.50) can be decomposed into

$$\begin{bmatrix} \dot{x}_a \\ \dot{x}_b \end{bmatrix} = \begin{bmatrix} A_{aa} & A_{bb} \\ A_{ba} & A_{bb} \end{bmatrix} \begin{bmatrix} x_a \\ x_b \end{bmatrix} + \begin{bmatrix} 0 \\ B_b \end{bmatrix} u + \begin{bmatrix} B_{va} \\ B_{vb} \end{bmatrix} v + \begin{bmatrix} B_{wa} \\ B_{wb} \end{bmatrix} w \tag{2.57}$$

The main point of this decomposition is that the control vector u acts directly only upon the partial vector x_b through the partial matrix B_b. To obtain this form, it is necessary to apply a linear transformation to the state equation (2.50).

In the same way we can decompose the switching law (2.44) into

$$-[K_a \quad K_b] \begin{bmatrix} x_a \\ x_b \end{bmatrix} + K_w w = 0 \tag{2.58}$$

K_b is an $m \times m$ matrix and must be nonsingular, but it can be freely chosen. Generally, it is set equal to the identity matrix. It is often judicious to choose K_b so that $K_b B_b$ assumes a diagonal form.

From (2.58) we obtain

$$x_b = -K_b^{-1} K_a x_a + K_b^{-1} K_w w \tag{2.59}$$

The state vector x_b of the second partial system is determined in the sliding mode by the state vector x_a of the first partial system and the reference vector w. For the first partial system, we can interpret the partial state vector x_b as the control vector. In fact, the decomposition of (2.57) gives

$$\dot{\boldsymbol{x}}_a = \boldsymbol{A}_{aa}\boldsymbol{x}_a + \boldsymbol{A}_{ab}\boldsymbol{x}_b + \boldsymbol{B}_{va}\boldsymbol{v} + \boldsymbol{B}_{wa}\boldsymbol{w} \qquad (2.60)$$

The equation (2.59) represents a continuous state variable feedback with the matrix $\boldsymbol{K}_b^{-1}\boldsymbol{K}_a$ on the first partial system. We can interpret the equivalent system S_a as a comprising only the first partial system together with a state variable feedback (Fig. 2.14).

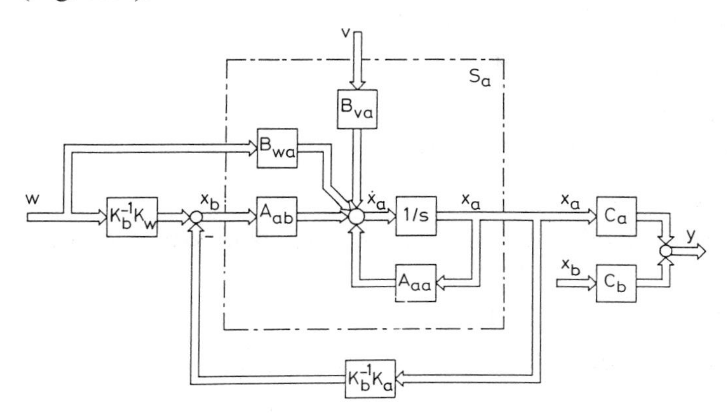

Fig. 2.14 *Equivalent system for a multivariable control working in the sliding mode*

Substituting in (2.60) for $\boldsymbol{x}_b$ from (2.59) we obtain the state equation in the sliding mode

$$\dot{\boldsymbol{x}}_a = (\boldsymbol{A}_{aa} - \boldsymbol{A}_{ab}\boldsymbol{K}_b^{-1}\boldsymbol{K}_a)\boldsymbol{x}_a + \boldsymbol{B}_{va}\boldsymbol{v} + (\boldsymbol{B}_{wa} + \boldsymbol{A}_{ab}\boldsymbol{K}_b^{-1}\boldsymbol{K}_w)\boldsymbol{w} \qquad (2.61)$$

which determines the trajectory of the partial vector $\boldsymbol{x}_a$. The trajectory of the partial vector $\boldsymbol{x}_b$ is also given by the linear equation (2.59).

2.4.4 Pole assignment

The equivalent system in Fig. 2.14 shows that the partial system S_a is fed by a continuous action through the matrix

$$\boldsymbol{M} = \boldsymbol{K}_b^{-1}\boldsymbol{K}_a \qquad (2.62)$$

In this $m \times (n - m)$ matrix there are $m(n - m)$ elements. For this equivalent system we can impose $n_a = n - m$ poles. This number is not sufficient to determine all the elements of the matrix $\boldsymbol{M}$. This problem is well known in the design of multivariable continuous feedback systems. There are several design methods. In the following we present concisely a method which normally gives very good results. For a detailed development see Bühler (1983)

For the equivalent system S_a the matrix $\boldsymbol{A}_{aa}$ is the system matrix and $\boldsymbol{A}_{ab}$ the input matrix. First, we have to form the controllability matrix

$$\boldsymbol{Q}_c = [\boldsymbol{A}_{ab} \quad \boldsymbol{A}_{aa}\boldsymbol{A}_{ab} \ldots \boldsymbol{A}_{aa}^{k-1}\boldsymbol{A}_{ab}] \qquad (2.63)$$

and determine the controllability index q according to

$$q = \min k : \text{rank } \boldsymbol{Q}_c = n_a = n - m \tag{2.64}$$

with the condition $q \leqslant n_a$. That means that the equivalent system S_a must be controllable.

The equivalent system will be decomposed into m subsystems with older n_1. This order must be chosen near n_a/m and must be an integer. One of these values must be equal to q. Furthermore the condition

$$n_1 + n_2 + \ldots + n_m = n_a \tag{2.65}$$

must be respected.

For each subsystem we form the relation

$$\boldsymbol{e}_i^T \boldsymbol{Q}_{ci} = d_i^T \tag{2.66}$$

In the commandability matrix $\boldsymbol{Q}_{ci}$, we have to introduce $k = n_i$. The row (mn_i)—vector $\boldsymbol{d}_i^T$ is given by

$$\boldsymbol{d}_i^T = [\boldsymbol{0}^T \quad \boldsymbol{0}^T \ldots \boldsymbol{h}_i^T] \tag{2.67}$$

The partial row vectors $\boldsymbol{0}^T$ and $\boldsymbol{h}_i^T$ are row m-vectors, where

$$\boldsymbol{h}_i^T = [0 \ldots 0 \quad h_{i,i} \quad h_{i,i+1} \ldots h_{i,m}] \tag{2.68}$$

with $h_{i,i} = 1$. The other elements $h_{i,i+1} \ldots h_{i,m}$ will be determined later.

With the relation (2.66) the row n_a-vector $\boldsymbol{e}_i^T$ can be calculated. One has to distinguish three cases:

$$n_i = n_a/m; \qquad n_i > n_a/m \quad \text{and} \quad n_i < n_a/m.$$

For $n_i = n_a/m$ the matrix $\boldsymbol{Q}_{ci}$ is a nonsingular $n_a \times n_a$ matrix. We obtain immediately

$$e_i^T = \boldsymbol{s}_i^T \boldsymbol{Q}_{ci}^{-1} \tag{2.69}$$

In this case we can choose $h_{i,i+1} \ldots h_{i,m} = 0$.

For $n_i > n_a/m$, the matrix $\boldsymbol{Q}_{ci}$ is an $n_a \times mn_i$ matrix and is decomposed as

$$\boldsymbol{Q}_{ci} = [\boldsymbol{Q}_{cn} \quad \boldsymbol{Q}_{ce}] \tag{2.70}$$

where $\boldsymbol{Q}_{cn}$ must be a nonsingular $n_a \times n_a$ matrix. Likewise, we decompose the row vector $\boldsymbol{d}_i^T$ as

$$\boldsymbol{d}_i^T = [\boldsymbol{d}_n^T \quad \boldsymbol{d}_e^T] \tag{2.71}$$

where $\boldsymbol{d}_n^T$ is a row n_a-vector. This partial row vector must contain at least the element $h_{i,i} = 1$ of $\boldsymbol{h}_i^T$. A certain number of elements of $\boldsymbol{h}_i^T$ which appear in $\boldsymbol{d}_n^T$ can be put equal to 0. The other elements are also included in $\boldsymbol{d}_e^T$. These elements cannot be chosen freely. If necessary the columns of $\boldsymbol{Q}_{ci}$ and $\boldsymbol{d}_n^T$ must be permuted in order to respect all these conditions. So

$$\boldsymbol{e}_i^T = \boldsymbol{d}_n^T \boldsymbol{Q}_{cn}^{-1} \tag{2.72}$$

and

$$d_e^T = d_n^T Q_{cn}^{-1} Q_{ce} \tag{2.73}$$

The row vector d_e^T also determines the elements $h_{i,k}$, which we could not choose preliminarily.

Finally, for $n_i < n_a/m$, the matrix Q_{ci} is an $n_a \times mn_i$ matrix which can be decomposed as

$$Q_{ci} = \begin{bmatrix} Q_{cn} \\ Q_{ce} \end{bmatrix} \tag{2.74}$$

Q_{cn} must be a nonsingular $mn_i \times mn_i$ matrix. We decompose the row vector e_i^T as

$$e_i^T = [e_n^T \quad e_e^T] \tag{2.75}$$

and obtain

$$e_n^T = (d_i^T - e_e^T Q_{ce}) Q_{cn}^{-1} \tag{2.76}$$

In d_i^T the elements $h_{i,i+1} \ldots h_{i,m}$ are set equal to 0. Normally, we can choose $e_e^T = 0^T$.

After the determination of the row vectors e_i^T one can assign eigenvalues (poles) to each subsystem (including the state feedback). This yields the characteristic polynomial

$$P_i(s) = s^{n_i} + \alpha_{i,n_i-1} s^{n_i-1} + \ldots + \alpha_{i,1} s + \alpha_{i,0} \tag{2.77}$$

All the eigenvalues are imposed on the subsystems from the eigenvalues of the equivalent system S_a provided with the feedback by the matrix $M = K_b^{-1} K_a$ (see Fig. 2.14). The other m poles of the complete system are zero valued because of the sliding mode.

With the coefficients $\alpha_{i,n_i-1}, \ldots, \alpha_{i,0}$ and the row vectors e_i^T for each subsystem the row vector

$$g_i^T = e_i^T (\alpha_{i,0} 1 + \alpha_{i,1} A_{aa} + \ldots + \alpha_{i,n_i-1} A_{aa}^{n_i-1} + A_{aa}^{n_j}) \tag{2.78}$$

can be formed.

Finally, the matrix $M = K_b^{-1} K_a$ of the state feedback comes from

$$M = H_u^{-1} G \tag{2.79}$$

with

$$H_u = \begin{bmatrix} h_1^T \\ h_2^T \\ \vdots \\ h_m^T \end{bmatrix}; \qquad G = \begin{bmatrix} g_1^T \\ g_2^T \\ \vdots \\ g_m^T \end{bmatrix} \tag{2.80}$$

The matrix $\boldsymbol{H}_u$ is composed of the row vectors $\boldsymbol{h}_i^T$. Since $h_{i,i} = 1$, the matrix is always nonsingular and its inversion is possible.

With the feedback matrix $\boldsymbol{M}$ for the equivalent system from (2.62) the feedback matrix $\boldsymbol{K}$ for the original system is

$$\boldsymbol{K} = [\boldsymbol{K}_a \quad \boldsymbol{K}_b] = \boldsymbol{K}_b[\boldsymbol{M} \quad \boldsymbol{1}] \tag{2.81}$$

The nonsingular matrix $\boldsymbol{K}_b$ can be chosen freely. It plays the same role as the coefficient c in (2.19) for the case of a scalar system.

2.4.5 Sliding mode domain

The sliding mode domain is given by the condition $\dot{\boldsymbol{s}}(\boldsymbol{x}) = \boldsymbol{0}$ and implies that each component u_{keq} of the equivalent control vector $\boldsymbol{u}_{eq}$, in accordance with (2.55), is situated within the limits u_{kmin} and u_{kmax} [see (2.56)]. Similarly to the scalar system there is a domain on the intersection of the hyperplanes $\boldsymbol{s}(\boldsymbol{x}) = \boldsymbol{0}$ in which the sliding mode exists.

For the determination of the sliding mode domain in the state space we must introduce in (2.55) for $\boldsymbol{u}_{eq}$ the vector $\boldsymbol{u}_{lim}$ whose elements are equal to u_{kmax} or u_{kmin}. From (2.55) the state vector $\hat{\boldsymbol{x}}$ on these limits is given by

$$\boldsymbol{K}(\boldsymbol{A}\hat{\boldsymbol{x}} + \boldsymbol{B}\boldsymbol{u}_{lim} + \boldsymbol{B}_v\boldsymbol{v} + \boldsymbol{B}_w\boldsymbol{w}) - \boldsymbol{K}_w\dot{\boldsymbol{w}} = \boldsymbol{0} \tag{2.82}$$

Of course the switching strategy

$$s(\hat{\boldsymbol{x}}) = -\boldsymbol{K}\hat{\boldsymbol{x}} + \boldsymbol{K}_w\boldsymbol{w} = \boldsymbol{0} \tag{2.83}$$

has to be satisfied.

2.4.6 Robustness

It is well known that systems in the sliding mode are robust in relation to parameter variations. The conditions for absolute invariance can be easily deduced with the help of the equivalent system represented by Fig. 2.14. In fact only the parameters appearing in the partial matrix $\boldsymbol{A}_{aa}$ and $\boldsymbol{A}_{ab}$ have an influence on the dynamic behaviour in (total) sliding mode operation. On the contrary all the parameters appearing in the partial matrixes $\boldsymbol{A}_{ba}$, $\boldsymbol{A}_{bb}$ and $\boldsymbol{B}_b$ [see decomposition in (2.57)] have no influence on the dynamic behaviour in the sliding mode; i.e. a variation of these parameters has no influence on the eigenvalues (poles) imposed for the sliding mode operation. It should be noted that variations of all the parameters in $\boldsymbol{A}$ and $\boldsymbol{B}$ (i.e. in $\boldsymbol{A}_{aa}$ and $\boldsymbol{A}_{ab}$ and also in $\boldsymbol{A}_{ba}$, $\boldsymbol{A}_{bb}$ and $\boldsymbol{B}_b$) have an influence on the sliding mode domain (Bühler, 1986).

2.5 Example of a multivariable system sliding mode control

2.5.1 Basic relations

To illustrate the above theory of the sliding mode control of multivariable systems and as evidence of some of its particulars, a theoretical example will be presented.

For this purpose a third order system ($n = 3$) with two inputs ($m = 2$) and two outputs will be considered. The matrices are

$$\boldsymbol{A} = \begin{bmatrix} -3 & | & 2 & 1 \\ \hline 0 & | & -1 & 1 \\ 0 & | & 1 & -2 \end{bmatrix}; \qquad \boldsymbol{B} = \begin{bmatrix} 0 & 0 \\ \hline 1 & 0 \\ 0 & 2 \end{bmatrix};$$

$$\boldsymbol{C} = \begin{bmatrix} 1 & | & 0 & 0 \\ 0 & | & 1 & 0 \end{bmatrix} \tag{2.84}$$

In this system a sliding mode control is applied without an integrator regulator. Thus the above matrices are also valid for the complete system. The matrix $\boldsymbol{B}$ is already in the particular form allowing decomposition into two subsystems according to the equation (2.57).

2.5.2 State feedback design (pole assignment)

With $n = 3$ and $m = 2$ the partial state vector $\boldsymbol{x}_a$ is a scalar ($\boldsymbol{x}_a = x_1$) and $\boldsymbol{x}_b = [x_2 \quad x_3]^T$. The controllability matrix (2.63) becomes

$$\boldsymbol{Q}_c = \boldsymbol{A}_{ab} = [2 \quad 1] \tag{2.85}$$

The rank is 1, equal to the controllability index q. The equivalent system S_a is also controllable. Corresponding to the Section 2.4.4, this equivalent system must be decomposed into $m = 2$ subsystems.

The first subsystems has the order $n_1 = q = 1 \quad n_a/m = 1/2$. In relation to (2.70)–(2.73), we have

$$\boldsymbol{e}_1^T = \boldsymbol{e}_{1,1} = 0{\cdot}5; \qquad \boldsymbol{h}_1^T = [1 \quad 0{\cdot}5] \tag{2.86}$$

Assigning a pole $p_1 = -10$, the coefficient of the characteristic polynomial is $\alpha_{1,0} = 10$ and in accordance with (2.78), we have

$$\boldsymbol{g}_1^T = g_{1,1} = \boldsymbol{e}_1^T(\alpha_{1,0}\boldsymbol{I} + \boldsymbol{A}_{aa}) = 3{\cdot}5 \tag{2.87}$$

The second subsystem is zero order, i.e. $n_2 = 0$. The controllability matrix $\boldsymbol{Q}_{c2}$ is also not defined. In this particular case we must set

$$\boldsymbol{h}_2^T = [0 \quad 1]; \qquad \boldsymbol{g}_2^T = g_{2,1} = 0 \tag{2.88}$$

The matrix $\boldsymbol{M}$ for the state feedback of the equivalent system becomes

$$\boldsymbol{M} = \boldsymbol{H}_u^{-1}\boldsymbol{G} = \begin{bmatrix} 1 & 0{\cdot}5 \\ 0 & 1 \end{bmatrix}^{-1} \begin{bmatrix} 3{\cdot}5 \\ 0 \end{bmatrix} = \begin{bmatrix} 3{\cdot}5 \\ 0 \end{bmatrix} \tag{2.89}$$

Finally, choosing $\boldsymbol{K}_b = \boldsymbol{1}$ the state feedback matrix for the original system will be

$$\boldsymbol{K} = v\boldsymbol{K}_b[\boldsymbol{M} \quad \boldsymbol{1}] = \left[\begin{array}{c|cc} 3{\cdot}5 & 1 & 0 \\ & & \\ 0 & 0 & 1 \end{array}\right] \tag{2.90}$$

2.5.3 Feedforward matrix design

Because the sliding mode control is not provided with an integrator regulator, the feedforward matrix $\boldsymbol{K}_w$ must be designed to obtain the output vector $\boldsymbol{y}$ equal to the reference vector $\boldsymbol{w}$ in steady state condition without the influence of the perturbation vector v. In a general manner we have the relation

$$\boldsymbol{y} = \boldsymbol{C}\boldsymbol{x} = \boldsymbol{C}_a\boldsymbol{x}_a + \boldsymbol{C}_b\boldsymbol{x}_b = (\boldsymbol{C}_a - \boldsymbol{C}_b\boldsymbol{K}_b^{-1}\boldsymbol{K}_a)\boldsymbol{x}_a + \boldsymbol{C}_b\boldsymbol{K}_b^{-1}\boldsymbol{K}_w\boldsymbol{w} \tag{2.91}$$

Here, $\boldsymbol{x}_b$ was substituted by (2.59). The partial vector $\boldsymbol{x}_a$ can be determined by (2.61). With $\dot{\boldsymbol{x}}_a = \boldsymbol{0}$, and $v = 0$, we obtain

$$\boldsymbol{x}_a = -(\boldsymbol{A}_{aa} - \boldsymbol{A}_{ab}\boldsymbol{K}_b^{-1}\boldsymbol{K}_a)^{-1}\boldsymbol{A}_{ab}\boldsymbol{K}_b^{-1}\boldsymbol{K}_w\boldsymbol{w} \tag{2.92}$$

Note that $\boldsymbol{B}_{wa} = \boldsymbol{0}$ without integrator regulation. Finally, substituting $\boldsymbol{x}_a$ in (2.92) by $\boldsymbol{y} = \boldsymbol{w}$, the feedforward matrix becomes

$$\boldsymbol{K}_w = \boldsymbol{K}_b[\boldsymbol{C}_b - (\boldsymbol{C}_b\boldsymbol{K}_b^{-1}\boldsymbol{K}_a)(\boldsymbol{A}_{aa} - \boldsymbol{A}_{ab}\boldsymbol{K}_b^{-1}\boldsymbol{K}_a)^{-1}\boldsymbol{A}_{ab}]^{-1} \tag{2.93}$$

For this example the numerical values are

$$\boldsymbol{K}_w = \begin{bmatrix} 3{\cdot}5 & 1 \\ 3 & -2 \end{bmatrix} \tag{2.94}$$

2.5.4 Switching planes

According to (2.44), the switching law is

$$-\begin{bmatrix} 3{\cdot}5 & 1 & 0 \\ 0 & 0 & 1 \end{bmatrix}\begin{bmatrix} x_1 \\ x_2 \\ x_3 \end{bmatrix} + \begin{bmatrix} 3{\cdot}5 & 1 \\ 3 & -2 \end{bmatrix}\begin{bmatrix} w_1 \\ w_2 \end{bmatrix} = \begin{bmatrix} 0 \\ 0 \end{bmatrix} \tag{2.95}$$

With a decomposition, we obtain two scalar equations

$$x_2 = 3{\cdot}5w_1 + w_2 - 3{\cdot}5x_1 \tag{2.96}$$

$$x_3 = 3w_1 - 2w_2 \tag{2.97}$$

The first relation (2.96) is independent of x_3. It describes a vertical switching plan S_1. The second relation (2.97) is independent of x_1 and x_2. The switching plane S_2 is horizontal. Fig. 2.15 shows these two switching planes for $w_1 = 1$ and $w_2 = 0{\cdot}5$.

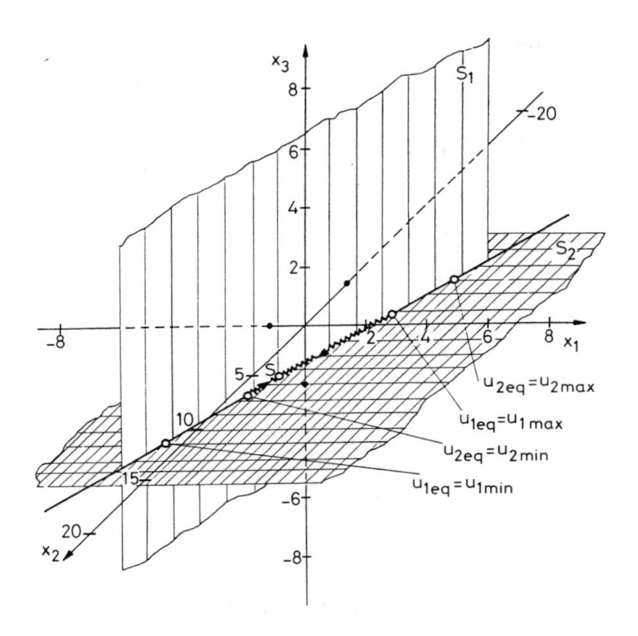

Fig. 2.15 *Representation in the state space of the switching planes and the trajectory in the sliding mode*

2.5.5 Trajectory in the sliding mode

The state equation in the sliding mode for the partial vector $\boldsymbol{x}_a$ becomes from (2.61)

$$\dot{\boldsymbol{x}}_a = \dot{x}_1 = -10x_1 + [10 \quad 0]\begin{bmatrix} w_1 \\ w_2 \end{bmatrix} = -10x_1 + 10w_1 \qquad (2.98)$$

It has the solution for a constant reference variable w_1

$$x_1 = (1 - e^{-10t})w_1 + e^{-10t}x_1(0) \qquad (2.99)$$

Evidently the initial condition $x_1(0)$ must be situated in the sliding mode domain (see Section 2.5.6).

The state variable x_1 evolves exponentially with a time constant equal to 0·1 s, in accordance with the assigned pole $p_1 = -10$.

The evolution for the other state variables x_2 and x_3 is given by (2.59)

$$\boldsymbol{x}_b = \begin{bmatrix} x_2 \\ x_3 \end{bmatrix} = -\begin{bmatrix} 3{\cdot}5 \\ 0 \end{bmatrix} x_1 + \begin{bmatrix} 3{\cdot}5 & 1 \\ 3 & -2 \end{bmatrix}\begin{bmatrix} w_1 \\ w_2 \end{bmatrix} \qquad (2.100)$$

or after a decomposition

$$x_2 = -3{\cdot}5x_1 + 3{\cdot}5w_1 + w_2$$
$$= 3{\cdot}5e^{-10t}w_1 - 3{\cdot}5e^{-10t}x_1(0) + w_2 \quad (2.101)$$

$$x_3 = 3w_1 - 2w_2 \quad (2.102)$$

The state variable x_3 remains constant and the trajectory in complete sliding mode is identical to the intersection of the switching planes S_1 and S_2.

2.5.6 Sliding mode domain

The limits of the sliding mode domain in the state space are given by the matrix equation (2.82). For this example we obtain for a constant reference vector ($\dot{\boldsymbol{w}} = 0$) the two scalar equations

$$-10{\cdot}5\hat{x}_1 + 6\hat{x}_2 + 4{\cdot}5\hat{x}_3 + u_{1lim} = 0 \quad (2.103)$$
$$\hat{x}_2 - 2\hat{x}_3 + 2u_{2lim} = 0$$

(Note that in this case $\boldsymbol{B}_v = 0$ and $\boldsymbol{B}_w = \boldsymbol{0}$.)

The switching strategy (2.83) gives the two scalar equations

$$-3{\cdot}5\hat{x}_1 - \hat{x}_2 + 3{\cdot}5w_1 + w_2 = 0 \quad (2.104)$$
$$-\hat{x}_3 + 3w_1 - 2w_2 = 0$$

It permits us to eliminate $\hat{x}_2$ and $\hat{x}_3$ and to obtain for the limits of $\hat{x}_1$ two relations:

$$-31{\cdot}5\hat{x}_1 + 34{\cdot}5w_1 - 3w_2 + u_{1lim} = 0 \quad (2.105)$$

$$-3{\cdot}5\hat{x}_1 - 2{\cdot}5w_1 + 5w_2 + 2u_{2lim} = 0 \quad (2.106)$$

Assuming that the limits of the control variables are $u_{1lim} = \pm 50$ and $u_{2lim} = \pm 2{\cdot}5$, with $w_1 = -1$ and $w_2 = -0{\cdot}5$, the relation (2.105) gives the limits $\hat{x}_{1max} = 0{\cdot}540$ and $\hat{x}_{1min} = -2{\cdot}635$. On the other hand, the relation (2.106) gives $\hat{x}_{1max} = 1{\cdot}428$ and $\hat{x}_{1min} = -1{\cdot}428$. Of course, the total sliding mode is valid in the limits

$$\hat{x}_{1max} = 0{\cdot}540 \quad \text{and} \quad \hat{x}_{1min} = -1{\cdot}428 \quad (2.107)$$

These limits are also indicated in Fig. 2.15. Starting from these limits, the transient motion converges to the steady state point s.

2.6 References

ACKERMANN, J. (1977): 'Entwurf durch Polvorgabe', *Regelungstechnik*, **25**, pp. 173–179 and 209–215

ACKERMANN, J. (1985): 'Sampled-data control systems' (Springer Verlag, Berlin)

BILALOVIC, F., SABANOVIC, A., MUSIC, O., and IZOSIMOV, D. B. (1983): 'Current inverter in the sliding mode for induction motor', Proc. IFAC Symposium, Lausanne, Control in power electronics and electrical drives (Pergamon Press, Oxford) pp. 139–144

BÜHLER, H. (1986): 'Rélage par mode de glissement' (Presses Polytechniques Romandes, Lausanne)

BÜHLER, H. (1985): 'Regelung mit Gleitzuständen. 59: Tagung der SGA' (Schweizerische Gesellschaft für Automatik, Zürich)

BÜHLER, H. (1983): 'Réglages échantillonnés. Vol. 2: Traitement dans l'espace d'état' (Presses Polytechniques Romandes, Lausanne)

FELLER, P., and BENZ, U. (1987): 'Sliding mode position control of a DC motor'. 10th World Congress in Automatic Control, IFAC, Munich, Preprints, Vol. 3, pp. 325–330

SABANOVIC, A., IZOSIMOV, D. B., BILALOVIC, F., and MUSIC, O. (1983): 'Sliding mode in controlled motor drives', Proc. IFAC Symposium Lausanne, Control in power electronics and electrical drives (Pergamon Press, Oxford) pp. 133–138

UTKIN, V. I. (1977): 'Variable structure systems with sliding modes', *IEEE Trans.,* **AC-22,** pp. 212–222

UTKIN, V. I. (1987): 'Discontinuous control systems: State of the art in theory and applications' 10th World Conference in Automatic Control, IFAC, Munich, Preprints, Vol. 1, pp. 75–94

Chapter 3

Hyperplane design and CAD of variable structure control systems

A. S. I. Zinober

University of Sheffield

C. M. Dorling

Topexpress Ltd.

3.1 Introduction

The control of uncertain dynamical systems in the presence of extraneous disturbances has been the subject of considerable study in recent years. If the statistics of the internal uncertainties (such as parameter variations) and the noise are available to the designer, then the stochastic approach to control design may be used. If, however, these statistics are not available but bounds on the uncertainties and disturbances are known, a deterministic approach can be employed. One approach to deterministic control design for a class of uncertain non-linear systems is by means of the variable structure controller. It has long been known (Drăzenović, 1969) that, during a certain mode of the transient motion, the response of a variable structure control system is unaffected by a class of parameter variations and disturbances occurring within a particular subspace of the state space. Ryan (1983) and Ryan and Corless (1984) have shown that variable structure control may be used to establish 'almost certain' convergence to a neighbourhood of the origin for a class of uncertain systems. The design approach is based on a nominal constant linearisation of the system, with the time-varying, non-linear and uncertain elements of the system and the extraneous disturbances grouped into an 'unknown' function.

3.2 Variable structure control systems

The essential feature of a variable structure control (VSC) system is that the generally non-linear feedback control has a discontinuity on one or more manifolds in the state space. Thus the structure of the feedback system is altered or switched as its state crosses each discontinuity surface.

The central feature of a VSC system is sliding motion. This occurs when the system state repeatedly crosses and immediately re-crosses a switching manifold, because all motion in the neighbourhood of the manifold is directed inwards (i.e. towards the manifold). In the sliding mode the motion of the system is effectively constrained to lie within a certain subspace of the full state space, and the system is thus formally equivalent to a system of lower order, termed the equivalent system. The transient motion of the system therefore consists of two independent stages: a (preferably rapid) motion bringing the state of the system to the maniford in which sliding occurs; and a slower sliding motion, in which the state slides towards the state-space origin while remaining in the sliding subspace.

Here we shall be mainly concerned with the design of the sliding mode for a nominal linear model of the uncertain system. This is therefore complementary to the work of Ryan (1983) and Ryan and Corless (1984). As will be shown in the next section, this stage of the VSC design may be completed independently of the form of the control functions selected for the actual uncertain system. Section 3.3 will also define the problem to be solved, and state the assumptions to be made about the system.

Some of the options available in the design of a stable sliding mode are considered. In Section 3.4 the canonical form that is employed throughout the design of the hyperplane matrix is introduced. This canonical form is also of use in designing the control structure. The sliding hyperplanes may be selected by quadratic minimisation, eigenstructure assignment or eigenvalue selection with robust eigenvector construction. Section 3.5 outlines these main approaches to the design of the sliding mode and discusses the points common to the methods, before the approaches are described in detail in Sections 3.6–3.8 respectively. A brief discussion of a control scheme which ensures that the previously designed sliding mode is attained appears in Section 3.9, while Section 3.10 presents examples of the design procedure. All the design approaches are available in the CAD design package VASSYD developed by the authors (see Dorling, 1985).

3.3 The system

The system we shall consider here has the form

$$\dot{x}(t) = [A + \Delta A(t)]x(t) + [B + \Delta B(t)]u(t) \tag{3.1}$$

where x is an n-vector of states and u is an m-vector of controls. It is assumed that $n > m$, that B is of full rank m and that the pair (A,B) is controllable. The matrices ΔA and ΔB represent the variations and uncertainties in the plant parameters and the control interface respectively. It is assumed further that the control interface uncertainties occur only on the control channels; i.e. $R(B) = R([B|\Delta B])$ (where $R(.)$ denotes the range space); and that rank $[B + \Delta B(t)] = m$ for all $t \geqslant 0$.

The overall aim of a VSC system design is to drive the system state from an initial condition $x(0) = x^0$ to the state space origin as $t \rightarrow \infty$. The jth component u_j ($j = 1, \ldots m$) of the state feedback control vector $u(x)$ has a discontinuity on the jth switching surface, which is a hyperplane M_j passing through the state origin. Defining the hyperplanes by

$$M_j = \{x : c^j x = 0\} \qquad (j = 1,2, \ldots m) \tag{3.2}$$

(where c^j is a row n-vector), the sliding mode occurs when the state lies in M_j for each $j = 1, \ldots m$; i.e. in $M = \bigcap_{j=1}^{m} M_j$ defined by

$$M = \{x : Cx = 0\} \tag{3.3}$$

Geometrically speaking, the sliding mode M is the null space or kernal of C, denoted $N(C)$. If C is chosen such that the product CB is non-singular then $N(C) \cap R(B) = \{0\}$ (Dorling and Zinober, 1986). Since the control inputs are effective only within the range space of B, motion constrained to $N(C)$ is independent of the controls.

Consider firstly the ideal nominal system ($\Delta A \equiv 0$, $\Delta B \equiv 0$)

$$\dot{x}(t) = Ax(t) + Bu(t) \tag{3.4}$$

The defining condition of the sliding mode is

$$Cx(t) \equiv 0 \qquad t \geqslant t_s \tag{3.5}$$

where t_s is the time of arrival at the sliding model. Differentiating and inserting (3.4),

$$C\dot{x}(t) = CAx(t) + CB\mathrm{u}(t) \equiv 0 \qquad (t > t_s) \tag{3.6}$$

from which the equivalent control u_{eq} may be determined as

$$u_{eq}(t) = -Kx(t) = -(CB)^{-1}CAx(t) \tag{3.7}$$

The motion of the equivalent system is thus described by

$$\begin{aligned} \dot{x}(t) &= A_{eq}x(t) = (A - BK)x(t) \\ &= [I - B(CB)^{-1}C]Ax(t) \qquad (t > t_s) \end{aligned} \tag{3.8}$$

and the existence problem reduces to finding the matrix K making (3.8) asymptotically stable. Clearly determination of K is equivalent to selection of the hyperplanes matrix C.

3.4 A canonical form for VSC design

The first task is to specify a particular canonical form for the system in order to simplify the development of the design scheme. This form, which is closely related to the controllability canonical form for a multivariable linear system (Kwakernaak and Sivan, 1972), is similar to that used by Utkin and Yang (1978) is an earlier paper on hyperplane design.

By assumption, the matrix B has full rank m, so that there exists an orthogonal $n \times n$ transformation matrix T such that

$$TB = \begin{bmatrix} 0 \\ B_2 \end{bmatrix} \tag{3.9}$$

where B_2 is $m \times m$ and non-singular. The orthogonality restriction is imposed on T for reasons of numerical stability and to remove the problem of inverting T when transforming back to the original system. A suitable method for determining T is the QU factorisation, whereby B is decomposed into the form

$$B = Q\begin{bmatrix} U \\ 0 \end{bmatrix} \tag{3.10}$$

with Q $n \times n$ and orthogonal, and U $m \times m$, non-singular and upper triangular; T is then determined by rearranging the rows of Q^T.

The transformed state variable $y = Tx$ is now defined, in terms of which the state equation (3.1) becomes

$$\dot{y}(t) = TAT^{\mathrm{T}}y(t) + TBu(t) \tag{3.11}$$

and the sliding condition (3.6) is

$$CT^{T}y(t) \equiv 0 \tag{3.12}$$

If the transformed state y is now partitioned as

$$y^T = [y_1^T y_2^T],\ y_1 \,\varepsilon\, R^{n-m},\ y_2 \,\varepsilon\, R^m \tag{3.13}$$

and the matrices TAT^T, TB and CT^T are partitioned accordingly, then (3.11)–(3.12) may be rewritten in the form

$$\dot{y}_1(t) = A_{11}y_1(t) + A_{12}y_2(t) \tag{3.14a}$$

$$\dot{y}_2(t) = A_{21}y_1(t) + A_{22}y_2(t) + B_2u(t) \tag{3.14b}$$

and

$$C_1y_1(t) + C_2y_2(t) \equiv 0 \tag{3.15}$$

where

$$TAT^T = \begin{bmatrix} A_{11} & A_{12} \\ A_{21} & A_{22} \end{bmatrix}; \qquad CT^{\mathrm{T}} = [C_1\ C_2] \tag{3.16}$$

The canonical form (3.14) is central to the hyperplane design methods to be described here. It also plays a significant role in the solution of the reachability problem; that is the determination of the control form ensuring the hitting of the sliding mode (see Section 3.9).

3.5 Methods for hyperplane design

The assumption that the product matrix CB is non-singular implies that the $m \times m$ matrix C_2 in (3.16) must also be non-singular, since

$$|C_2|\,|B_2| = |C_2 B_2| = |CT^T TB| = |CB| \neq 0 \tag{3.17}$$

and therefore $|C_2| \neq 0$. Condition (3.15) defining the sliding mode may now be written as

$$y_2(t) \equiv -Fy_1(t) \tag{3.18a}$$

where the $m \times (n - m)$ matrix F is defined by

$$F = C_2^{-1} C_1 \tag{3.18b}$$

This indicates that the evolution of y_2 in the sliding mode is related linearly to that of y_1. The relationship (3.18) may also be established by transforming (3.9) to show that $\dot{y}_2(t) = F\dot{y}_1(t)$ during sliding motion, equivalent to the requirement that $s(t) \equiv 0$ during sliding. The ideal sliding mode is therefore governed by the equations

$$\dot{y}_1(t) = A_{11} y_1(t) + A_{12} y_2(t) \tag{3.19a}$$

$$y_2(t) = -Fy_1(t) \tag{3.19b}$$

an $(n - m)$th order system in which y_2 plays the role of a state feedback control. Closing the loop in (3.19) gives

$$\dot{y}_1(t) = (A_{11} - A_{12}F)y_1(t) \tag{3.20}$$

which indicates that the design of a stable sliding mode ($y \to 0$ as $t \to \infty$) requires the determination of the gain matrix F such that $A_{11} - A_{12}F$ has $n - m$ left-hand half-plane eigenvalues. This may be achieved by using a modified form of any standard design method giving a linear feedback controller for a linear dynamical system. In particular, the methods which will be considered here are those based on the minimisation of an integral cost functional with quadratic integrand (Section 3.6) and two eigenstructure assignment methods (Sections 3.7 and 3.8). However, before discussing these methods in more detail, we note that, whichever scheme is chosen for the design, fixing F does not uniquely determine C, since m^2 degrees of freedom (DOF) remain in the relationship

$$C_2 F = C_1 \tag{3.21}$$

Two possibilities for determining the hyperplane matrix C, assuming that F has been fixed, are as follows.

In certain control schemes, such as hierarchical methods (Utkin, 1978; Ryan, 1983), the control design may be considerably simplified if the product $R = CB$ is diagonal or, at the very least, diagonally dominant. To this end it would be useful to be able to specify this product matrix. Suppose, therefore, that the $m \times m$ matrix R has been selected. Then

$$R = CB = C_2 B_2 \tag{3.22a}$$

so that

$$C_2 = RB_2^{-1} \tag{3.22b}$$

and

$$C = RB_2^{-1} [F \; I_m] T \tag{3.23}$$

Clearly, C_2 should be found in practice by solving (3.22*a*) as a set of m equations in m unknowns with m right-hand sides.

If, however, the product CB is immaterial to the designer, the simplest method of determining C from F is that the employed by Utkin and Yang (1978); namely, letting $C_2 = I_m$. This is equivalent to specifying $R = B_2$ in (3.22) – (3.23), giving

$$C = [F \; I_m]\mathrm{T} \tag{3.24}$$

This approach has the merit of minimising the amount of calculation in proceeding from F to C, and hence reduces the possibility of numerical errors.

3.6 Hyperplane design by quadratic minimisation

The design of switching hyperplanes by minimising a cost functional in which the integrand is a quadratic function of the state $x(.)$ was proposed by Utkin and Yang (1978). Since the method was described in that work, it will only be discussed briefly here.

Let t_s denote the time at which the sliding mode starts; and define the cost functional

$$J(\mathrm{u}) = \tfrac{1}{2} \int_{t_s}^{\infty} x^{\mathrm{T}}(t) \mathrm{Q} x(t) \mathrm{d}t \tag{3.25}$$

where the matrix $Q > 0$ is constant and symmetric. The aim is to minimize J subject to the system equation (3.19*a*), assuming that $x(t_s)$ is a known initial condition, and such that $x(t) \to 0$ as $t \to \infty$. Partitioning the product

$$TQT^T = \begin{bmatrix} Q_{11} & Q_{12} \\ Q_{21} & Q_{22} \end{bmatrix} \tag{3.26}$$

compatibly with y, and defining

$$Q^* = Q_{11} - Q_{12}Q_{22}^{-1}Q_{21}; \qquad A^* = A_{11} - A_{12}Q_{22}^{-1}Q_{21} \tag{3.27a}$$

$$v(t) = y_2(t) + Q_{22}^{-1}Q_{21}y_1(t) \tag{3.27b}$$

this problem may be restated as: minimise

$$J(v) = \tfrac{1}{2}\int_{t_s}^{\infty} \{y_1^T(t)Q^*y_1(t) + v^T(t)Q_{22}v(t)\}dt \tag{3.28a}$$

subject to

$$\dot{y}_1(t) = A^*y_1(t) + A_{12}v(t) \tag{3.28b}$$

which has the form of the standard linear quadratic optimal regulator problem.

The controllability of (A, B) is sufficient to ensure the controllability of (A^*, A_{12}) (Utkin and Yang, 1978). Moreover, the positivity condition on Q ensures that $Q_{22} > 0$ (so that Q_{22}^{-1} exists) and that $Q^* > 0$. Thus a positive-definite unique solution P is guaranteed for the algebraic matrix Riccati equation

$$PA^* + A^{*T}P - PA_{12}Q_{22}^{-1}A_{12}^TP + Q^* = 0 \tag{3.29}$$

associated with the problem (3.28), and the optimal 'control' v minimising (3.28a) is given by

$$v(t) = -Q_{22}^{-1}A_{12}^TPy_1(t) \tag{3.30}$$

Using (3.27*b*), this may be transformed to give

$$y_2(t) = -Q_{22}^{-1}[Q_{21} + A_{12}^TP]y_1(t) = -Fy_1(t) \tag{3.31}$$

and F has been determined as required.

3.7 Hyperplane design by eigenstructure assignment

The majority of VSC designs appearing in the literature to date have suggested eigenvalue assignment methods in the design of the sliding mode. Since these problems have generally been scalar-controlled examples with a single switching hyperplane, specification of the $n - 1$ eigenvalues to be associated with the sliding mode completely determines the feedback matrix F which in this case is an $n - 1$ element row vector.

For the multiple input case Utkin and Yang (1978) show that the pair (A_{11}, A_{12}) is controllable and that eigenvalue assignment in (3.20) is therefore feasible. It is well known, however, that the assignment of eigenvalues of an nth order m-input system requires only n of the nm DOF available in choosing the feedback gain matrix (Srinathkumar and Rhoten, 1975; Shah *et al.*, 1975). The remaining $n(m - 1)$ DOF may be utilised in some other way; particular, in partially assigning the eigenvectors. Two complementary approaches to the specification of the eigenvectors have been presented throughout the literature, as will be discussed below. The capability of partially assigning the eigenvectors may be used to shape the response of a closed-loop system and this motivates the present work.

In the present context, the aim is to assign the eigenvectors of (3.20) directly in order to shape the system response during sliding. For convenience it is assumed throughout the sequel that the non-zero sliding-mode eigenvalues are distinct from each other and from the eigenvalues of A_{11}.

Suppose that the sliding mode has commenced on $N(C)$. Then

$$\dot{x}(t) = (A - B\mathrm{K})x(t) \tag{3.32}$$

where K is defined by (3.7). During the sliding mode, $x(.)$ must remain in $N(C)$, so that

$$C[A - BK] = 0 \Leftrightarrow R(A - BK) \subseteq CN(C) \tag{3.33}$$

Let $\{\lambda_i\colon i = 1, \ldots n\}$ be the eigenvalues of $A - BK$ with corresponding eigenvectors v_i. Then (3.33) implies that

$$C[A - BK]v_i = \lambda_i C v_i = 0 \tag{3.34}$$

so that either λ_i is zero or $v_i \in N(C)$. Now $A - BK = A_{\mathrm{eq}}$ has precisely m zero-valued eigenvalues, so let $\{\lambda_i\colon i = 1 \ldots n - m\}$ be the non-zero eigenvalues (distinct by assumption). Then specifying the corresponding eigenvectors $\{v_i\colon i = 1, \ldots n - m)$ fixes the null space of C, since $\dim[N(C)] = n - m$. However, C is not uniquely determined, because the equation

$$CV = 0, \qquad V = [v_1 \ldots v_{n-m}] \tag{3.35}$$

has m^2 DOF. This may be seen if we define

$$W = \begin{bmatrix} W_1 \\ W_2 \end{bmatrix} = TV \tag{3.36}$$

where the partitioning of W is compatible with that of y; then (3.35) becomes

$$0 = CT^T TV = [C_1 C_2]\begin{bmatrix} W_1 \\ W_2 \end{bmatrix} = C_2[F2I_m]\begin{bmatrix} W_1 \\ W_2 \end{bmatrix} \tag{3.37a}$$

giving the equation

$$FW_1 = - W_2 \tag{3.37b}$$

Thus, providing that W_1 is non-singular, the matrix F is uniquely determined by (3.37*b*), but the m^2 degrees of freedom associated with (3.21) still remain.

The drawback to this approach is that the eigenvectors v_i of $A - BK$ are not, in general, freely assignable. Shah *et al.* (1975) show that at most m elements of an eigenvector may be assigned arbitrarily, after which the remaining $n - m$ elements are fully determined by the assigned elements. Thus one approach to eigenvector assignment is to select m elements according to some scheme and accept the remaining elements as determined. This might allow a degree of adjustment to be carried out by inspection.

The alternative method of eigenvector assignment is by consideration of the assignable subspace corresponding to a given eigenvalue (Moore, 1976; Klein and Moore, 1977). It is shown by Sinswat and Fallside (1977) that this assignable subspace for an eigenvalue λ_i may be characterised as the null space of the $n \times n$ matrix $H(\lambda_i)$ defined by

$$H(\lambda_i) = (I_n - BB^+)(A - \lambda_i I_n), \qquad B^+ = (B^T B)^{-1} B^T \tag{3.38}$$

which follows from the requirement that $(A - \lambda_i I_n)v_i$ must lie in $R(B)$. Now, the transformation matrix T is non-singular, so that

$$H(\lambda_i)v = 0 \Leftrightarrow TH(\lambda_i)T^T \cdot Tv = 0 \tag{3.39}$$

while the canonical form (3.11) – (3.16) may be used to give

$$\begin{aligned} TH(\lambda_i)T^T &= (I - TBB^+T^T)(TAT^T - \lambda_i I_n) \\ &= \left(I_n - \begin{bmatrix} 0 \\ B_2 \end{bmatrix} [0 \ B_2^{-1}] \right) \left(\begin{bmatrix} A_{11} & A_{12} \\ A_{21} & A_{22} \end{bmatrix} - \lambda_i I_n \right) \\ &= \begin{bmatrix} A_{11} - \lambda_i I_{n-m} & A_{12} \\ 0 & 0 \end{bmatrix} \end{aligned} \tag{3.40}$$

Eqns. (3.39) – (3.40) show clearly that an arbitrary vector v lies within $N[H(\lambda_i)]$ if and only if $w = Tv \in N[H^*(\lambda_i)]$, where

$$H^*(\lambda_i) = [A_{11} - \lambda_i I_{n-m} \quad A_{12}] \tag{3.41}$$

Note that $H^*(\lambda_i)$ has dimensions $(n - m) \times n$, and therefore requires less storage space than the original matrix $H(\lambda_i)$ of dimension $n \times n$. Moreover, $H^*(\lambda_i)$ provides clarification of the number of DOF available in assigning the eigenvector corresponding to λ_i; for if $w = Tv$ is partitioned compatibly with y then $w \, \varepsilon \, H^*(\lambda_i)$ implies

$$(A_{11} - \lambda_i I_{n-m})w_1 = - A_{12} w_2 \tag{3.42}$$

from which it is clear that fixing the m elements of w_2 uniquely determines w_1, and hence fixes v. Note also that the condition that w_1 must have linearly independent columns is a further restriction on the assignable eigenvectors arising from the requirement that the reduced-order system (3.20) should have distinct eigenvalues and hence linearly independent eigenvectors.

The concept of assignable subspaces is applied to slidingmode design as follows. The designer selects the desired elements of the closed-loop eigenvector v_i corresponding to a non-zero sliding mode eigenvalue λ_i. If r elements of v_i are specified $(1 \leqslant r \leqslant m)$, the remaining $n - r$ elements are determined directly by solving $H^*(\lambda_i)Tv_i = 0$, taking the minimum norm solution if $r < m$. If all of the n elements of v_i are specified, the assignability of the vector is tested by transforming it and applying (3.41). If this result is non-zero, v_i must be modified to give the 'closest' assignable eigenvector, which is found by projection into the current assignable subspace $N(H^*(\lambda_i)T)$. Finally, if more than m but less than n elements are specified, a least-squares approach is employed to determine an assignable eigenvector which gives as close a fit to the specified elements as is possible and then minimises the remaining elements.

3.8 Hyperplane design by robust eigenvalue assignment

For the non-ideal system, the perturbations $\Delta A(\cdot)$ and $\Delta B(\cdot)$ may also be transformed:

$$T[\Delta A(\cdot)]T^T = \begin{bmatrix} \Delta A_{11}(\cdot) & \Delta A_{12}(\cdot) \\ \Delta A_{12}(\cdot) & \Delta A_{12}(\cdot) \end{bmatrix} \tag{3.43a}$$

$$T[\Delta B(\cdot)] = \begin{bmatrix} 0 \\ \Delta B_2(\cdot) \end{bmatrix} \tag{3.43b}$$

Note that (3.43b) follows from the assumption on the control interface uncertainties. The closed loop system in the transformed state space then becomes

$$\dot{y}_1 = (A_{11} - A_{12}F)y_1 + (\Delta A_{11} - \Delta A_{12}F)y_1 \tag{3.44}$$

(where the explicit time dependence of y, ΔA_{11} and ΔA_{12} has been dropped). The second term on the right hand side of (3.44) represents the perturbation of the nominal sliding mode system due to parameter variations, and will generally be time-varying and unknown. The aim of a robust controller design is to minimise the effect of this perturbation on the overall performance of the sliding regime; and, in particular, to ensure that stability is maintained. Since the variations in the parameters are unknown, the design technique must be centred on the nominal system (3.4).

During the sliding mode the nominal system has a set of eigenvalues comprising the $n - m$ members of $\sigma(A_{11} - A_{12}F)$, the spectrum of the 'closed loop' matrix $[A_{11} - A_{12}F]$, plus the value zero repeated m times. Now the purpose of the control functions u_j during sliding is to maintain the motion within the null space of C, which amounts to keeping the m zero-valued eigenvalues at zero. One of the aims of robust control function design is therefore to minimise the effect on the sliding mode behaviour of parameter variations within the range space of B. Methods of achieving this aim are described by Ryan (1983) and Ryan and Corless (1984) in some detail, and are therefore not repeated here in full. One such method is similar to that used in Section 3.9.

The problem considered in detail here is the assignment of the non-zero sliding mode eigenvalues; and, in particular, how to select the eigenvectors in such a way that the eigenvalues are maximally robust, in the sence that they are as insensitive as possible to parameter variations not lying within $R(B)$.

3.8.1 Redundancy in the sliding mode system

The system (3.4) represents an $(n - m)$th order linear system with m control inputs. The interface matrix A_{12} has dimensions $(n - m) \times m$ and rank p where

$$1 \leqslant p \leqslant \min\{n - m, m\}$$

(note that $p = 0$ would imply that the pair (A_{11}, A_{12}) is uncontrollable, which would contradict the controllability of the original system. If $p < m$, then the sliding mode system has redundant inputs which may be removed in the following mannuer. Let P be an $m \times m$ permutation matrix which reorders the columns of A_{12} such that

$$A_{12}P = [\tilde{A}_{12} \quad \hat{A}_{12}] \tag{3.45}$$

where $\tilde{A}_{12}$ has p linearly independent columns, while the columns of $\hat{A}_{12}$ depend linearly on those of $\tilde{A}_{12}$. (This operation amounts to a renumbering of the 'control' inputs in y_2.) Then setting

$$[\Delta\tilde{A}_{12} \quad \Delta\hat{A}_{12}] = (\Delta A_{12})P \tag{3.46}$$

and

$$P^{T}F = \begin{bmatrix} \tilde{F} \\ \hat{F} \end{bmatrix} \tag{3.47}$$

the perturbation effect of the second term in (3.44) becomes

$$(\Delta A_{11} - \Delta\tilde{A}_{12}\tilde{F} - \Delta\hat{A}_{12}\hat{F})y_1 \tag{3.48}$$

In the absence of any detailed information it seems appropriate to set $\hat{F} = 0$, removing the effects of the uncertain elements in A_{12}. Eqn. (3.44) then becomes

$$\dot{y}_1 = (A_{11} - \tilde{A}_{12}\tilde{F})y_1 + (\Delta A_{11} - \Delta\tilde{A}_{12}\tilde{F})y_1 \tag{3.49}$$

The remaining feedback matrix $\tilde{F}$ may then be determined by applying the algorithm of the next section to the pair $(A_{11}, \tilde{A}_{12})$, and the full feedback matrix F is found by permuting the rows of the partitioned matrix

$$F = P\begin{bmatrix} \tilde{F} \\ \hat{F} \end{bmatrix}$$

3.8.2 Hyperplane design by robust eigenvalue assignment

An algorithm for hyperplane design by robust eigenvalue assignment is now described for the case $p = m$: the general case of $p < m$ is readily derived by suitable notational changes.

Suppose that $\lambda_i \in \sigma\,(A_{11} - A_{12}F)$, with corresponding right eigenvector $w_i \in R^{n-m}$ and left eigenvector $r_i \in R^{n-m}$. Then it is well known (Wilkinson, 1965) that the sensitivity of λ_i to perturbations in A_{11}, A_{12} and F depends on the angle between w_i and r_i. More precisely the sensitivity of λ_i is, to first order, inversely proportional to the cosine c_i of the angle between w_i and r_i:

$$c_i = \frac{|r_i{}^{\mathrm{T}} w_i|}{\|r_i\| \, \|w_i\|}. \tag{3.50}$$

Since c_i is a cosine, we have $1/c_i \geqslant 1$. In the case of A_{12} being square (equal number of 'states' and 'inputs') and non-singular we have complete freedom in assigning the eigenvectors of $A_{11} - A_{12}F$, and can do no better than selecting an orthogonal eigenvector matrix $W = [w_1 w_2 \ldots w_{n-m}]$ giving $1/c_i = 1$ ($i = 1, \ldots n - m$). In particular, we may take $W = I_{n-m}$. More generally, A_{12} is non-square ($1 < m < n - m$) and we should aim to make $1/c_i$ as small as possible for each $i = 1, \ldots n - m$.

Now, it is also well known (Wilkinson, 1965) that an upper bound on each of the sensitivities is the spectral condition number $\kappa(W)$ defined by

$$\kappa(W) = \|W\| \, \|W^{-1}\| \tag{3.51}$$

where the spectral matrix norm ($\|M\| = \max \{\sqrt{\mu}: \mu \varepsilon \sigma(M^T M)\}$ is employed. Thus minimising $\kappa(W)$ should ensure that the sensitivities of all the assigned eigenvalues is acceptable. Moreover, minimising $\kappa(W)$ ensures that W is well-conditioned with respect to inversion, so that the solution process for the feedback matrix F is stable and the bounds on the magntidues of $\|F\|$ and the transient response are minimised. It should be noted, however, that we are not free to choose our eigenvector matrix W arbitrarily if $m < n - m$, and that the minimum attainable condition number $\kappa(W)$ is therefore not necessarily unity for a given set of eigenvalues.

Four methods for determining a suitable set of eigenvectors for a specified spectrum in the usual linear state feedback system are described by Kautsky and Nichols (1983) and by Kautsky, Nichols and Van Dooren (1985). The four methods each consist of three steps, labelled A, X and F; steps A and F are identical in all cases with the crucial step X differing from method to method. We have adapted one of these methods to the restricted problem of assigning the eigenvalues of the sliding mode system, as outline below. This particular method was selected as being the most efficient, in general. Fuller details of the algorithm may be found in the work of Kautsky *et al.* (1983, 1985).

As shown in Section 3.7, a criterion for the assignability of an $n - m$ vector w_i as the eigenvector corresponding to eigenvalue λ_i is that w_i should lie in the null space of the matrix $H^*(\lambda_i)$ given by (3.41). Clearly this condition must still be satisfied in our present algorithm. We are therefore only interested in looking at vectors which lie in $N[H^*(\lambda_i)]$. With this in mind we outline the algorithm:

Step A: Construct a basis for the null space of $H^*(\lambda_i)$ ($i = 1, \ldots n - m$) as the columns of an $n \times m$ matrix $\begin{vmatrix} W_i \\ S_i \end{vmatrix}$ such that $S_i^{\mathrm{T}} S_i = I_{n-m}$.

Step X: Determine linearly independent vectors w_i in $R(S_i)$ such that $\|w_i\| = 1$ ($i = 1, \ldots n - m$); and set $W = [w_1 \ldots w_{n-m}]$.

Step F: Define $x_i = S_i^T w_i (i = 1, \ldots n - m)$, set $X = [W_1 x_1 \ldots W_{n-m} x_{n-m}]$ and solve $FW = X$ for the feedback gain matrix F.

Note that the method of construction of the vector W_i used here ensures that it lies in $N[H^*(\lambda_i)]$.

In the method we have chosen to implement, step X is further divided. Firstly, a set of orthonormal vectors $\tilde{v}_i$ $(i = 1, \ldots n - m)$ giving a 'best fit' to the subspaces $R(S_i)$ is determined. In the ideal situation of freely assignable eigenvectors, $\tilde{v}_i$ would lie within $R(S_i)$ $(i = 1, \ldots n - m)$; but as this is not generally the case the second stage of step X is the projection of $\tilde{v}_i$ into $R(S_i)$ to give v_i.

The technique used for determining the 'best fit' set of vectors $\{\tilde{v}_i : i = 1, \ldots n - m\}$ aims to minimise the sum of magnitudes of the angles between the vectors $\tilde{v}_i$ and their respective subspaces $R(S_i)$. Equivalently, the sum of magnitudes of the angles between $\tilde{v}_i$ and the orthogonal complement of $R(S_i)$ may be *maximised*; this is the procedure adopted, as it is more efficient for systems in which the number of states is at least twice the number of controls (as is usually the case for large systems). This sum of angles is maximised by rotating the vectors according to an iterative scheme, wherein the increment in the sum is maximised at each step. Since each term of the sum is clearly bounded above (by a right angle), the sum must eventually approach a limit, and the process may be terminated when the increments in the sum drop below a specified tolerance.

The approach described here does not produce the minimum value for $\kappa(V)$, but does minimise a conditioning measure which closely approximates to the optimal conditioning. This is because the method does not pick the eigenvectors directly from the spaces $R(S_i)$, but rather selects vectors that are orthogonal to each other and at the same time close to the corresponding $R(S_i)$ and then projects each one into its desired space. The nearness of each vector thus obtained to its corresponding subspace determines how close the conditioning is to the optimal value. A lower bound κ_0 on the minimum attainable condition number for the selected set of eigenvalues λ_i $(i = 1, \ldots n - m)$ may be determined from the matrices S_i calculated at step A. The value of this lower bound is found by forming the partitioned matrix $S = [S_1 \ S_2 \ldots S_{n-m}]$; κ_0 is then giving by

$$\kappa_0 = \kappa(S)/\sqrt{(n - m)}$$

where $\kappa(S)$ is determined as the ratio of the largest to smallest of the singular values of S. The closeness of κ_0 to unity gives an indication of the suitability for assignment of the chosen spectrum $\sigma(A_{11} - A_{12}F)$. Note, however, that this bound is not necessarily attainable.

To date the technique has been implemented only for the case of real eigenvalues in the CAD design package VASSYD, because of additional complications in the complex case (see Kautsky and Nichols, 1983, Appendix IV). In the problems tackled the algorithm has given rapid convergence and a condition number close to the minimum has been attained. Mudge and Patton (1988) have recently proposed an effective algorithm for the complex case.

3.9 Control scheme designs

Once the VSC existence problem has been solved (that is, C has been fixed) attention must be turned to solving the reachability problem. This involves the selection of a state feedback control function $u: R^n \rightarrow R^m$ which will drive the state x into $N(C)$ and thereafter maintain it within this subspace. In theory, there is a virtually unlimited number of possible forms for this control function, the only essential features of the form chosen being discontinuity on one or more subspaces containing $N(C)$. However, only a few control forms have been studied in depth to date.

In general the variable structure control law consists of two parts: a linear control law u^L and a non-linear part u^N, which are added to form u. The linear control is merely a state feedback controller:

$$u^L(x) = Lx \tag{3.52}$$

while the non-linear feedback controller u^N incorporates the discontinuous elements of the control law. These non-linearities may include:

(*a*) Relays with constant gains:

$$u_j^N(x) = m_j \operatorname{sgn}(c^j x), \quad m_j > 0 \tag{3.53a}$$

(*b*) Relays with state-dependent gains:

$$u_j^N(x) = m_j(x) \operatorname{sgn}(c^j x), \quad m_j(\cdot) > 0 \tag{3.53b}$$

(*c*) Linear feedback with switched gains:

$$u^N(x) = \Psi x; \quad \Psi = [\psi_{ij}]; \quad \psi_{ij} = \begin{cases} \alpha_{ij}, & s_i x_j > 0 \\ \beta_{ij}, & s_i x_j < 0 \end{cases} \tag{3.53c}$$

(*d*) Unit vector non-linearity (with scale factor):

$$u^N(x) = \frac{\varrho}{\|Cx\|} Cx, \quad \varrho > 0 \tag{3.53d}$$

In forms (*a*) – (*c*) the control functions u_j^N are discontinuous on the individual switching planes. Sliding conditions are derived separately for each hyperplane, which often leads to a waste of control effort. In particular, there are situations in which the controller drives the state away from an unattained hyperplane as it attempts to force the state into another surface. In case (*d*), the individual controls are continuous except on the final intersection $N(C)$ of the hyperplanes, where all the controls are discontinuous together. This method is based on a design technique which ensures that the motion is always towards the final target $N(C)$, and the control is, moreover, easier to implement because of its simpler structure. Attention is therefore focused on case (d) here: see Itkis (1976), Utkin (1978), Ryan (1983) and Young (1977) for details of types (a) – (c).

3.9.1 Unit vector control

The control structure to be described here is based on that of Ryan and Corless (1984) and has the form

$$u(x) = Lx + \frac{\varrho}{\|Mx\|} Nx \tag{3.54}$$

where the null spaces of N, M and C are coincident: $N(N) = N(M) = N(C)$. Starting from the transformed state y, we form a second transformation $T_2 : R^n \rightarrow R^n$ such that

$$z = T_2 y \tag{3.55a}$$

where

$$T_2 = \begin{bmatrix} I_{n-m} & 0 \\ F & I_m \end{bmatrix} \tag{3.55b}$$

The matrix T_2 is clearly non-singular, with inverse

$$T_2^{-1} = \begin{bmatrix} I_{n-m} & 0 \\ -F & I_m \end{bmatrix} \tag{3.56}$$

Partitioning $z^T = [z_1^T \; z_2^T]$ with $z_1 \in R^{n-m}$ and $z_2 \in R^m$

$$z_1 = y_1; \quad z_2 = Fy_1 + y_2 \tag{3.57}$$

from which it is clear that the conditions $s \equiv 0$ and $z_2 \equiv 0$ are equivalent (in the sense that the points of the state space at which $s = 0$ are precisely the points at which $z_2 = 0$). The transformed system equations become

$$\dot{z}_1 = Gz_1 + A_{12}z_2 \tag{3.58a}$$

$$\dot{z}_2 = Hz_1 + \Phi z_2 + B_2 u \tag{3.58b}$$

where

$$G = A_{11} - A_{12}F; \quad H = FG - A_{22}F + A_{21}; \quad \Phi = FA_{12} + A_{22} \tag{3.58c}$$

In order to attain the sliding mode, it is necessary to force z_2 and $\dot{z}_2$ to become identically zero. To this end the linear part of the control is defined to be

$$u^L(z) = -B_2^{-1}\{Hz_1 + (\Phi - \Phi_*)z_2\} \tag{3.59}$$

where Φ_* is any $m \times m$ matrix with left-hand half-plane eigenvalues. In particular, given a spectrum $\{\mu_i : \mathrm{Re}(\mu_i) < 0, i = , \ldots m\}$, we may set $\Phi_* = \mathrm{diag}\{\mu_i : i = 1, \ldots m\}$. Transforming back into the original state space (x-space) gives

$$L = -B_2^{-1}[H \quad \Phi - \Phi_*]T_2 T \tag{3.60}$$

The linear control law u^L serves only to drive the state component z_2 to zero asymptotically; to attain $N(C)$ in finite time, the non-linear control component u^N is required. This non-linear control must be discontinuous whenever $z_2 = 0$, and continuous elsewhere. Letting P_2 denote the positive-definite unique solution of the Lyapunov equation

$$P_2\Phi_* + \Phi_*^T P_2 + I_m = 0 \tag{3.61}$$

then $P_2 z_2 = 0$ if and only if $z_2 = 0$, and we may take

$$u^N(z) = \frac{-\varrho}{\|P_2 z_2\|} B_2^{-1} P_2 z_2, \quad z_2 \neq 0 \tag{3.62}$$

where $\varrho > 0$ is a scalar parameter to be selected by the designer. When $z_2 = 0$, u^N may be arbitrarily defined as any function satisfying $\|u^N\| \leqslant \varrho$. The control $u = u^L + u^N$ defined by (3.59) and (3.62) drives an arbitrary initial state z^0 to the sliding manifold in a time

$$t \leqslant \frac{1}{\varrho}\sqrt{\frac{\langle z_2^0, P_2 z_2^0\rangle}{\sigma_{min}(P_2)}} \tag{3.63}$$

where $\sigma_{min}(P_2)$ denotes the minimum eigenvalue of P_2 and $\langle \cdot, \cdot \rangle$ is the usual Euclidean inner product on R^m. Finally, expressing the control in x-space, we have

$$N = B_2^{-1}[0 \; P_2]T_2 T \tag{3.64a}$$

$$M = [0 \; P_2]T_2 T \tag{3.64b}$$

For the more general system in which disturbances and uncertainties are present, a similar control structure may be employed. However, in this case, the scalar ϱ of (3.53) is replaced by a time-varying state-dependent function incorporating two design parameters γ_1, γ_2, upon which the time to $N(C)$ also depends. For details of this case, refer to Ryan and Corless (1984).

3.9.2 Practical implementation of VSC

The control structures described so far in this section are designed to drive the system state into the ideal sliding mode. In order to achieve this they rely on certain idealisation which, because of delays in switching, hysteresis and other dynamic non-idealities, cannot be achieved in any practical system. Thus the ideal sliding mode as defined by (3.5) will, in general, be unattainable. Instead, the state will remain only within a neighbourhood of $N(C)$, continually passing backwards and forwards through the sliding manifold as the control switches repeatedly between two (or more) distinct values. This behaviour is known as chattering and is the real sliding mode, of which the ideal sliding mode is a theoretical abstraction. Chattering is particularly undesirable in any physical situation in which the actuator mechanisms may be damaged by rapid switching.

The simplest way of overcoming this major disadvantage of VSC implementation is to 'soften' the action of the non-linearity by substituting a continuous approximation to the discontinuous part of the control. For example, a relay might be replaced by a saturating amplifier, giving a small region of unsaturated control effort around the switching surface. We shall refer to a control incorporating a softened non-linearity as a smoothing control.

The most rigorous way to introduce this smoothing control is to design the system under the assumption of the required smoothing control, as proposed for the unit vector method by Ryan and Corless (1984). An alternative approach is to design the system using VSC methods under the assumption of discontinuous control, but to actually implement an approximating smoothing control. For example, the unit vector non-linearity (3.53*d*) may be replaced by a control of the form

$$u^N(x) = \frac{\varrho \, Cx}{\|Cx\| + \delta}, \qquad \varrho > 0 \tag{3.65}$$

where $\delta > 0$ is a small positive constant. Such an approach has been examined by Burton and Zinober (1986), Slotine (1985) and others. Suitable choices of δ vary from one application to another; systematic methods for fixing δ are currently being sought.

3.10 Examples

3.10.1 Example 1

To illustrate the modal design technique described in Section 3.7, consider the VSTOL aircraft autostabilisation problem outlined by Porter and Crossley (1972). This is a fourth-order problem, the states x_1 to x_4 being the sideslip angle, the roll rate, the yaw rate and the bank angle respectively. Two inputs are available to control these states: the rudder angle u_1 and the aileron angle u_2. The system is linearised about a nominal operating point equivalent to an airspeed of 80 knots at sea level. The linearised system, of the form (3.4) with matrices

$$A = \begin{bmatrix} -0{\cdot}0506 & 0 & -1 & 0{\cdot}2380 \\ -0{\cdot}7374 & -1{\cdot}3345 & 0{\cdot}3696 & 0 \\ 0{\cdot}0100 & 0{\cdot}1074 & -0{\cdot}3320 & 0 \\ 0 & 1 & 0 & 0 \end{bmatrix} \tag{3.66a}$$

$$B = \begin{bmatrix} 0{\cdot}0409 & 0 \\ 1{\cdot}2714 & -20{\cdot}3106 \\ -2{\cdot}0625 & 1{\cdot}3350 \\ 0 & 0 \end{bmatrix} \tag{3.66b}$$

is readily found to be controllable, with B of full rank 2. The transformation matrix T is then calculated as

$$T = \begin{bmatrix} 0 & 0 & 0 & 1 \\ 0{\cdot}998 & 0{\cdot}0014 & 0{\cdot}0207 & 0 \\ 0{\cdot}0120 & -0{\cdot}8513 & -0{\cdot}5245 & 0 \\ -0{\cdot}0169 & -0{\cdot}5247 & 0{\cdot}8511 & 0 \end{bmatrix} \tag{3.67}$$

A sliding mode for this system is to be designed by the eigenvector assignment method of Section 3.7. Following Porter and Crossley, it is desired to assign the complex pair $-0{\cdot}9 \pm 2i$ as the Dutch roll mode. Ideally this mode should be isolated from the states x_2 and x_4. This could be achieved if the corresponding pair of eigenvectors w_1 and w_2 could be assigned with zero entries in the second and fourth positions. However, this leads to a singular matrix W_1 in (3.37*b*), so it is necessary to choose non-zero entries of small magnitude in these positions. Accordingly, let the desired eigenvectors have real and imaginary parts

$$w_d^R = [1 \quad 0{\cdot}1 \quad * \quad 0{\cdot}1] \tag{3.68}$$

$$w_d^I = [* \quad 0{\cdot}1 \quad 1 \quad 0{\cdot}1]$$

where the * indicates that the size of the element in that position is immaterial. This choice of eigenvectors will hopefully ensure that the Dutch role mode will show up very weakly in the states x_2 and x_4 once sliding commences. Following the procedure outlined in Section 3.7, the normalised assignable eigenvectors providing the 'best fit' to (3.54) are found to be

$$w_1 = w_R + iw_I; \; w_2 = w_R - iw_I = \bar{w}_1 \tag{3.69a}$$

$$w_R = [0{\cdot}1012 \quad 0{\cdot}0033 \quad 0{\cdot}9046 \quad 0{\cdot}0037] \tag{3.69b}$$

$$w_I = [0{\cdot}4013 \quad 0{\cdot}0103 \quad 0{\cdot}1012 \quad -0{\cdot}0033]$$

Applying (3.35) – (3.37) to this pair of eigenvectors, and letting $C_2 = I_2$ (as in (3.24)), the resulting hyperplane matrix is found to be

$$C = \begin{bmatrix} 0{\cdot}9650 & -0{\cdot}8500 & -0{\cdot}5048 & 98{\cdot}9266 \\ -1{\cdot}5162 & -0{\cdot}5267 & 0{\cdot}8201 & -160{\cdot}4030 \end{bmatrix} \tag{3.70}$$

which can be shown to give the desired eigenvalues and eigenvectors when A_{eq} is formed.

In order to show that the desired behaviour in the sliding mode is obtained, the control (3.54), (3.60), (3.64) is implemented, with

$$\Phi_* = \operatorname{diag}\{-1 \quad -1\}, \quad \varrho = 10 \tag{3.71}$$

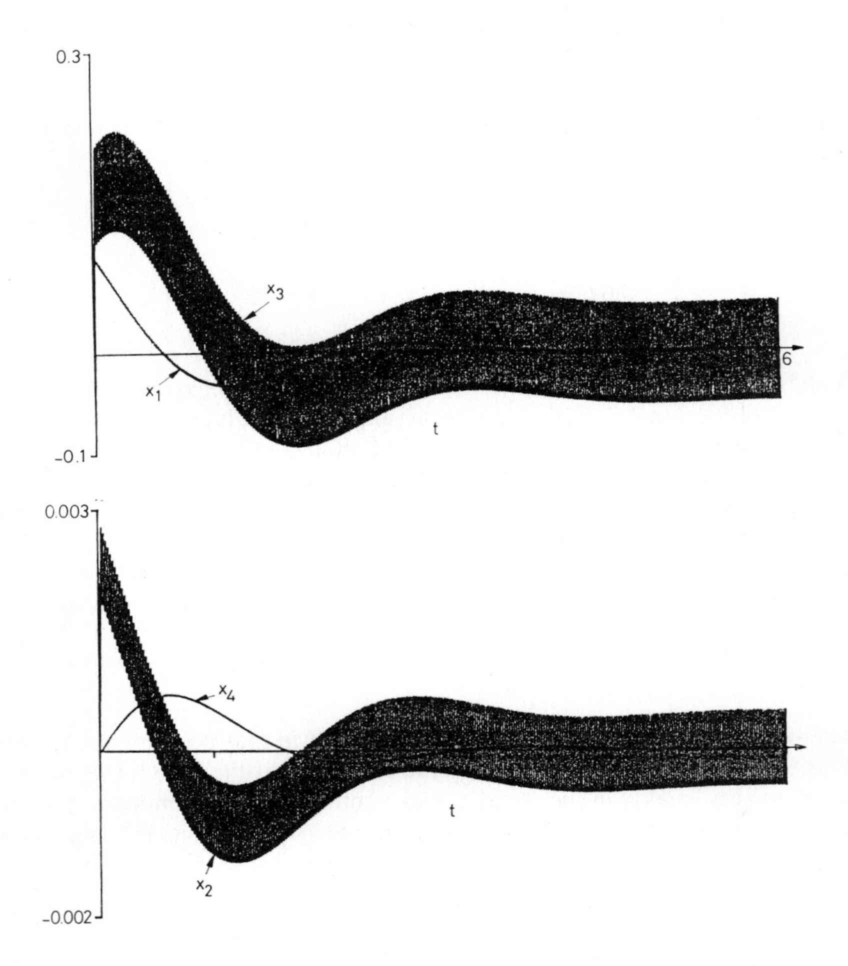

Fig. 3.1 *Response of the system under control* (3.71) *to the initial condition* $x(0) = [0{\cdot}1 \; 0{\cdot}0 \; -0{\cdot}1 \; 0{\cdot}0]^T$

The response of the system under this control to the initial condition $x(0) = [0{\cdot}1 \quad 0{\cdot}0 \quad -0{\cdot}1 \quad 0{\cdot}0]^T$ is shown in Fig. 3.1, while Fig. 3.2 displays the control functions. It will be noticed from Fig. 3.2 that a very large chattering component is present in the controls, particularly in the rudder angle u_1. This chatter also shows up in the states of Fig. 3.1. Since chattering is not really practicable in this aircraft control system, the smoothing control (3.65) described in Section 3.9.2

is substituted, using a value of $\delta = 0{\cdot}2$. The resulting responses and controls are shown in Figs. 3.3 and 3.4. From these figures it is clear that the sideslip angle and yaw rate have the required shape (damping coefficient 0·9, period π) once the sliding mode is approached. The roll rate and bank angle also have the same shape, but on a much smaller scale. In practice, the rapid growth of x_2 and x_3

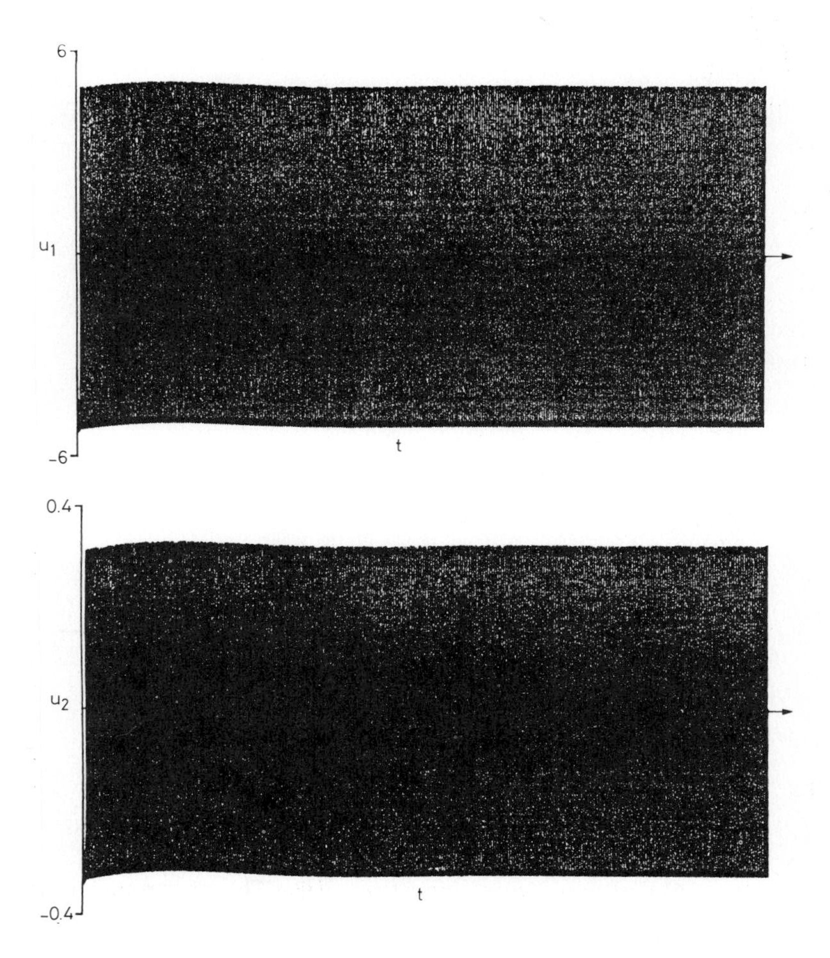

Fig. 3.2 *Control functions for Fig. 3.1*

before the onset of sliding is likely to be unacceptable and modifications to the control function design would be needed in a practical implementation.

3.10.2 Example 2

As further example of the design procedures consider the following system with

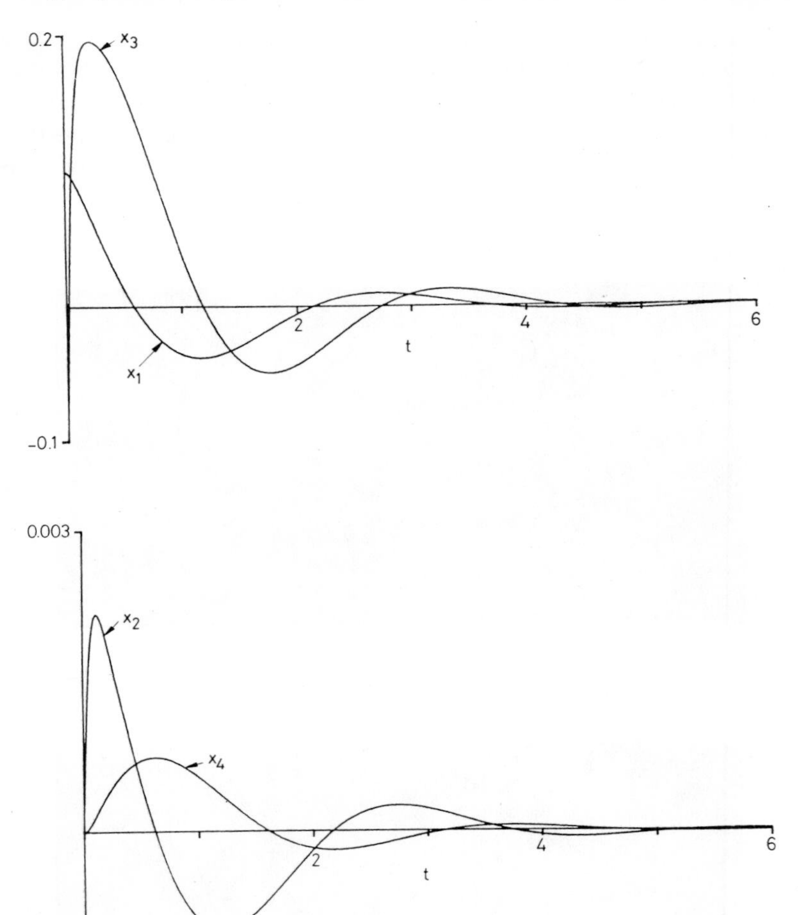

Fig. 3.3 *Response of the system with smoothing control* (3.65) *substituted*

five states ($n = 5$) and two control inputs ($m = 2$). The system matrix A and interface matrix B are:

$$A = \begin{bmatrix} -1 & 1 & 0 & 0 & 0 \\ 0 & -2 & 1 & 0 & 0 \\ 0 & 0 & 0 & 1 & 0 \\ 0 & 0 & 0 & 0 & 1 \\ 0 & 0 & 0 & 0 & 0 \end{bmatrix}, \tag{3.72}$$

$$B = \begin{bmatrix} 0 & 0 \\ 0 & 0 \\ 0 & 1 \\ 0 & 1 \\ 1 & 0 \end{bmatrix}. \tag{3.73}$$

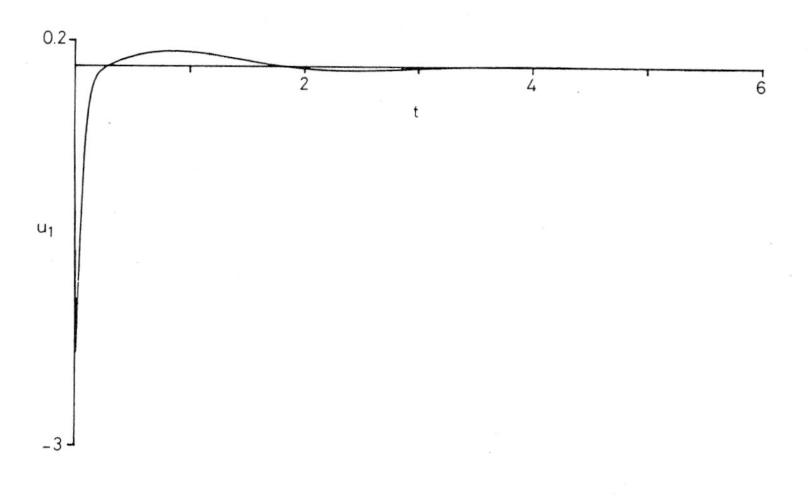

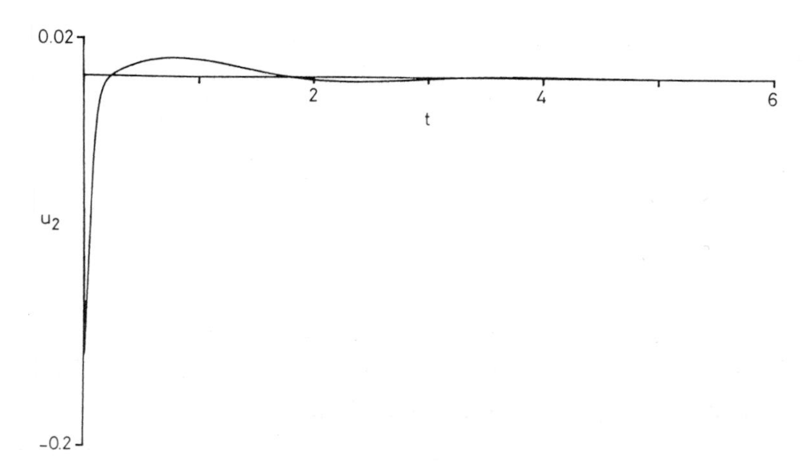

Fig. 3.4 *Control functions for Fig. 3.3*

The transformation matrix for this problem is

$$T = -\tfrac{1}{2}\begin{bmatrix} 2 & 0 & 0 & 0 & 0 \\ 0 & \sqrt{2} & 1 & -1 & 0 \\ 0 & \sqrt{2} & -1 & 1 & 0 \\ 0 & 0 & \sqrt{2} & \sqrt{2} & 0 \\ 0 & 0 & 0 & 0 & 2 \end{bmatrix}, \tag{3.74}$$

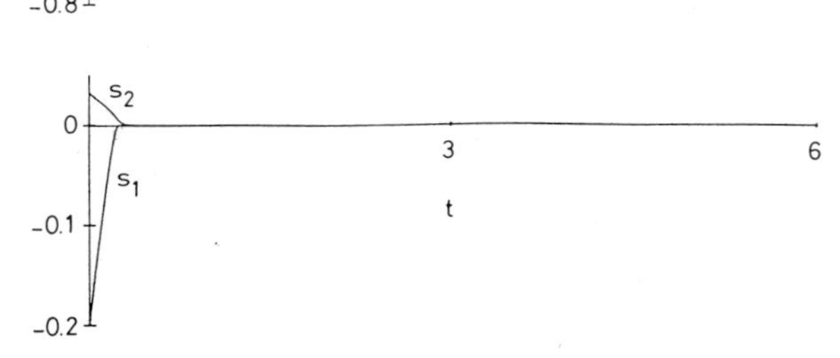

Fig. 3.5 *Simulation of the normal system under control* (3.79) *for* $x_0 = (0, 0, 0{\cdot}5, 0, 0)^T$

and the desired sliding mode spectrum is defined to be

$$\Lambda = \{-2{\cdot}5, \quad -3{\cdot}5, \quad -4{\cdot}5\} \tag{3.75}$$

Partitioning the product TAT^{T} gives

$$A_{11} = \frac{1}{4}\begin{bmatrix} -4 & 2\sqrt{2} & 2\sqrt{2} \\ 0 & -5+\sqrt{2} & -3-\sqrt{2} \\ 0 & -3+\sqrt{2} & -5-\sqrt{2} \end{bmatrix} \tag{3.76a}$$

$$A_{12} = \frac{1}{4}\begin{bmatrix} 0 & 0 \\ 2+\sqrt{2} & -2 \\ 2-\sqrt{2} & 2 \end{bmatrix} \tag{3.76b}$$

The matrix (3.76b) has full rank, so that $p = m = 2$. The eigenvector matrix associated with the robust assignment of the spectrum (3.75) to the closed loop system (3.52) is (to 4 DP.)

$$V = \begin{bmatrix} -0{\cdot}5547 & 0{\cdot}0000 & -0{\cdot}2747 \\ 0{\cdot}5884 & -0{\cdot}7071 & 0{\cdot}6800 \\ 0{\cdot}5884 & 0{\cdot}7071 & 0{\cdot}6800 \end{bmatrix} \tag{3.77}$$

which has a condition number

$$\kappa(V) = 6{\cdot}4061 \tag{3.78}$$

This compares with a minimum attainable condition number of 4·4995 for this system. The sensitivities of the individual eigenvalues can be calculated from (3.50) and are shown in Table 3.1.

Continuing with step F of the assignment algorithm and specifying the product matrix $CB = I_2$, the resulting hyperplane matrix is

$$C = \begin{bmatrix} -7{\cdot}4246 & 5{\cdot}6569 & 1{\cdot}4142 & 0 & 0 \\ -5{\cdot}2498 & -4{\cdot}0000 & 2{\cdot}5000 & -2{\cdot}5000 & -1{\cdot}0000 \end{bmatrix} \tag{3.79}$$

The following matrices are obtained for range space eigenvalues $-0{\cdot}5$, $-1{\cdot}5$:

$$L = \begin{bmatrix} -20 & -15{\cdot}5 & -5{\cdot}5 & -7{\cdot}5 & -6{\cdot}5 \\ -12 & -10{\cdot}5 & -7{\cdot}5 & -1 & 0 \end{bmatrix}$$

$$M = \begin{bmatrix} -2{\cdot}2627 & -1{\cdot}4142 & -0{\cdot}2828 & 0 & -0 \\ -1{\cdot}1429 & -0{\cdot}7143 & 0{\cdot}4286 & -0{\cdot}4286 & -0{\cdot}1429 \end{bmatrix}$$

$$N = \begin{bmatrix} -1{\cdot}1429 & -0{\cdot}7143 & 0{\cdot}4286 & -0{\cdot}4286 & -0{\cdot}1429 \\ -1{\cdot}6 & -1 & -0{\cdot}2 & 0 & 0 \end{bmatrix}$$

A simulation of the normal system under this control is given in Fig. 3.5 for $x(0) = (0, 0, 0{\cdot}5, 0, 0)^T$. In implementing this example a smoothed unit vector non-linearity of the form (3.65) with $\delta = 0{\cdot}01$ has been used in order to remove the chattering motion. Note the attainment of the sliding mode.

Table 13.1 *Sensitivities of sliding mode eigenvalues in system (3.76) using eigenvector matrix (3.77)*

Eigenvalue	Sensitivity
−2·5	3·2811
−3·5	1·0000
−4·5	3·2811

Alternatively the quadratic performance index approach may be preferred by the designer to specify the sliding hyperplanes. Suppose

$$Q = \text{diag}\,(1 \quad 1 \quad 1 \quad 10 \quad 10) \tag{3.80}$$

which corresponds to the performance index

$$\begin{aligned} J &= \tfrac{1}{2}\int_{t_s}^{\infty} x^T\, Q(t)\, x(t)\, dt \\ &= \tfrac{1}{2}\int_{t_s}^{\infty} (x_1^2 + x_2^2 + x_3^2 + 10\, x_4^2 + 10\, x_5^2) dt \end{aligned} \tag{3.81}$$

This weights the system states in a particular desired manner. Using the VASSYD CAD package the hyperplane matrix is calculated to be

$$C = \begin{bmatrix} -0{\cdot}0310 & -0{\cdot}0582 & -0{\cdot}3972 & -1{\cdot}0170 & 0 \\ 0{\cdot}0079 & 0{\cdot}0126 & 0{\cdot}1963 & -0{\cdot}1963 & -1 \end{bmatrix} \tag{3.82}$$

In the sliding mode A_{eq} has the eigenvalues 0, 0, −0·5314, −0·9789, −2·0080. Specifying the range space eigenvalues to be −0·5 and −1·5 one obtains

$$L = \begin{bmatrix} 0{\cdot}0039 & 0{\cdot}0016 & 0{\cdot}3071 & -0{\cdot}0981 & -1{\cdot}6963 \\ 0{\cdot}0110 & 0{\cdot}0398 & -0{\cdot}1816 & -0{\cdot}6404 & -0{\cdot}7191 \end{bmatrix}$$

$$M = \begin{bmatrix} -0{\cdot}0310 & -0{\cdot}0582 & -0{\cdot}3972 & -1{\cdot}0170 & 0{\cdot}0000 \\ 0{\cdot}0026 & 0{\cdot}0042 & 0{\cdot}0654 & -0{\cdot}0654 & -0{\cdot}3333 \end{bmatrix}$$

$$N = \begin{bmatrix} 0{\cdot}0026 & 0{\cdot}0042 & 0{\cdot}0654 & -0{\cdot}0654 & -0{\cdot}3333 \\ -0{\cdot}0219 & -0{\cdot}0411 & -0{\cdot}2809 & -0{\cdot}7191 & 0{\cdot}0000 \end{bmatrix}$$

Simulation results are presented in Fig. 3.6.

An alternate control design is a modal approach to specify the closed-loop eigenvalues as in Section 3.7. This technique has yielded condition numbers $\kappa(V)$ at least four times larger than the robust design for the above system. In many examples $\kappa(V)$ may be excessively large.

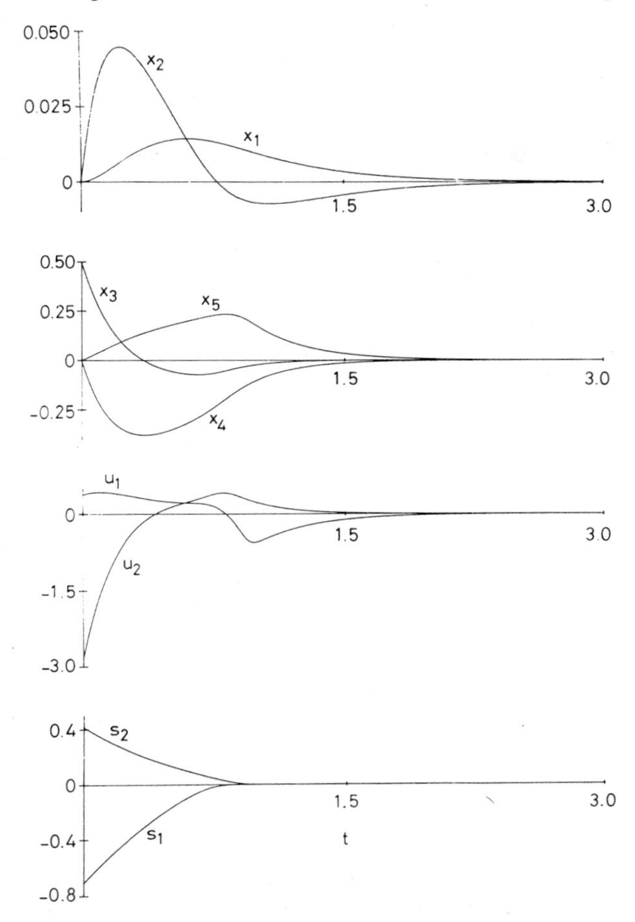

Fig. 3.6 *Simulation results*

3.11 Conclusion

In this chapter we have outlined an approach to hyperplane design in variable structure systems based on the assignment of the eigenstructure in the sliding mode. In order to clarify the number of degrees of freedom in this process and to simplify the numerical implementation of the design procedures, a particular canonical form for the system has been proposed. An assignability criterion that requires that the assignable eigenvectors lie in the null space of an $n \times n$ matrix has been shown to be equivalent to a criterion requiring only that an $(n - m) \times n$ matrix be checked. The partial freedom to assign the eigenvectors of the sliding mode whenever $m > 1$ may be exploited in order to maximise the robustness of the sliding mode eigenvalues. A control scheme to drive the state into the sliding mode has been outlined, along with a method of reducing the chattering component of the control. A computer-aided design package, VASSYD, has been developed by the authors to assist the designer. This package has been used for the examples in this chapter and applications in the fields of robotics and flight control; see, for example, Zinober *et al.* (1988), Zinober (1988) and Spurgeon *et al.*, (1989). It allows the user to carry out a complete design of hyperplane matrix and controls, including the case when the system parameters are varying within knowns band or the system is subject to noise. It also provides simple simulation facilities, graphical display and facilities for the storage of designs. The package runs interactively with an accompanying manual (Dorling, 1985) providing the underlying theory and a detailed description of the runtime procedure.

3.12 References

BURTON, J. A., and ZINOBER, A. S. I. (1986): 'Continuous approximation of variable structure control', *Int. J. Systems Sci.,* **17,** pp. 876–885

DORLING, C. M. (1985). 'The design of variable structure control systems: a manual for VASSYD package. Available from A. S. I. Zinober, University of Sheffield

DORLING, C. M., and ZINOBER, A. S. I. (1986): 'Two approaches to hyperplane design in multivariable variable structure control systems', *Int. J. Control,* **44,** pp. 65–82

DRAŽENOVIĆ, B. (1969): 'The invariance conditions in variable structure systems', *Automatica,* **5,** pp. 287–295

EL-GHEZAWI, O. M. E., ZINOBER, A. S. I., and BILLINGS, S. A. (1983): 'Analysis and design of variable structure systems using a geometric approach', *Int. J. Control,* **38,** pp. 657–671

ITKIS, U., (1976): 'Control systems of variable structure' (Wiley, NY)

KAUTSKY, J., and NICHOLS, N. K., (1983): 'Robust eigenstructure assignment in state feedback controls', University of Reading, Dept of Mathematics, Numerical Analysis Report NA/2/83

KAUTSKY, J., NICHOLS, N. K., and VAN DOOREN, P. (1985): 'Robust pole assignment in linear state feedback', *Int. J. Control,* **41,** pp. 1129–1155

KLEIN, G., and MOORE, B. C. (1977): 'Eigenvalue-generalised eigenvector assignment with state feedback', *IEEE Trans.,d* **AC-22,** pp. 140–141

KWAKERNAAK, H., and SIVAN, R. (1972): 'Linear optimal control system' (Wiley, NY)

MOORE, B. C., (1976): 'On the flexibility offered by state feedback in multivariable systems beyond closed loop eigenvalue assignment,' *IEEE Trans.,* **AC-21,** pp. 689–692

MUDGE, S. K., and PATTON, R. J. (1988): 'Analysis of the technique of robust eigenstructure assignment with application to aircraft control', *Proc. IEE,* **135D**, pp. 275–281

PORTER, B., and CROSSLEY, R. (1972): 'Modal control: Theory and applications' (Taylor & Francis, London)

RYAN, E. P. (1983): 'A variable structure approach to feedback regulation of uncertain dynamical systems', *Int. J. Control,* **38,** pp. 1121–1134

RYAN, E. P., and CORLESS, M. (1984): 'Ultimate boundedness and asymptotic stability of a class of uncertain dynamical systems via continuous and discontinuous feedback control', *IMA J. Math. Control Information,* **1,** pp. 223–242

SHAH, S. L., FISHER, D. G., and SEBORG, D. E., (1975): 'Eigenvalue/eigenvector assignment for multivariable systems and further results for output feedback control', *Electron. Lett.,* **11,** pp. 388–389

SINSWAT, V., and FALLSIDE, F., (1977): 'Eigenvalue/eigenvector assignment by state-feedback', *Int. J. Control,* **26,** pp. 389–403

SLOTINE J. -J. E. (1985): 'The robust control of robot manipulators', *Int. J. Robotic Res.*, **4,** pp. 49–64

SPURGEON, S. K. *et al.*, Chapter 5 in this book

SRINATHKUMAR, S., and RHOTEN, R. P. (1975): 'Eigenvalue/eigenvector assignment for multivariable systems', *Electron. Lett.*, **11,** pp. 124–125

UTKIN, V. I. (1978): 'Sliding modes and their application in variable structure systems (MIR, Moscow)

UTKIN, V. I., and YANG, K. D. (1978): 'Methods for constructing discontinuity planes in multidimensional variable structure systems', *Autom. Remote Control,* **39,** pp. 1466–1470

WILKINSON, J. H. (1965): 'The algebraic eigenvalue problem' (Clarendon Press, Oxford)

YOUNG, K. -K. D. (1977): 'Asymptotic stability of model reference systems with variable structure control', *IEEE Trans.*, **AC-22, 2,** pp. 279–281

YOUNG, K. -K. D. (1978): 'Design of variable structure model-following control systems', *IEEE Trans.*, **AC-23,** pp. 1079–1085

ZINOBER, A. S. I. (1988): 'Variable structure model-following control of a robot manipulator' Proc. 4th IFAC Symposium on Computer Aided Design in Control Systems, CADCS '88, Beijing, PR China, pp. 175–180

ZINOBER, A. S. I. (1988): 'Continuous self-adaptive control of a time-varying nonlinear crane system.' Proc 8th IFAC/IFORS Symposium on Identification and System Parameter Estimation, Beijing, PR China, pp. 611–616

ZINOBER, A. S. I., YEW, M. K., and PATTON, R. J. (1988): 'Variable structure control of flight dynamics', Proc. IEE Int. Conf., Control 88, Oxford, pp. 707–711

Chapter 4

Subspace attractivity and invariance: Ultimate attainment of prescribed dynamic behaviour

E.P. Ryan

University of Bath.

4.1 Introduction

Consider a dynamical system of the form

$$\dot{x}(t) \quad = \quad \mathrm{M}(t, x(t), u(t)), \ x(t) \in \mathscr{R}^n, \ u(t) \in \ \mathscr{R}^m \tag{4.1}$$

If (4.1) is a mathematical description of some 'real-world' process, then, inevitably, some imprecision or uncertainty will have been encountered during the modelling procedure; the function M is, at best, a 'reasonable' approximation of the actual controlled vector field. In many cases, the determination of an acceptably accurate model of form (4.1) may prove impossible. For example, it may only be possible to determine a model structure with unknown parameters (with values in a known set); furthermore, realistic processes are generally subject to unknown extraneous disturbance (possibly with known bounds). In order to handle such system uncertainty, we adopt the following approach in the present Chapter. For each t, we assume that the right hand side of (4.1) is unknown, but that the set of all possible right hand sides is known, i.e. we replace (4.1) by the controlled differential inclusion

$$\dot{x}(t) \in \mathscr{M}(t, x(t), u(t)), \ x(t) \in \mathscr{R}^n, \ u(t) \in \mathscr{R}^m \tag{4.2}$$

where $(t, x, u) \mapsto \mathscr{M}(t, x, u)$ is a known set-valued map, or multifunction, on which additional structure will be imposed later. In essence, $\mathscr{M}(t, x(t), u(t))$ is the set of all possible 'velocities' $\dot{x}(t)$ of the uncertain system at time t.

The adoption of a controlled differential inclusion formulation of the underlying system provides an additional bonus from the analytic viewpoint. Given that control *synthesis* is an objective, discontinuous feedback is a natural candidate in many contexts. Within a framework of controlled differential *equations* (e.g. model (4.1)), this gives rise to many analytical difficulties (since the hypotheses of the classical theory do not hold). However, as we shall demonstrate, such difficulties can be circumvented by invoking the theory of differential inclusions (see also Leitmann, 1979; Gutman, 1979; and Gutman & Palmor, 1983).

The outline of the Chapter is as follows. We begin with a brief resumé of relevant concepts and results from the theory of differential inclusions. The class of uncertain systems to be considered in then made precise. We continue with a treatment of the variable structure systems concept of an invariant subspace $\mathscr{S}$ (with prescribed dynamic behaviour therein) and construct a discontinuous feedback strategy which renders $\mathscr{S}$ globally finite-time attractive (thereby ensuring ultimate attainment of prescribed dynamic behaviour). The approach is essentially that of Ryan & Corless (1984) (with origins in Corless & Leitmann, 1981), subsequently recast in a differential inclusion setting by Goodall & Ryan (1986, 1988). Finally, our results are extended to problems of tracking and model following.

4.2 Background concepts

Before describing the class of systems to be considered and posing the basic problem of stabilisation by feedback, we introduce some notation and catalogue some properties of multifuctions to be invoked later (see Aubin & Cellina, 1984, for details).

Let Z_1 and Z_2 be real Banach spaces. A multifunction $\mathscr{Z}: Z_1 \rightrightarrows Z_2$ is a mapping from Z_1 into the subsets of Z_2. Here, we will only be concerned with *compact-valued* multifunctions. The continuity concept best suited to our requirements is that of *upper semi-continuity*:

Definition 1: A compact-valued multifunction $\mathscr{Z}: Z_1 \rightrightarrows Z_2$ is upper semi-continuous at $z_1 \in Z_1$ if, for each $\varepsilon > 0$, there exists $\delta > 0$ such that $\mathscr{Z}(z) \subset \mathscr{Z}(z_1) + \varepsilon\mathscr{B}_{Z_2}$ for all $z \in z_1 + \delta\mathscr{B}_{\mathscr{Z}_1}$.
$\mathscr{Z}$ is said to be upper semi-continuous if it is upper semi-continuous at each $z_1 \in Z_1$.

In the above definition the following notation is used. $\mathscr{B}_Z$ denotes the open unit ball in Banach space Z, its closure will be denoted $\bar{\mathscr{B}}_Z$. For $z \in Z$ and $\mathscr{S}_1$, $\mathscr{S}_2 \subset Z$, $z + \mathscr{S}_1$ denotes the set $\{z + s_1: s_1 \in \mathscr{S}_1\}$ and $\mathscr{S}_1 + \mathscr{S}_2$ denotes the set $\{s_1 + s_2: s_1 \in \mathscr{S}_1; s_2 \in \mathscr{S}_2\}$. Note that, in the case of a singleton-valued multifunction (i.e. a function in the usual sense), upper semi-continuity is equivalent to continuity in the usual sense.

Proposition 1: Let $K \subset Z_1$ be compact and let $\mathscr{Z}: Z_1 \rightrightarrows Z_2$ be upper semi-continuous with compact values. Then $\mathscr{Z}(K) \subset Z_2$ is compact.

Proposition 2: Let $\mathscr{Z}_1: Z_1 \rightrightarrows Z_2$ and $\mathscr{Z}_2: Z_2 \rightrightarrows Z_3$ have non-empty values. Define the composition $\mathscr{Z}_2 \circ \mathscr{Z}_1 \colon Z_1 \rightrightarrows Z_3$ by

$$z \mapsto (\mathscr{Z}_2 \circ \mathscr{Z}_1)(z) := \bigcup_{v \in \mathscr{Z}_1(z)} \mathscr{Z}_2(v)$$

If $\mathscr{Z}_1$ and $\mathscr{Z}_2$ are upper semi-continuous, then $\mathscr{Z}_2 \circ \mathscr{Z}_1$ is upper semi-continuous.

Proposition 3: If $f\colon Z_1 \to \mathscr{R}$ is continuous and $\mathscr{Z}\colon Z_1 \rightrightarrows Z_2$ is upper semi-continuous with compact values, then $f\mathscr{Z}\colon Z_1 \rightrightarrows Z_2$, $z \mapsto f(z)\mathscr{Z}(z)$, is upper semi-continuous with compact values.

Throughout this Chapter, multifunctions on $\mathscr{R} \times X \times X$ or on $\mathscr{R} \times X \times U$ or on $\mathscr{R} \times X$ or on U will be considered, with $X := \mathscr{R}^n$ and $U := \mathscr{R}^m$ $(1 \leqslant m \leqslant n)$. The Euclidean inner product (on X or U as appropriate) and the induced norm will be denoted by $\langle , \rangle$ and $\| \ \|$, respectively. For a linear map L, $\|L\| = [\max \sigma(L^T L)]^{1/2}$, where σ denotes spectrum. For a non-empty compact subset K of X or U, $\xi(K) := \max\{\|v\| \colon v \in K\}$ and $\xi(\varnothing) := 0$.

Finally, we state a fundamental existence result (Aubin & Cellina, 1984; Filippov, 1988) for differential inclusions of the form

$$\dot{x}(t) \in \mathscr{W}(t, x(t)), \; x(t) \in X \tag{4.3a}$$

$$x(t_0) = x^0. \tag{4.3b}$$

By a (local) solution of (4.3) we mean an absolutely continuous function $x\colon [t_0, t_1) \to X$ satisfying (4.3*b*) and (4.3*a*) almost everywhere.

Lemma 1: If $\mathscr{W}\colon \mathscr{R} \times X \rightrightarrows X$ is upper semi-continuous with convex and compact values, then, for each $(t_0, x^0) \in \mathscr{R} \times X$, the initial-value problem (4.3) admits a local solution.

4.3 The system

The system to be considered is of the form

$$\dot{x}(t) \in \mathscr{M}(t, x(t), u(t)), \; x(t) \in X, \; u(t) \in U \tag{4.4a}$$

where the multifunction $\mathscr{M}$ has the following structure:

$$\mathscr{M}(t, x, u) = Ax + B[u + \mathscr{G}(t, x) + \beta(t)\mathscr{G}_c(u)] \tag{4.4b}$$

and where the upper semi-continuous (in fact, continuous) multifunction $\mathscr{G}_c$ has the simple form

$$\mathscr{G}_c(u) = \|u\|\bar{\mathscr{B}}_U \tag{4.4c}$$

Furthermore, the following assumptions are made:
Assumption 1:

(i) The pair (A, B) is controllable, and B has full rank m;
(ii) $\mathscr{G}\colon \mathscr{R} \times X \rightrightarrows U$ is upper semi-continuous with convex and compact values

(iii) β: $\mathscr{R} \to [0, \kappa_0]$ is continuous, with $\kappa_0 < 1$.

The multifunction $(t, x, u) \mapsto \mathscr{G}(t, x) + \beta(t)\|u\|\bar{\mathscr{B}}_U$ may be interpreted as modelling uncertainty in the system, in which case we note that, in the terminology of Barmish & Leitmann (1982) and Barmish, Corless & Leitmann (1983), the uncertainty is *matched*. A simple class of matched uncertain systems which can be embedded into the more general class under consideration is typified by the following:

$$\dot{x}(t) = Ax(t) + B[u(t) + g(t, x(t), u(t))]$$

where g is an unknown Carathéodory function with

$$\|g(t, x, u)\| \leqslant \psi(t, x) + \beta(t)\|u\| \quad \forall \quad (t, x, u)$$

and where the bounding functions ψ: $\mathscr{R} \times X \to \mathscr{R}^+ := [0, \infty)$ and β: $\mathscr{R} \to [0, \kappa_0]$, $\kappa_0 < 1$, are known and continuous.

The first problem we study is that of stabilisation by feedback, viz. determine a (time-dependent) feedback strategy $(t, x) \mapsto \mathscr{F}(t, x)$ such that, loosely speaking, all trajectories of (4.4), with $u(t) \in \mathscr{F}(t, x(t))$, exhibit "stable" behaviour. As our class of admissible feedbacks, we take the class of upper semi-continuous multifunctions $\mathscr{F}$: $\mathscr{R} \times X \rightrightarrows X$ with non-empty, convex and compact values, henceforth referred to as *generalised feedbacks*. In the scalar case, a simple example of a generalised feedback is the multifunction $\mathscr{F} = \mathscr{sgn}$ given by

$$\mathscr{sgn}: x \mapsto \begin{cases} \{+1\}, & x > 0 \\ [-1, 1], & x = 0 \\ \{-1\}, & x < 0 \end{cases}$$

i.e. a relay-type control function which is permitted to take any value between (and including) its limits (± 1) for zero argument; in practice, parasitic inertias in the system may give rise to chattering and, in this sense, the physical system itself implicitly determines the appropriate control value.

4.4 Stabilisation by generalised feedback

4. 4.1 Problem formulation

We now state the first problem as: determine a generalised feedback $\mathscr{F}$ which renders $\{0\}$ a globally uniformly asymptotically stable state of the feedback-controlled differential inclusion

$$\dot{x}(t) \in \mathscr{M}_{\mathscr{F}}(t, x(t)) \tag{4.5a}$$

$$\mathscr{M}_{\mathscr{F}}(t, x) = Ax + B[\mathscr{F}(t, x) + \mathscr{G}(t, x) + \beta(t)(\mathscr{G}_c \circ \mathscr{F})(t, x)] \tag{4.5b}$$

in the following sense.

Definition 3: $\{0\}$ is a globally uniformly asymptotically stable state of (4.5) if:

(i) *Existence and continuation of solutions*: for each $(t_0, x^0) \in \mathscr{R} \times X$, there exists a (local) solution x: $[t_0, T) \to X$ with $x(t_0) = x^0$, and every such solution can be extended into a solution on $[t_0, \infty)$
(ii) *Uniform boundedness of solutions*: for each $\delta > 0$, there exists $d(\delta) > 0$ such that $x(t) \in d(\delta)\mathscr{B}_x$ for all t on every solution x: $[t_0, \infty) \to X$ with $x(t_0) \in \delta\mathscr{B}_x$
(iii) *Uniform stability of* $\{0\}$: for each $\delta > 0$, there exists $D(\delta) > 0$ such that $x(t) \in \delta\mathscr{B}_x$ for all t on every solution x: $[t_0, \infty) \to X$ with $x(t_0) \in D(\delta)\mathscr{B}_x$
(iv) *Uniform attractivity of* $\{0\}$: for each $\delta > 0$ and $\varepsilon > 0$, there exists $\tau(\delta, \varepsilon) \geq 0$ such that $x(t) \in \varepsilon\mathscr{B}_x$ for all $t \geq t_0 + \tau(\delta, \varepsilon)$ on every solution x: $[t_0, \infty) \to X$ with $x(t_0) \in \delta\mathscr{B}_x$.

4.4.2 Outline of approach

The method to be followed in the design of a generalised feedback $\mathscr{F}$ draws on concepts from (*a*) variable structure systems theory and (*b*) Lyapunov-based theory. In particular, we adopt the concept of an invariant $(n - m)$-dimensional subspace $\mathscr{S} \subset X$ from variable structure systems theory; this subspace is implicitly determined by the selection of a set $\Lambda = \{\lambda_1, \lambda_2, \ldots, \lambda_{n-m}\} \subset \mathscr{C}^-$ (the open left half complex plane) of ideal model eigenvalues. The generalised feedback $\mathscr{F}$ is then designed (via Lyapunov based analysis) to ensure that: (i) $\mathscr{S}$ is positively invariant under $\mathscr{M}_{\mathscr{F}}$ and the flow on $\mathscr{S}$ coincides with that of a linear system with prescribed spectrum Λ; (ii) $\mathscr{S}$ is globally finite-time attractive in the sense that $\mathscr{S}$ is ultimately (i.e. in finite time) attained on every solution of (4.5).

4.4.3 The subspace $\mathscr{S}$

Let $L_1 \in \mathscr{R}^{(n-m)\times n}$ be such that ker L_1 = im B. Define $L_2 := (B^T B)^{-1} B^T$ and

$$L := \begin{bmatrix} L_1 \\ L_2 \end{bmatrix}$$

with inverse $L^{-1} = R = [R_1 \vdots B]$, then

$$LAR = \begin{bmatrix} L_1 AR_1 & L_1 AB \\ L_2 AR_l & L_2 \mathrm{AB} \end{bmatrix}; \quad LB = \begin{bmatrix} O \\ I \end{bmatrix}$$

Let $\Lambda = \{\lambda_1, \lambda_2, \ldots, \lambda_{n-m}\} \subset \mathscr{C}^-$ be any selected ideal model spectrum. Since (by Assumption 1(i)) the pair (A, B) is controllable, we may conclude that $(L_1 AR_1, L_1 AB)$ is also a controllable pair. Hence there exists $S \in \mathscr{R}^{m\times(n-m)}$ such that

$$\sigma(L_1 AR_1 + L_1 ABS) = \sigma(L_1 AS^*) = \Lambda$$

where, for notational convenience, we have introduced $S^* := R_1 + BS$. The $(n - m)$-dimensional subspace $\mathscr{S} \subset X$ is now defined as

$$\mathscr{S} := \ker \bar{S}, \text{ where } S := L_2 - SL_1 \tag{4.6}$$

Lemma 2: Let $C \in \mathscr{R}^{m \times m}$ be such that $\sigma(C) \subset \mathscr{C}^-$ and define $F := C\bar{S} \quad \bar{S}A$. Then (i) $\sigma(A + BF) \subset \mathscr{C}^-$ and (ii) $\mathscr{S} \subset \mathscr{R}^n$ is an $(A + BF)$-invariant subspace.
Proof:

$$\text{Let } T = \begin{bmatrix} L_1 \\ \bar{S} \end{bmatrix} = \begin{bmatrix} L_1 \\ L_2 - SL_1 \end{bmatrix}$$

with inverse

$$T^{-1} = [R_1 + BS \,\vdots\, B],$$

then

$$T(A + BF)T^{-1} = \begin{bmatrix} L_1 AS^* & L_1 AB \\ O & C \end{bmatrix}$$

whence

$$\begin{aligned} \sigma(A + BF) &= \sigma(T(A + BF)T^{-1}) = \sigma(L_1 AS^*) \cup \sigma(C) \\ &= \Lambda \cup \sigma(C) \subset \mathscr{C}^- \end{aligned}$$

which establishes (i).

Now, noting that $\bar{S}BF = F = C\bar{S} \quad \bar{S}A$, we may conclude that $\bar{S}(A + BF)x = 0$ for every $x \in \mathscr{S} = \ker \bar{S}$, thereby proving (ii).

The above lemma ensures that, in the absence of uncertainty (i.e. when $\mathscr{G} \equiv \{0\}$ and $\beta \equiv 0$), the feedback law $u(t) = Fx(t)$ asymptotically stabilises system (4.4) and renders the subspace $\mathscr{S}$ invariant. Moreover, trajectories of the controlled system in $\mathscr{S}$ are precisely those trajectories x expressible as $x = S^*w$, where w is a solution of the linear system $\dot{w} = (L_1 AS^*)w$ with prescribed spectrum $\sigma(L_1 AS^*) = \Lambda$.

In the next Section, the linear feedback operator F is augmented by an appropriate multifunction to yield a generalised feedback $\mathscr{F}$ which preserves the property of asymptotic stability of the origin in the presence of uncertainty.

4.4.4 Generalised feedback $\mathscr{F}$

Recalling that $\sigma(L_1 AS^*) = \Lambda \subset \mathscr{C}^-$ and $\sigma(C) \subset \mathscr{C}^-$, let $P_1 \in \mathscr{R}^{(n-m)\times(n-m)}$ and $P_2 \in \mathscr{R}^{m \times m}$ denote the unique positive-definite symmetric solutions of the Lyapunov equations

$$P_1(L_1 AS^*) + (L_1 AS^*)^T P_1 + I = 0 \tag{4.7}$$

and

$$P_2 C + C^T P_2 + I = 0. \tag{4.8}$$

Define multifunctions $\mathscr{P}_1: X \rightrightarrows \mathscr{R}^{n-m}$ and $\mathscr{P}_2: X \rightrightarrows U$ by

$$x \mapsto \mathcal{P}_1(x) := \{v \in \mathcal{R}^{n-m}: \langle v, P_1 L_1 x \rangle \geqslant 0\}$$

and

$$x \mapsto \mathcal{P}_2(x) := \{u \in U: \langle u, P_2 \bar{S} x \rangle \geqslant 0\}.$$

$\mathcal{P}_1$ plays a role only in the discussion in Section 4.7
We now introduce the upper semi-continuous multifunction $\mathcal{U}$: $U \rightrightarrows U$ (an n-dimensional analogue of the multifunction *sgn* of Section 4.3), defined by

$$u \mapsto \mathcal{U}(u) := \begin{cases} \|u\|^{-1}u; & u \neq 0 \\ \bar{\mathcal{B}}_U; & u = 0 \end{cases} \tag{4.9}$$

Then, the proposed generalised feedback is given by

$$(t, x) \mapsto \mathcal{F}(t, x) := Fx - \varrho(t, x)\mathcal{U}(P_2\bar{S}x) \tag{4.10a}$$

where F is defined as in Lemma 2 and ϱ is any continuous function $\mathcal{R} \times X \to [0, \infty)$ satisfying

$$\varrho(t, x) \geqslant \varrho_0(t, x) := (1 - \beta(t))^{-1} [\beta(t)\|Fx\| + \xi(\mathcal{G}(t, x) \cap \mathcal{P}_2(x)) + \gamma] \tag{4.10b}$$

and where $\gamma > 0$ is a design parameter. Clearly, $\mathcal{F}$ takes convex and compact values; moreover, the continuity of ϱ and the upper semi-continuity of $\mathcal{U}$ ensure, by Proposition 3, that $\mathcal{F}$ is also upper semi-continuous. Thus, $\mathcal{F}$ qualifies as a generalised feedback. We note that $\mathcal{F}$ is singleton-valued except on the set $\Gamma_{\mathcal{F}} = \mathcal{R} \times \mathcal{S}$.

4.4.5 *Global uniform asymptotic stability of the zero state of the feedback system*

To establish existence of local solutions, it suffices to show that the feedback system (4.5), with $\mathcal{F}$ given by (4.10), satisfies the hypotheses of Lemma 1. Since $\beta\mathcal{G}_c$ and $\mathcal{F}$ are upper semi-continuous with compact values, it follows from Propositions 1 and 2 that $(t, x) \mapsto \beta(t)(\mathcal{G}_c \circ \mathcal{F})(t, x)$ is also upper semi-continuous with compact values; thus, $\mathcal{M}_{\mathcal{F}}$ is the sum of upper semi-continuous compact-valued multifunctions and hence is itself upper semi-continuous with compact values. It remains only to show that $\mathcal{M}_{\mathcal{F}}$ is convex-valued. Now, $\beta(t)(\mathcal{G}_c \circ \mathcal{F})(t, x) = \beta(t)\xi(\mathcal{F}(t, x))\bar{\mathcal{B}}_U$, and hence is convex; it immediately follows that $\mathcal{M}_{\mathcal{F}}(t, x)$ is convex, being the sum of convex sets. For each (t_0, x^0), the existence of a local solution x: $[t_0, \tau) \to X$ of (4.5) with $x(t_0) = x^0$ is assured by Lemma 1.

To establish that every such local solution can be extended into a solution on $[t_0, \infty)$, we argue as follows. Consider the behaviour, along local solutions, of the function V: $X \to [0, \infty)$ given by

$$V(x) := \tfrac{1}{2}\langle x, Q_\zeta x\rangle \tag{4.11a}$$

where

$$Q_\zeta := L_1^T P_1 L_1 + \zeta \bar{S}^T P_2 \bar{S} \tag{4.11b}$$

and ζ is positive real number to be specified later. In particular, along each local solution x: $[t_0, t_1) \to X$, we have

$$\dot{V}(x(t)) \in \mathscr{L}(t, x(t)) \qquad \text{for almost all } t \tag{4.12}$$

where

$$\mathscr{L}(t, x) := \{\langle Q_\zeta x, v\rangle : v \in \mathscr{M}_{\mathscr{F}}(t, x)\} \tag{4.12b}$$

Writing $\mathscr{L}_1(t, x) = \{\langle P_2\bar{S}x, \bar{S}v\rangle : v \in \mathscr{M}_{\mathscr{F}}(t, x)\}$, then a straightforward calculation (using (4.7)) reveals that

$$\begin{aligned}\mathscr{L}(t, x) &= \{\langle P_1 L_1 x, L_1 v\rangle + \zeta\langle P_2\bar{S}x, \bar{S}v\rangle : v \in \mathscr{M}_{\mathscr{F}}(t, x)\}\\ &= \langle P_1 L_1 x, L_1 Ax\rangle + \zeta\mathscr{L}_1(t, x)\\ &= \langle P_1 L_1 x, (L_1 AS^*)L_1 x\rangle + \langle P_1 L_1 x, L_1 \bar{A}BSx\rangle + \zeta\mathscr{L}_1(t, x)\\ &= -\tfrac{1}{2}\|L_1 x\|^2 + \langle P_1 L_1 x, L_1 \bar{A}BSx\rangle + \zeta\mathscr{L}_1(t, x) \end{aligned}\tag{4.13}$$

Moreover, using (4.8) and (4.10a), we have

$$\begin{aligned}\mathscr{L}_1(t, x) &= -\tfrac{1}{2}\|\bar{S}x\|^2 - \rho(t, x)\|P_2\bar{S}x\|\\ &\quad + \{\langle v, P_2\bar{S}x\rangle : v \in \mathscr{G}(t, x) + \beta(t)(\mathscr{G}_c \circ \mathscr{F})(t, x)\}\end{aligned}$$

whence, in view of (4.10b),

$$\max \mathscr{L}_1(t, x) \leqslant -\tfrac{1}{2}\|\bar{S}x\|^2$$

which, when combined with (4.12b) and (4.13), yields

$$\max \mathscr{L}(t, x) \leqslant -\tfrac{1}{2}\left\langle \begin{bmatrix}\|L_1 x\|\\ \|\bar{S}x\|\end{bmatrix}, E_\zeta \begin{bmatrix}\|L_1 x\|\\ \|\bar{S}x\|\end{bmatrix}\right\rangle \tag{4.14a}$$

where

$$E_\zeta := \begin{bmatrix} 1 & -\|P_1 L_1 AB\|\\ -\|P_1 L_1 AB\| & \zeta\end{bmatrix}. \tag{4.14b}$$

Thus, choosing $\zeta > \|P_1 L_1 AB\|^2$, (4.12a) and (4.14) ensure that every local solution x: $[t_0, t_1) \to X$ evolves within the compact set $\{x: V(x) \leqslant V(x(t_0))\} \subset X$ and hence can be continued indefinitely. Moreover, the properties of uniform boundedness of solutions and uniform stability and attractivity of $\{0\}$ readily follow.

We summarise the above analysis as:

Lemma 3: $\{0\}$ is a globally uniformly asymptotically stable state of the feedback controlled differential inclusion system (4.5) and (4.10).

4.4.6 Positive invariance and finite-time attractivity of $\mathscr{S}$

To establish that every solution of the feedback controlled differential inclusion attains the subspace $\mathscr{S}$ in finite time and thereafter remains in $\mathscr{S}$, we consider the behaviour of the function $V_2: X \to [0, \infty)$, $x \mapsto \frac{1}{2}\langle \bar{S}x, P_2\bar{S}x\rangle$ along solutions. Note that $V_2(x) = 0$ if and only if $x \in \mathscr{S}$.

Lemma 4: For each $(t_0, x^0) \in \mathscr{R} \times X$, the subspace $\mathscr{S}$ is attained in finite time

$$t_f \;<\; \gamma^{-1}[2\|P_2\|V_2(x^0)]^{1/2} \tag{4.15}$$

on every solution $x: [t_0, \infty) \to X$ of the feedback controlled system (4.5) and (4.10) with $x(t_0) = x^0$; moreover, $x(t) \in \mathscr{S}$ for all $t \geqslant t_0 + t_f$.

Proof: Along every solution $x: [t_0, \infty) \to X$ with $x(t_0) = x^0$ the following holds almost everywhere:

$$\dot{V}(x(t)) \leqslant -\gamma\|P_2\bar{S}x(t)\| \;\leqslant\; -\gamma[2\|P_2^{-1}\|^{-1}V_2(x(t))]^{1/2} \tag{4.16}$$

which, on integration, ensures that $V_2(x(t_0 + t_f)) = 0$ (and hence $x(t_0 + t_f) \in \mathscr{S}$) for some t_f satisfying (4.15). Furthermore, (4.16) ensures that $V_2(x(t)) = 0$ (and hence $x(t) \in \mathscr{S}$) for all $t \geqslant t_f + t_0$. Therefore, the subspace $\mathscr{S}$ is positively $\mathscr{M}_{\mathscr{F}}$-invariant.

We note that the upper bound in (4.15) on the time to attain $\mathscr{S}$ is inversely proportional to the value of the design parameter $\gamma > 0$, and hence, in principle, can be made arbitrarily small (of course, this is achieved at the expense of high control gains).

4.4.7 Linear motion in $\mathscr{S}$

In view of the $\mathscr{M}_{\mathscr{F}}$-invariance of $\mathscr{S}$, $\bar{S}x(\cdot) \equiv 0$ on every solution $x: [t_0, \infty) \to X$ of the feedback system with $x(t_0) = x^0 \in \mathscr{S}$. Thus motion in $\mathscr{S}$ is governed by the asymptotically stable linear ideal model (with prescribed spectrum Λ)

$$\dot{w}(t) \;=\; L_1AS^*w(t), \qquad w(t) \in \mathscr{R}^{n-m} \tag{4.17a}$$

$$w(t_0) \;=\; L_1x^0 \tag{4.17b}$$

in the sense of the following lemma:

Lemma 5: For each $(t_0, x^0) \in \mathscr{R} \times \mathscr{S}$, the feedback controlled differential inclusion (4.5) and (4.10) admits the solution $x: [t_0, \infty) \to \mathscr{S}$, with $x(t_0) = x^0$, given by

$$t \mapsto x(t) \;=\; S^* \exp\,[L_1AS^*(t-t_0)]L_1x^0$$

Proof: In view of the $\mathscr{M}_{\mathscr{F}}$-invariance of $\mathscr{S}$, solutions of (4.5) and (4.10) which satisfy $x(t_0) = x^0 \in \mathscr{S}$ are precisely those functions $x(\cdot)$ which also satisfy the following almost everywhere:

$$L_1\dot{x}(t) \in L_1\mathscr{M}_{\mathscr{F}}(t, x(t))$$

$$\bar{S}\dot{x}(t) \;=\; 0.$$

Now, $L_1 \mathcal{M}_{\mathcal{F}}(t, x) = \{L_1 Ax\} = \{(L_1 AS^*)L_1 x\}$ for all $x \in \mathcal{S}$. Thus, solutions of (4.5) and (4.10) which satisfy $x(t_0) = x^0 \in \mathcal{S}$ are precisely those functions expressible by

$$\begin{bmatrix} L_1 \\ \bar{S} \end{bmatrix} x(\cdot) = \begin{bmatrix} w(\cdot) \\ 0 \end{bmatrix}$$

where $w(\cdot)$ solves the linear initial-value problem (4.17). Since $w(\cdot)$ is uniquely determined by (4.17) as $t \mapsto w(t) = \exp\,[L_1 AS^*(t - t_0)]L_1 x^0$, we deduce that

$$x(\cdot) = \begin{bmatrix} L_1 \\ \bar{S} \end{bmatrix}^{-1} \begin{bmatrix} w(\cdot) \\ 0 \end{bmatrix} = [S^* \quad B] \begin{bmatrix} w(\cdot) \\ 0 \end{bmatrix} = S^* w(\cdot)$$

is the unique solution of (4.5) and (4.10) emanating from $x^0 \in \mathcal{S}$ at time t_0.

4.4.8 Summary of results

The following theorem summarises the results of this Section and is a direct consequence of Lemmas 3, 4 and 5.

Theorem 1: The generalised feedback $\mathcal{F}$, given by (4.10), renders the zero state of the differential inclusion system (4.5) globally uniformly asymptotically stable. Moreover, the dynamic behaviour of the feedback controlled system ultimately corresponds to that of a linear system with prescribed spectrum in the sense that $x(t) = S^* w(t)$ for all $t \geq t_0 + \gamma^{-1}[\|P_2^{-1}\|\,\|P_2\|]^{1/2}\delta$ on every solution $x(\cdot)$ with $\|\bar{S}x(t_0)\| \leq \delta$, where $w(\cdot)$ is a solution of the linear system $\dot{w} = L_1 AS^* w$ with prescribed spectrum $\sigma(L_1 AS^*) = \Lambda \subset \mathscr{C}^-$.

4.5 Tracking in the presence of uncertainty

In this Section, we extend the results of the previous Section to the problem of tracking a prescribed function $x^*(\cdot)$. The approach is similar in concept to that of Corless, Leitmann & Ryan (1985). Our underlying system is given by (4.5) and all assumptions of Section 4.4 remain in force.

To ensure *feasibility* of the motion x^* to be tracked, we make the following assumption:

Assumption 2: There exists a function $v^* \in L^\infty(\mathscr{R};\ \mathscr{R}^m)$ such that $\dot{x}^*(t) = Ax^*(t) + Bv^*(t)$ almost everywhere.

Now, let $v: \mathscr{R} \to \mathscr{R}^m$ be any bounded continuous function, which approximates v^* in some appropriate sense, and let $\gamma^*: \mathscr{R} \to [0, \infty)$ be a bounded continuous function satisfying

$$\gamma^*(t) \geq \|v(t) - v^*(t)\| \qquad \text{for almost all } t$$

Clearly, if v^* is continuous, then $v = v^*$ and $\gamma^* = 0$ suffice. Let S, $\bar{S}$, S^*, F, P_1 and P_2 be defined as in Section 4.4. Define multifunction $\mathscr{P}_2^*: \mathscr{R} \times X \rightrightarrows U$ by

$$(t, x) \mapsto \mathscr{P}_2^*(t, x) := \{u \in U: \langle u, P_2\bar{S}[x - x^*(t)]\rangle \geqslant 0\}$$

Then the proposed generalised feedback for tracking x^* is given by

$$(t, x) \mapsto \mathscr{F}(t, x) := F[x - x^*(t)] + v(t) - \varrho^*(t, x)\mathscr{U}(P_2\bar{S}[x - x^*(t)]) \tag{4.18a}$$

where $\mathscr{U}$ is defined by (4.9) and ϱ^* is any continuous function $\mathscr{R} \times X \to [0, \infty)$ satisfying

$$\varrho^*(t, x) \geqslant \varrho_0^*(t, x) := (1 - \beta(t))^{-1}[\beta(t)\|F[x - x^*(t)] + v(t)\| + \xi(\mathscr{G}(t, x) \cap \mathscr{P}_2^*(t, x)) + \gamma + \gamma^*(t)] \tag{4.18b}$$

with $\gamma > 0$. We shall now demonstrate that the feedback controlled system (4.5) and (4.18) asymptotically tracks x^* in the sense that, on every solution x, $\|x(t) - x^*(t)\| \to 0$ as $t \to \infty$ and moreover the convergence is (ultimately) exponential; furthermore, if $x(t_0) = x^*(t_0)$ then $x(t) = x^*(t)$ for all $t \geqslant t_0$. These results will be established by proving that $\{0\}$ is a globally uniformly asymptotically stable state (in the sense of Definition 3) of a differential inclusion system, the solutions of which contain the tracking error functions $x(\cdot) - x^*(\cdot)$.

4.5.1 *The tracking error system*

Introducing the transformed state (tracking error) $e(t) = x(t) - x^*(t)$, then in view of (4.5) and Assumption 2 we have

$$\dot{e}(t) \in Ae(t) + B[\mathscr{F}(t, x(t)) + \mathscr{G}(t, x(t)) + \beta(t)(\mathscr{G}_c \circ \mathscr{F})(t, x(t)) - v^*(t)] \tag{4.19}$$

Defining multifunctions $\mathscr{F}^*$, $\mathscr{G}^*$, and $\mathscr{M}^*_{\mathscr{F}^*}$ as

$$(t, e) \mapsto \mathscr{F}^*(t, e) := \mathscr{F}(t, e + x^*(t)) \tag{4.20}$$

(where $\mathscr{F}$ is given by (4.18)),

$$(t, e) \mapsto \mathscr{G}^*(t, e) := \mathscr{G}(t, e + x^*(t)) \tag{4.21}$$

and

$$(t, e) \mapsto \mathscr{M}^*_{\mathscr{F}^*}(t, e) := Ae + B[\mathscr{F}^*(t, e) + \mathscr{G}^*(t, e) + \beta(t)(\mathscr{G}_c \circ \mathscr{F}^*)(t, e) + \gamma^*(t)\bar{\mathscr{B}}_U - v(t)] \tag{4.22}$$

Noting that, for each t, the right hand side of (4.19) is contained in $\mathscr{M}^*_{\mathscr{F}^*}(t, e(t))$, we may conclude that stability properties of the differential inclusion system

$$\dot{e}(t) \in \mathscr{M}^*_{\mathscr{F}^*}(t, e(t)) \tag{4.23}$$

hold *a fortiori* for system (4.19). Thus, asymptotic tracking of x^* is assured if it can be shown that $\{0\}$ is a globally uniformly asymptotically stable state of system (4.23).

4.5.2 Asymptotic tracking of x^*

Noting that $\mathcal{M}^*_{\mathcal{F}^*}$ is upper semi-continuous with convex and compact values, an argument similar to that used in the proof of Theorem 1 (via Lemmas 3, 4 and 5) yields the following result:

Theorem 2: The generalized feedback $\mathcal{F}^*$, given by (4.18) and (4.20) renders $\{0\}$ a globally uniformly asymptotically stable state of (4.23). Moreover, the dynamic behaviour of (4.23) ultimately corresponds to that of a linear system with prescribed spectrum in the sense that $e(t) = S^*w(t)$ for all $t \geqslant t_0 + \gamma^{-1}[\|P_2^{-1}\|\|P_2\|]^{1/2}\delta$ on every solution $e(\cdot)$ with $\|\bar{S}e(t_0)\| \leqslant \delta$, where $w(\cdot)$ is a solution of the linear system $\dot{w} = L_1AS^*w$ with prescribed spectrum $\sigma(L_1AS^*) = \Lambda \subset \mathscr{C}^-$.

Thus asymptotic tracking of x^* is assured by the generalised feedback (4.18); moreover, the tracking error ultimately decays exponentially with rate determined *a priori* by the prescribed spectrum $\Lambda \subset \mathscr{C}^-$.

4.6 Model following in the presence of uncertainty

We now extend the results of Section 4.4 to a model-following problem (similar problems are considered by, for example, Erzberger, 1968; Chen, 1973; Shaked, 1977; Landau, 1979; Balestrino, DeMaria & Zinober, 1984; Zinober, 1985) in which a nonlinear system to be emulated by (4.5) is specified *a priori* as

$$\begin{aligned} \dot{z}(t) &= A^\# z(t) + B^\# v(t) + f^\#(t, z(t)), \quad z(t) \in \mathscr{R}^n, \\ v(t) &\in \mathscr{V}^\#(t) \subset \mathscr{R}^p. \end{aligned} \tag{4.24}$$

It is assumed that the triple $(A^\#, B^\#, f^\#)$ is known and that $f^\#$ is continuous. Furthermore, $t \mapsto \mathscr{V}^\#(t)$ is assumed to be a known continuous multifunction with non-empty convex and compact values in $\mathscr{R}^p$. The only additional model information available at time t to the controller is the state $z(t)$. The availability, to the controller, of $\mathscr{V}^\#$ rather than exact model input information distinguishes the problem under consideration from that considered by Corless, Goodall, Leitmann & Ryan (1985). All assumptions of Section 4.4 relating to the underlying system (4.5) remain in force. To ensure feasibility of the model, we make the following assumptions.

Assumption 3:

(i) *Global existence and uniqueness of model solutions*: For each $(t_0, z^0, v) \in \mathscr{R} \times \mathscr{R}^n \times L^\infty(\mathscr{R}; \mathscr{R}^p)$, there exists a unique absolutely continuous function $z(\cdot) = Z(\cdot, t_0, z^0, v)$: $[t_0, \infty) \to \mathscr{R}^n$ satisfying (4.24) almost everywhere, with $z(t_0) = z^0$. (Z is the model state transition map.)
(ii) *Model-following conditions* (see, for example, Erzberger, 1968; and Chen, 1973):
(a) $\mathrm{im}(A^\# - A) \subset \mathrm{im}\, B$

(*b*) $\operatorname{im} B^{\#} \subset \operatorname{im} B$
(*c*) $\operatorname{im} f^{\#} \subset \operatorname{im} B$

In relation to the continuous multifunction $\mathscr{V}^{\#}$, we introduce some further concepts (Aubin & Cellina, 1984). The *Chebishev radius* $r_c(t)$ of $\mathscr{V}^{\#}(t) \subset \mathscr{R}^p$ is first defined as

$$r_c(t) := \inf\{r: \mathscr{V}^{\#}(t) \subset v + r\mathscr{B}_{\mathscr{R}^p} \text{ for some } v \in \mathscr{R}^p\}$$

then it can be shown (Aubin & Cellina, 1984) that the set

$$\bigcap_{r > r_c(t)} \{v: \mathscr{V}^{\#}(t) \subset v + r\mathscr{B}_{\mathscr{R}^p}\}$$

consists of a single point $c(\mathscr{V}^{\#}(t))$, called the *Chebishev centre* of $\mathscr{V}^{\#}(t)$ and, moreover, $\mathscr{V}^{\#}(t) \subset c(\mathscr{V}^{\#}(t)) + r_c(t)\mathscr{B}_{\mathscr{R}^p}$. Furthermore, the functions $r_c(\cdot)$ and $c(\mathscr{V}^{\#}(\cdot))$ are continuous. In essence, $c(\mathscr{V}^{\#}(t))$ will play the role of an estimate of the true (but unknown) model input $v(t) \in \mathscr{V}^{\#}(t)$; $r_c(t)$ then provides a bound for the input error $\tilde{v}(t) = v(t) - c(\mathscr{V}^{\#}(t))$, i.e. $\|\tilde{v}(t)\| \leqslant r_c(t)$.

Let L_1, L_2, S, $\bar{S}$, S^*, F, P_1 and P_2 be defined as in Section 4.4. Then, in view of Assumption 3(ii),

$$A^{\#} - A = BK^{\#}, \qquad \text{where } K^{\#} := L_2(A^{\#} - A) \tag{4.25a}$$

$$B^{\#} = BL^{\#}, \qquad \text{where } L^{\#} := L_2 B^{\#} \tag{4.25b}$$

$$f^{\#} = Bg^{\#}, \qquad \text{where } g^{\#} := L_2 f^{\#} \tag{4.25c}$$

The proposed feedback for model following is now given as

$$(t, x, z) \mapsto \mathscr{F}^{\#}(t, x, z) := F[x - z] + K^{\#} z + L^{\#} c(\mathscr{V}^{\#}(t))$$
$$+ g^{\#}(t, z) - \varrho^{\#}(t, x, z)\mathscr{U}(P_2\bar{S}[x - z]) \tag{4.26a}$$

where $\mathscr{U}$ is defined by (4.9) and $\varrho^{\#}$ is any continuous function $\mathscr{R} \times X \times X \to [0, \infty)$ satisfying

$$\varrho^{\#}(t, x, z) \geqslant \varrho_0^{\#}(t, x, z) := (1 - \beta(t))^{-1}[\beta(t)\|F[x - z]$$
$$+ K^{\#} z + L^{\#} c(\mathscr{V}^{\#}(t)) + g^{\#}(t, z)\|$$
$$+ \xi(\mathscr{G}(t, x)) \cap \mathscr{P}_2(x - z)) + \|L^{\#}\| r_c(t) + \gamma] \tag{4.26b}$$

with $\mathscr{P}_2$ defined as in Section 4.4 and $\gamma > 0$.

It remains to show that the feedback controlled system (4.5) and (4.26) asymptotically follows system (4.24) in the sense that, for each $(t_0, x^0, z^0, v) \in \mathscr{R} \times X \times X \times L^{\infty}(\mathscr{R}; \mathscr{R}^p)$ with $v(t) \in \mathscr{V}^{\#}(t)$ for all t, $\|x(t) - z(t)\| \to 0$ as $t \to \infty$ on every solution x of the feedback system with $x(t_0) = x^0$, where $z(\cdot) = Z(\cdot, t_0, z^0, v)$; furthermore, if $x^0 = z^0$, then $x(t) = z(t)$ for all $t \geqslant t_0$.

As in the previous Section, these results will be established by considering a differential inclusion system, the solutions of which contain the model-following error functions.

4.6.1 The model-following error system
Introducing the error $e(t) = x(t) - z(t) = x(t) - Z(t, t_0, z^0, v)$, then

$$\dot{e}(t) \in Ae(t) + B[\mathcal{F}^{\#}(t, x, z) + \mathcal{G}(t, x) + \beta(t)(\mathcal{G} \circ \mathcal{F}^{\#})(t, x, z) - K^{\#} z - L^{\#} v(t) - g^{\#}(t, z)] \tag{4.27}$$

Now, for each (t_0, z^0, v), define the multifunction

$$(t, e) \mapsto \mathcal{M}^{\#}_{(t_0, z^0, v)}(t, e) := Ae + B[\![Fe + \mathcal{G}(t, e + z(t)) + \beta(t)(\mathcal{G} \circ \mathcal{F}^{\#})(t, e + z(t), z(t)) + \|L^*\| r_c(t)\bar{\mathcal{B}}_U - \varrho^{\#}(t, e + z(t), z(t))\mathcal{U}(P_2 \bar{S} e)] \tag{4.28}$$

where $z(t) = Z(t, t_0, z^0, v)$. That (4.28) is upper semi-continuous with convex and compact values is readily seen (recall that $r_c(\cdot)$ is continuous). Moreover, for each t, the right hand side of (4.27) is contained in $\mathcal{M}^{\#}_{(t_0, z^0, v)}(t, e(t))$. Thus, all stability properties of solutions $e: [t_0, \infty) \to \mathcal{R}^n$ of the system

$$\dot{e}(t) \in \mathcal{M}^{\#}_{(t_0, z^0, v)}(t, e(t)) \tag{4.29}$$

hold *a fortiori* for solutions of (4.27). Thus, if it can be established that $\{0\}$ is an asymptoically stable state (4.29), then asymptotic tracking of (4.24) is assured by the feedback system (4.5) and (4.26).

4.6.2 Asymptotic model-following
Again, an argument similar to that used in establishing Theorem 1 (via Lemmas 3, 4 and 5) yields the following result.

Theorem 3: $\{0\}$ is a globally uniformly asymptotically stable state of system (4.29). Moreover, the dynamic behaviour of (4.29) ultimately corresponds to that of a linear system with prescribed spectrum $\Lambda \subset \mathcal{C}^-$ (in the same sense as in Theorems 1 and 2).

Thus asymptotic model following of (4.24) is assured for the feedback system (4.5) and (4.26); moreover, the model-following error ultimately decays exponentially with rate determined *a priori* by the spectrum $\Lambda \subset \mathcal{C}^-$.

4.7 Discussion

Underlying the analysis of this Chapter is a hypothesis of matched uncertainty. We conclude with some brief remarks on relaxing this structural hypothesis. Loosely speaking, some unmatched or residual uncertainty (i.e. functions/multifunctions of (t, x) with values not contained in im B) can be tolerated provided that it is bounded in a rather specific manner (see Ryan & Corless, 1984; and Goodall & Ryan, 1988 for details). In particular, unmatched or residual uncertainty modelled by a multifunction $(t, x) \mapsto \mathcal{G}_r(t, x) \subset (\text{im } B)^{\perp}$ is admissible provided that $\mathcal{G}_r$ is upper semi-continuous with convex and compact values, and that there exist constants $\kappa_1 < \frac{1}{2}\|P_1\|^{-1}, \kappa_2, \kappa_3$ such that, for all (t, x),

$$\xi(L_1 \mathscr{G}_r(t, x) \cap \mathscr{P}_1(x)) \leqslant \kappa_1 \|L_1 x\| + \kappa_2 \|\bar{S}x\| + \kappa_3$$

The introduction of $\mathscr{G}_r$ necessitates the following modification to the generalised feedback (4.10): the term $\xi(\mathscr{G}(t, x) \cap \mathscr{P}_2(x))$ in the definition of the function ϱ_0 in (4.10*b*) is replaced by $\xi((\mathscr{G}(t, x) - SL_1 \mathscr{G}_r(t, x)) \cap \mathscr{P}_2(x))$.

In the context of the stabilisation problem of Section 4.4, the effect of such unmatched uncertainty is as follows: positive invariance and global finite-time attractivity of $\mathscr{S}$ are preserved; however, if $\kappa_3 > 0$, then asymptotic stability of $\{0\}$ is lost and replaced by global uniform asymptotic stability of a calculable compact set Σ ($\{0\} \subset \Sigma \subset \mathscr{S}$), the diameter of which increases with increasing values of κ_1 and κ_3 (and $\Sigma = \{0\}$ if $\kappa_3 = 0$); moreover, prescribed linear motion in $\mathscr{S}$ cannot be guaranteed but is approximated with calculable error determined by κ_1 and κ_3.

In the contexts of the tracking problem of Section 4.5 and the model-following problem of Section 4.6, the effects of unmatched uncertainty are such that asymptotic tracking/model-following cannot be guaranteed; instead, tracking/model-following with error ultimately bounded within a calculable set is assured and again the diameter of this set increases with increasing values of κ_1 and κ_3. Finally, we remark that there is some limited scope for relaxing the feasibility Assumption 2 (for tracking) and the model-following Assumption 3(ii) and refer the reader to Corless, Leitmann & Ryan (1985), and Coreless, Goodall, Leitmann & Ryan (1985), respectively, for details.

4.8 References

AUBIN, J. P., and CELLINA, A. (1984): 'Differential Inclusions', (Springer-Verlag, Berlin, Heidelberg, New York, Tokyo)

BALESTRINO, A., DEMARIA, G., and ZINOBER, A. S. I. (1984): 'Nonlinear adaptive model-following control', *Automatica,* **20,** pp. 559–568

BARMISH, B. R., CORLESS, M., and LEITMANN, G. (1983): 'A new class of stablizing controllers for uncertain dynamical systems', *SIAM J. Control & Optimiz.,* **21,** pp. 246–255

BARMISH, B. R., and LEITMANN, G. (1982): On ultimate boundedness control of uncertain systems in the absence of matching conditions, *IEEE Trans.* ,**AC-27,** pp. 153–157

CHEN, Y. T. (1973): 'Perfect model-following with real model', *Proc.* JACC, pp. 287–293

CORLESS, M., GOODALL, D. P., LEITMANN, G., and RYAN, E. P. (1985): 'Model-following controls for a class of uncertain dynamical systems' Proc. IFAC Symposium on Identification and System Parameter Estimation, pp. 1895–1899 (Pergamon Press, Oxford)

CORLESS, M., and LEITMANN, G. (1981): 'Continuous state feedback guaranteeing uniform ultimate boundedness for uncertain dynamic systems', *IEEE Trans.,* **AC-26,** 1139–1141

CORLESS, M., LEITMANN, G. and RYAN, E. P. (1985): 'Tracking in the presence of bounded uncertainties'. Proc. 4th IMA Inter. Conf. on Control Theory, pp. 347–361 (Academic Press, London)

ERZBERGER, H. (1968): 'Analysis and design of model following control systems by state space techniques'. Proc. JACC, pp. 572–581

FILIPPOV, A. F. (1988): 'Differential Equations with Discontinuous Righthand Sides', (Kluwer Academic Publishers, Dordrecht, Boston, London)

GOODALL, D. P., and RYAN, E. P. (1986): 'Controlled differential inclusions and stabilization of uncertain dynamical systems'. Proc. 25th IEEE Conf. on Decision & Control, Athens, pp. 31–33

GOODALL, D. P., and RYAN, E. P. (1988): 'Feedback controlled differential inclusions and stabilization of uncertain dynamical systems', *SIAM J. Control & Optimiz.*, 26, pp. 1431–1441

GUTMAN, S. (1979): 'Uncertain dynamical systems, a Lyapunov min-max approach', *IEEE Trans.* **AC-24,** pp. 437–443

GUTMAN, S., and PALMOR, Z. (1983): 'Properties of min-max controllers in uncertain dynamical systems', *SIAM J. Control & Optimiz.*, **20,** 850–861

LANDAU, I. D. (1979). 'Adaptive control: The model reference approach' (Marcel Dekker, New York)

LEITMANN, G. (1979): 'Guaranteed asymptotic stability for some linear systems with bounded uncertainties', *Trans. ASME, J. Dynamic Systems, Measurement & Control,* **101,** pp. 212–216

RYAN, E. P., and CORLESS, M. (1984): 'Ultimate boundedness and asymptotic stability of a class of uncertain dynamical systems via continuous and discontinuous feedback control', *IMA J. Math. Control & Information,* **1,** pp. 223–242

SHAKED, U. (1977): Design of general model following control systems, *Int. J. Control,* **25,** pp. 213–238

ZINOBER, A. S. I. (1985): Properties of adaptive discontinuous model-following control systems. Proc. 4th IMA Inter. Conf. on Control Theory, pp. 337–346, (Academic Press, London)

Chapter 5

Model-following control of time-varying and nonlinear avionics systems

S. K. Spurgeon

Loughborough University of Technology

M. K. Yew and A. S. I. Zinober

University of Sheffield

R. J. Patton

University of York

5.1 Introduction

The basic philosophy of variable structure control (VSC) design (Itkis, 1976; Utkin, 1977; Zinober, 1979) is that the structure of the feedback control is altered as the state crosses sliding surfaces in the state space. The control in the sliding mode remains on the intersection of these surfaces, yielding total invariance to a class of system parameter variations and disturbances. Precise closed-loop eigenvalue placement can be achieved in time-varying and uncertain systems.

The designer of a model-following VSC system may choose the stable eigenvalues of the closed-loop error system directly specifying the sliding hyperplanes. Eigenvectors may also be assigned. Alternatively, a cost functional can be minimised in order to enable desirable weightings to be placed upon particular elements of the error state when in the sliding mode. Thereafter, a suitable control law is computed to ensure that sliding motion is achieved sufficiently rapidly. This design technique ensures that the desired error transients are obtained.

Here the model-following VSC design approach using the CAD package, VASSYD, will be described for two different desired aircraft configurations. Firstly, a stability augmentation system will be designed for the lateral motion model of a light aircraft. This will illustrate the eigenstructure assignment option for hyperplane design. Secondly, a pitchpointing/vertical-translation manoeuvre for a conventional combat aircraft will be considered. This will involve incorporation of a pilot demand in the form of a model input. An ideal model can be defined using eigenstructure assignment to modally shape the aircraft's response. A fully nonlinear simulation incorporating wind tubulence will be used to assess the performance of the controller. The performance of the VSC system will be compared with the performance of the purely linear control used to define the model dynamics in order to demonstrate the inherent robustness of the VSC design technique. For both cases a continuous nonlinear controller is shown to yield trajectories arbitarily close to the sliding mode with the complete elimination of chattering motion, which is undersirable for the aircraft actuator system.

5.2 Model-following control systems

The technique of model reference adaptive control (MRAC) has been discussed in the literature by numerous authors, using two main approaches; Lyapunov and hyperstability theory (Landau, 1979). The actual plant is required to follow the dynamic behaviour of a specific model plant. The model is included as part of the system and it specifies the main design objectives. The adaptive controller should force the error between the model and the plant states to zero as time tends to infinity.

Consider the plant and model described by

$$\dot{x}(t) = A\,x(t) + B\,u(t) + f(x,u) \tag{5.1}$$

$$\dot{x}_m(t) = A_m\,x_m(t) + B_m\,r(t) \tag{5.2}$$

where x is the plant state n-vector, x_m is the model state n-vector, u is the plant control m-vector, f represents nonlinear additive terms and r is the model input. The tracking error vector is

$$e(t) = x_m(t) - x(t) \tag{5.3}$$

We shall assume that the pairs (A,B), (A_m, B_m) and (A_m, B) are stabilisable; i.e. any uncontrollable modes lie in the left-hand half of the s-plane (Balestrino, De Maria and Zinober, 1984; Young, 1979; Zinober, 1984). The model matrix A_m

is assumed to be stable. The plant matrices A and B, and the vector f may be uncertain and time-varying. The upper and lower magnitude bounds of the elements of these matrices and f are assumed to be known to the designer. It can be easily shown that

$$\dot{e}(t) = A_m e(t) + [(A_m - A)x(t) - f + B_m r(t)] - Bu(t) \qquad (5.4)$$

Matching conditions are necessary for perfect model-following to ensure that the equality $x = x_m$ can be attained. The error e and its derivative should be zero for any input r and plant state x, for $e(0) = 0$. To determine the necessary conditions, consider the linear control law

$$u = Ku + Rr + Pf \qquad (5.5)$$

Substituting (5.5) into (5.4) gives

$$\dot{e} = A_m e + (A_m - A - BK)x - (I + BP)f + (B_m - BR)r \quad (5.6)$$

We require that

$$(A_m - A - BK)x - (I + BP)f + (B_m - BR)r = 0$$

This can be achieved if the following matching conditions (Chan, 1968; Erzber ger, 1973) are satisfied:

$$\text{rank}[B\ B_m] = \text{rank}[B\ A - A_m] = \text{rank}\ B \qquad (5.7)$$

Equivalently G, f and $A - A_m$ should lie in the range space of B, $R(B)$. Then R and K will exist and have the values

$$R = B^* B_m,\ K = B^*(A_m - A) \qquad (5.8)$$

where $B^* = (B^{\mathrm{T}} B)^{-1} B^I$ is the pseudo-inverse of B.

5.3 Variable structure model-following control systems

Into the framework of the above we shall now add a variable structure nonlinear control which ensures that all error states are attracted to a particular sliding subspace in the error state space, and thereafter approach the origin (e = 0) on this subspace (Young, 1978; Zinober, 1984). Variable structure model-following control systems are characterised by control discontinuities on m switching hyperplanes, s_i, in the tracking error space where

$$s = (s_1 s_2 \ldots s_m)^T = Ce(t) \qquad (5.9)$$

The controller takes the form

$$u = Le + u_N(e) + Kx + Rr \qquad (5.10)$$

where u_N is a suitable discontinuous (or smoothed nonlinear) control to be described below. The linear feedback Le ensures rapid approach towards the sliding subspace, $s = 0$, and also maintains the error state in this subspace in the case of the nominal system. The nonlinear term is required to attract the state into the subspace and maintain it on $s = 0$ in the presence of uncertain disturbances and parameter variations (Ryan and Corless, 1984). The variable structure model-following design philosophy is similar to that of the multivariable VSC regulator (Dorling and Zinober, 1987; Young, 1978; Zinober, 1979). The analysis is, however, carried out in the error state space.

Idealised sliding occurs if, at a point on the sliding subspace $s(e) = 0$, the directions of motion along the error state trajectories on either side of the surface are not away from the switching surface. The state then slides and remains for some finite time on the surface $s(e) = 0$.

Before studying the nature of the control enforcing sliding for the case of constant C, i.e. fixed sliding hyperplanes, the behaviour of the system dynamics during sliding needs to be considered. Suppose that the sliding mode exists on all the hyperplanes. Then

$$s = Ce = 0 \text{ and } \dot{s} = C\dot{e} = 0 \qquad (5.11)$$

The equations governing the system dynamics may be obtained by substituting an equivalent control U for the original control u. Assuming CB is nonsingular, we have from (5.4) and (5.11)

$$\dot{s} = C[A_m e + (A_m - A)x - f + B_m r - Bu] = 0$$

and

$$U = (CB)^{-1} C[Fe + (F - A)x - f + B_m r] \qquad (5.12)$$

Substitution of $u = U$ into (5.4) yields an nth order system equation

$$\dot{e} = [I - B(CB)^{-1}C)\,[A_m e + (A_m - A)x - f + B_m r] \qquad (5.13)$$

For the perfect model-following case, since the matching conditions (5.7) correspond to the invariance conditions of VSC systems (Young, 1978) we obtain

$$\dot{e} = [I - B(CB)^{-1}C]\,A_m e = A_e e \qquad (5.14)$$

where A_e is the equivalent error system matrix. The system is formally equivalent to a system of lower order which has dimension $(n - m)$, since m state variables can be expressed in terms of the remaining $n - m$ variables. During sliding the state error vector e remains on the $(n - m)$-dimensional manifold of the intersection of the m sliding surfaces. It can be readily shown that m eigenvalues of the equivalent system are equal to zero. The error system response is then determined only by the other $n - m$ eigenvalues (El-Ghezawi, Zinober and Billings, 1982).

Since the matching conditions hold, we can design our system using simply

$$\dot{e} = A_m e - B\,u \qquad (5.15)$$

because the terms in [ ] in (5.4) lie in $R(B)$. These terms do, however, affect the gain requirements of the nonlinear control as will be mentioned below. Therefore, given A_m, B and assuming the pair (A_m, B) is stabilisable, a matrix C can be found such that the error tends to zero with increasing time. Sufficiently fast error decay in the sliding subspace can be ensured by placing $n - m$ eigenvalues deep in the left-hand half of the complex plane.

Eqn. (5.14) is independent of the actual values of the control u during the sliding mode and depends only on the choice of C. The function of the nonlinear control is to drive the error state into $s = 0$, and thereafter to maintain it within the sliding subspace. The determination of the matrix C may be completed without prior knowledge of the form of the control vector u (although the reverse is not true). The indepence of the sliding motion from the actual control, arises from the nonsingularity of CB. In geometrical terms, the condition that CB is nonsingular implies that the null space of C, $N(C)$, and the range splace of B, $R(B)$, are complementary subspaces; i.e. $N(C) \cap R(B) = 0$. If a system satisfies the perfect model-following conditions, the error state lies entirely within $N(C)$ during the sliding mode, and the behaviour of the ideal system in $N(C)$ is unaffected by the form of the control. On the other hand, if $|CB| = 0$, then $N(C)$ and $R(B)$ have points in common, and motion in $N(C)$ will no longer be independent of u. Since we are concerned with the selection of the hyperplane matrix C, we may reasonably demand that CB is nonsingular in our design process.

In the user-friendly CAD package VASSYD, (Dorling, 1985; Dorling and Zinober, 1986, 1988) three methods, namely direct eigenvalue/eigenvector assignment, robust eigenvalue assignment and quadratic minimisation methods, are available for the design of the hyperplane matrix C. Firstly, the designer can select some or all of the elements of the closed-loop eigenvectors corresponding to the nonzero sliding mode eigenvalues in order to achieve desired modal error responses during the sliding mode. Secondly, the robust eigenstructure assignment approach employed in VASSYD is based on a method proposed by Kautsky and Nichols (1983) in which the eigenvectors are rotated within the assignable subspaces to maximise the sum of the angles between each pair of vectors. For further details, see Dorling and Zinober (1988). Finally, the designer can minimise a cost functional in which the integrand is quadratic in terms of the error state.

In this study, both robust eigenstructure assignment and quadratic minimisation methods will be employed. To ensure that the error state approaches the sliding surface in a suitably short time, the m range space eigenvalues should be specified to determine the linear part of the control, Le. The time in which the error state reaches $N(C)$ depends on the values of these eigenvalues; i.e. the further left the eigenvalues are in the complex plane, the shorter the time in which the system will reach the sliding subspace $s = 0$, from an error state not on the subspace.

The equivalent control (5.12) is the effective control function which yields sliding motion. The value of U is in practice the average value of u which maintains the state on the surface $s = 0$. The actual control consists of a low-frequency (average) component and a high-frequency (chatter) component. The occurrence of chattering motion is due to the state continually passing backwards and forwards through the sliding manifold as the control switches repeatedly between two (or more) distinct values. This chattering is undesirable in many physical situations in which the actuation mechanism may be damaged by rapid switching. The simplest way of removing the high frequency chattering is to 'soften' the action of the nonlinearity by substituting a continuous nonlinear approximation for the discontinuous part of the control. A control incorporating a softened nonlinearity is termed a smoothing control and the 'unit-vector' form (Balestrino, De Maria and Sciavicco, 1982; Balestrino, De Maria and Zinober, 1984; Burton and Zinober, 1986) has been found to be easily implemented. Its form is

$$u_N(e) = \varrho\, Ne/(\| Me \| + \delta) \tag{5.16}$$

where $\rho = \text{diag}\,(\rho_1 \rho_2 \ldots \rho_m)$, the gains ρ_i, are positive, and δ is a small positive constant. The error state remains in the neighbourhood of $N(C)$ during the sliding mode. (Note that for $\delta = 0$ we obtain a control discontinuous on $s = 0$.) The details of the calculation of M, N and L can be found in Ryan and Corless (1984) and Dorling and Zinober (1986). C, M and N relate to the same sliding subspace.

5.4 Example 1: Lateral motion stability augmentation system design

The aircraft is a complex, highly nonlinear system whose aerodynamic parameters change considerably during flight, resulting in variations in the dynamic performance over the flight envelope. This presents a challenge to flight control systems design.

The equations of motion of the fixed wing aircraft are normally decoupled into two sub-systems each describing the longitudinal and the lateral motions. The longitudinal motions are concerned with forward velocity and pitching excursions whilst the lateral motions relate to roll, yaw and sideslip velocity. Here a controller will be designed for the lateral motion of a particular remotely piloted vehicle for which the equations of motion and the aerodynamic parameters are well defined (Aslin, 1985; Aslin, Patton and Dorling, 1985, Yew 1987).

A stick-fixed linearisation of the non-linear dynamic airframe system yields the following linear model:

$$A = \begin{bmatrix} y_v & 0 & y_r - U_0 & g & 0 & y_\zeta & 0 \\ l_v & l_p & l_r & 0 & 0 & l_\zeta & l_\xi \\ n_v & n_p & n_r & 0 & 0 & n_\zeta & n_\xi \\ 0 & 1 & 0 & 0 & 0 & 0 & 0 \\ 0 & 0 & 1 & 0 & 0 & 0 & 0 \\ 0 & 0 & 0 & 0 & 0 & \mu_1 & 0 \\ 0 & 0 & 0 & 0 & 0 & 0 & \mu_2 \end{bmatrix}$$

$$B = \begin{bmatrix} 0 & 0 \\ 0 & 0 \\ 0 & 0 \\ 0 & 0 \\ K_1 & 0 \\ 0 & K_2 \end{bmatrix}$$

where $x \in R^7$ and $u \in R^2$. The state vector for the lateral motion comprises the variables

$$x = \begin{bmatrix} v \\ p \\ r \\ \phi \\ \psi \\ \zeta \\ \xi \end{bmatrix} \begin{matrix} \text{sideslip velocity, m/s} \\ \text{roll rate, rad/s} \\ \text{yaw rate, rad/s} \\ \text{roll angle, rad} \\ \text{heading angle, rad} \\ \text{rudder angle, rad} \\ \text{aileron angle, rad} \end{matrix}$$

and the control vector comprises

$$u = \begin{bmatrix} \zeta_c \\ \xi_c \end{bmatrix} \begin{matrix} \text{rudder angle demand, rad} \\ \text{aileron angle demand, rad} \end{matrix}$$

The aerodynamic derivatives for the aircraft are denoted by y_v, n_v, ... , etc., U_0 is the nominal forward air speed and μ_1, K_1, μ_2, K_2 are the rudder and aileron actuator parameters. Also, in the design of autopilots it has been established that certain modes are associated with particular dynamic subsystems. Thus, the

Dutch roll mode is associated with the yaw subsystem states v and r, whilst the spiral mode dominates the roll response.

The design of a model-following control system consists of choosing C and u so that the plant variables will faithfully follow the model states. The plant is represented by the linearised equations of lateral motion of a light aircraft, and the model by the equations corresponding to the nominal aircraft model with ideal response characteristics provided by suitable linear feedback control. The unmanned aircraft under consideration has the following nominal trim flight linear lateral motion model matrices corresponding to a 33 ms^{-1} airspeed:

$$A = \begin{bmatrix} -0{\cdot}277 & 0 & -32{\cdot}9 & 9{\cdot}81 & 0 & -5{\cdot}432 & 0 \\ -0{\cdot}1033 & -8{\cdot}325 & 3{\cdot}75 & 0 & 0 & 0 & -28{\cdot}64 \\ 0{\cdot}3649 & 0 & -0{\cdot}639 & 0 & 0 & -9{\cdot}49 & 0 \\ 0 & 1 & 0 & 0 & 0 & 0 & 0 \\ 0 & 0 & 1 & 0 & 0 & 0 & 0 \\ 0 & 0 & 0 & 0 & 0 & -10 & 0 \\ 0 & 0 & 0 & 0 & 0 & 0 & -5 \end{bmatrix}$$

and

$$B = \begin{bmatrix} 0 & 0 \\ 0 & 0 \\ 0 & 0 \\ 0 & 0 \\ 0 & 0 \\ 20 & 0 \\ 0 & 10 \end{bmatrix}$$

The ideal model matrix A_m is given by $A_m = A + BF$, where F is a feedback matrix which gives a predetermined ideal response (see Mudge and Patton, 1988*a*).

$$F = \begin{bmatrix} 0{\cdot}0246 & 0{\cdot}0021 & 0{\cdot}0789 & -0{\cdot}0647 & -0{\cdot}00282 & -0{\cdot}349 & 0{\cdot}53 \\ 0{\cdot}0028 & -0{\cdot}0078 & 0{\cdot}0784 & 0{\cdot}181 & 0{\cdot}0292 & -0{\cdot}102 & -0{\cdot}082 \end{bmatrix}$$

$$A_m = \begin{bmatrix} -0{\cdot}277 & 0 & -32{\cdot}9 & 9{\cdot}81 & 0 & -5{\cdot}432 & 0 \\ -0{\cdot}1033 & -8{\cdot}325 & 3{\cdot}75 & 0 & 0 & 0 & -28{\cdot}64 \\ 0{\cdot}3649 & 0 & -0{\cdot}639 & 0 & 0 & -9{\cdot}49 & 0 \\ 0 & 1 & 0 & 0 & 0 & 0 & 0 \\ 0 & 0 & 1 & 0 & 0 & 0 & 0 \\ 0{\cdot}492 & 0{\cdot}0422 & 1{\cdot}578 & -1{\cdot}295 & 0{\cdot}056 & -16{\cdot}98 & 3{\cdot}06 \\ 0{\cdot}0285 & -0{\cdot}078 & 0{\cdot}784 & 1{\cdot}81 & 0{\cdot}292 & -1{\cdot}02 & -5{\cdot}829 \end{bmatrix}$$

and B_m takes the nominal value of B.

The eigenvalues of A are $-0{\cdot}5018 \pm i3{\cdot}509$, $-8{\cdot}359$, $0{\cdot}1217$, 0, -5 and -10. F is chosen so that A_m has eigenvalues $-2 \pm j$, $-1{\cdot}5 + j1{\cdot}5$, -15, -10 and $-0{\cdot}05$. By choosing these eigenvalues with appropriate eigenvectors the model will have the desired system responses, i.e. the spiral mode has a very slow response, the roll mode is fast and stable and the Dutch roll mode is relatively slow with damping ratio equal to 0·7. Also, from (5·6) it follows that rank $[B\ B_m]$ = rank $[B, A - A_m]$ = rank $B = 2$; i.e. the perfect model-following matching conditions are satisfied.

To design the hyperplanes, a set of five $(n - m)$ null space eigenvalues for the closed-loop modes is required to prescribe the rate of decay of the error vector. These eigenvalues correspond to the non-zero roots of the characteristic polynomial of A_e.

The robust eigenvalue assignment method (Dorling and Zinober, 1985*a*, *b*, *c*, 1986, 1988) is used interactively in order to choose an eigenvalue set which provides sufficiently rapid decay of the error vector whilst being sufficiently robust in the sense that it will be minimally sensitive to perturbations in the system matrices. Such a set of eigenvalues is given by $-1, -2, -2{\cdot}5, -9, -12$ and the following switching matrix C is obtained;

$$C = \begin{bmatrix} 0{\cdot}05 & 0{\cdot}1456 & 0{\cdot}2081 & 1{\cdot}1481 & 2{\cdot}5925 & 0 & -1 \\ -0{\cdot}0135 & -0{\cdot}0217 & 1.387 & -0{\cdot}3705 & 0.1832 & -1 & 0 \end{bmatrix}$$

The eigenvalues of the range space dynamics, linked with the design of the linear part of the control, are -14, -15 for rapid approach to the subspace $s = 0$. The control structure based on that of Ryan and Corless (1984) has the form (5·12), and the matrices L, N, M designed for the nominal system are:

$$L = \begin{bmatrix} 0{\cdot}0091 & 0{\cdot}0279 & -0{\cdot}9443 & 0{\cdot}2198 & -0{\cdot}1346 & 0{\cdot}5555 & 0{\cdot}1219 \\ -0{\cdot}0717 & -0{\cdot}2052 & -0{\cdot}3492 & -1{\cdot}4754 & -3{\cdot}6003 & 0{\cdot}1226 & 1{\cdot}2341 \end{bmatrix}$$

$$N = \begin{bmatrix} 0 & 0 & -0{\cdot}0023 & 0{\cdot}0006 & -0{\cdot}0003 & 0{\cdot}0017 & 0 \\ -0{\cdot}0002 & -0{\cdot}0005 & -0{\cdot}0007 & -0{\cdot}0041 & -0{\cdot}0093 & 0 & 0{\cdot}0036 \end{bmatrix}$$

and

$$M = \begin{bmatrix} 0{\cdot}0018 & 0{\cdot}0052 & 0{\cdot}0074 & 0{\cdot}0410 & 0{\cdot}0926 & 0 & -0{\cdot}0357 \\ -0{\cdot}0005 & -0{\cdot}0007 & 0{\cdot}0462 & -0{\cdot}0124 & 0{\cdot}0061 & -0{\cdot}0333 & 0 \end{bmatrix}$$

This control system is now tested using a nonlinear simulation of the light aircraft. The simulation incorporates both the full force and moment lateral and longitudinal dynamics together with cross-coupling effects. Fig. 5.1 ($\varrho = I$, $\delta = 0{\cdot}1$) shows time responses of the plant together with those of the ideal model. To demonstrate the effectiveness of this approach we use different initial conditions for the plant and the model, namely

$$x(0) = [0{\cdot}1 \quad 0{\cdot}5 \quad 0 \quad 0 \quad 0 \quad 0 \quad 0]^T$$

$$x_m(0) = [0 \quad 0 \quad 0 \quad 0 \quad 0 \quad 0 \quad 0]^T$$

In addition a pilot aileron command of 0·02 rad is made for time $t < 2{\cdot}5$ s. This is incorporated as an input to the model with $B_m = B$ and $r_2 = 0{\cdot}02$ for $t < 2{\cdot}5$ s. The result of this manoeuvre is to produce a rolling moment with little or no change in lift.

The result shows clearly that after an initial transient the switching functions remain close to zero and the plant state x follows the model state x_m.

5.5 Example 2: Pitchpointing/vertical translation control system design

The plant under consideration here is the longitudinal motion model of a conventional combat aircraft. The lateral and longitudinal motions are assumed decoupled for the control system design, and a linearisation at the trim flight condition corresponding to an altitude of 3048 m and 0·77 Mach number yields

$$x = \begin{bmatrix} \gamma \\ q \\ \alpha \\ \eta \\ \delta \end{bmatrix} \begin{matrix} \text{flight path angle, rad} \\ \text{pitch rate, rad/s} \\ \text{angle of attack, rad} \\ \text{elevator deflection, rad} \\ \text{flap deflection, rad} \end{matrix} \tag{5.17}$$

and

$$u = \begin{bmatrix} \eta_c \\ \delta_c \end{bmatrix} \begin{matrix} \text{elevator command, rad} \\ \text{flap command, rad} \end{matrix} \tag{5.18}$$

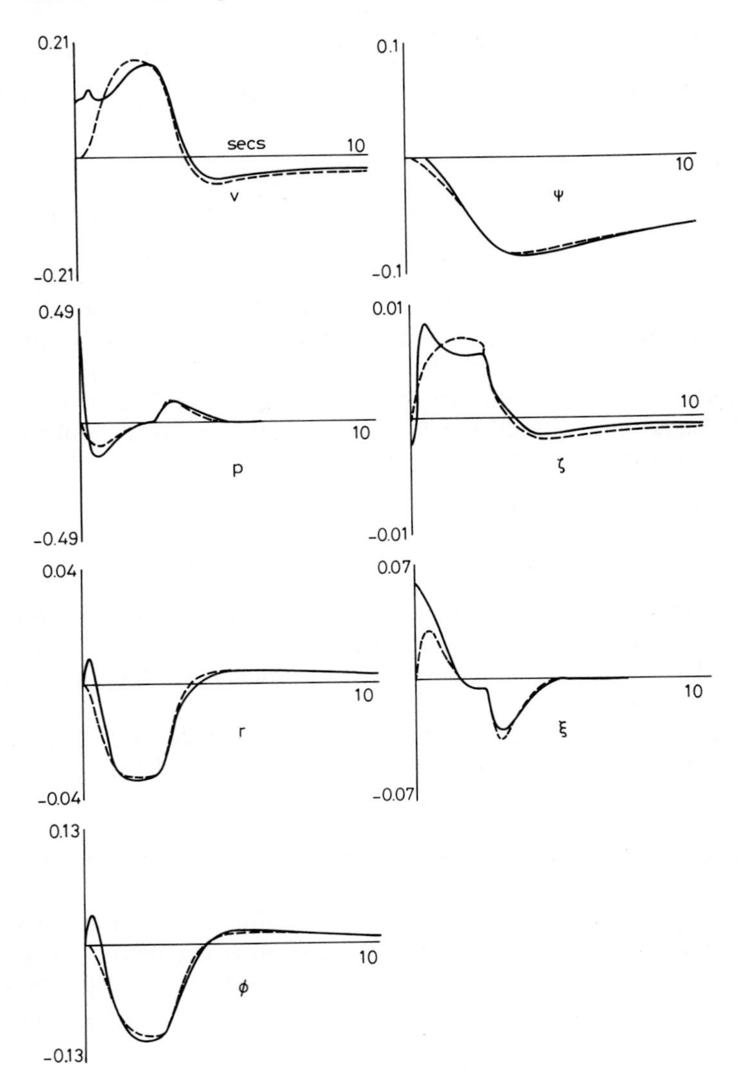

Fig. 5.1 *Model following state regulator system responses*

——— plant

– – – – model

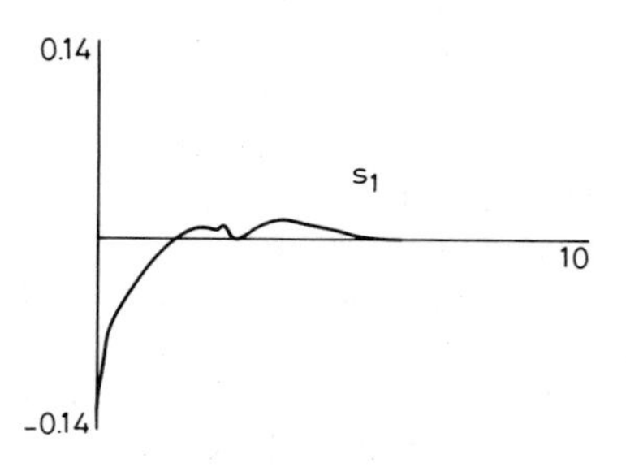

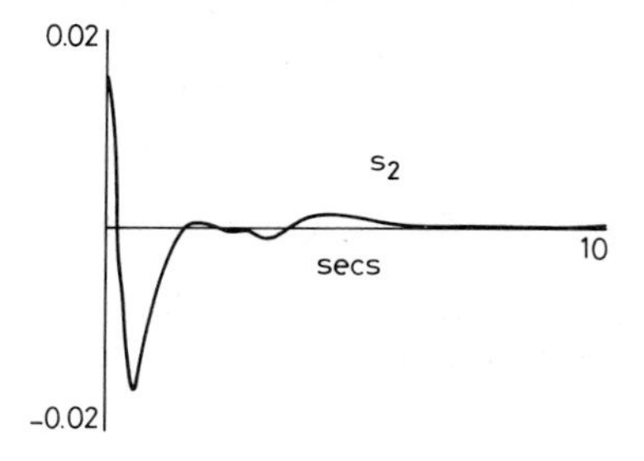

Fig. 5.1 *Continued*

with linearised plant matrices of the form:

$$A = \begin{bmatrix} 0 & 0 & -z_w & -z_\eta/U_0 & -z_\delta/U_0 \\ 0 & m_q & U_0 m_w & m_\eta & m_\delta \\ 0 & 1 & z_w & z_\eta/U_0 & z_\delta/U_0 \\ 0 & 0 & 0 & T_\eta & 0 \\ 0 & 0 & 0 & 0 & T_\delta \end{bmatrix} \tag{5.19}$$

$$B^T = \begin{bmatrix} 0 & 0 & 0 & K_\eta & 0 \\ 0 & 0 & 0 & 0 & K_\delta \end{bmatrix} \tag{5.20}$$

where z_w, z_η, ... etc. are the aerodynamic derivatives of the aircraft, U_0 is the nominal forward airspeed and T_η, K_η, T_δ, K_δ are the time constants and gains of the elevator and flap actuators, respectively, both of which are modelled as simple first order lags.

The model is represented by the equations corresponding to the nominal aircraft model together with ideal response characteristics provided by eigen-structure assignment (Sobel and Shapiro, 1985; Andry, Shapiro and Chung, 1983) in conjunction with a command tracker (O'Brien and Broussard, 1978).

The eigenvectors are chosen to obtain the desired decoupling and the eigen-values are chosen to obtain the desired damping and rise time. For pitchpointing control, the objective is to command the pitch angle whilst maintaining zero perturbation in the flight path angle. For a vertical translation controller, the converse is true; pilot command of flight path angle is required with no perturbation of pitch angle. From the result due to Srinathkumar (1978), five eigenvalues can be arbitrarily placed and a maximum of two entries in each corresponding eigenvector can be chosen; if more entries are chosen a corresponding achievable eigenvector can be found by projecting the desired eigenvector into the allowable subspace. In order to comply with MIL–F–8785C (1980) specifications for category A, level 1 flight which includes rapid manoeuvring and precise flight path control for mission objectives, the following eigenvalues are chosen:

$$\begin{aligned} \lambda_{1,2} &= -5{\cdot}6 \pm j4{\cdot}2 && \text{short period mode} \\ \lambda_3 &= -1{\cdot}0 && \text{flight path mode} \\ \lambda_4 &= -20 && \text{elevator actuator mode} \\ \lambda_5 &= -20 && \text{flap actuator mode} \end{aligned}$$

The corresponding eigenvectors are chosen to decouple the short period and flight path modes in order to minimise the coupling between pitch rate and flight path angle

$$V_d = [v_1^d, v_2^d, v_3^d, v_4^d, v_5^d] = \begin{bmatrix} 0 & 0 & 1 & x & x \\ 1 & x & 0 & x & x \\ x & 1 & 0 & x & x \\ x & x & x & 1 & x \\ x & x & x & x & 1 \end{bmatrix} \tag{5.21}$$

where x denotes that the magnitude of the particular element is unimportant. The achievable eigenvector matrix is found to be

$$V_a = \begin{bmatrix} 0 & 0 & 0{\cdot}67 & 0{\cdot}01 & 0{\cdot}3 \\ 1{\cdot}0 & -9{\cdot}5 & -0{\cdot}33 & 0{\cdot}97 & 0{\cdot}11 \\ -0{\cdot}93 & 1{\cdot}0 & -0{\cdot}33 & -0{\cdot}06 & 0{\cdot}03 \\ -1{\cdot}82 & -2{\cdot}24 & 0{\cdot}29 & 1{\cdot}0 & -0{\cdot}11 \\ 2{\cdot}97 & -2{\cdot}62 & -0{\cdot}18 & -0{\cdot}21 & 1{\cdot}0 \end{bmatrix} \tag{5.22}$$

This yields the following feedback matrix:

$$K = \begin{bmatrix} 1{\cdot}93 & 0{\cdot}57 & 2{\cdot}11 & -0{\cdot}48 & 0{\cdot}11 \\ -1{\cdot}94 & -0{\cdot}22 & -3{\cdot}12 & 0{\cdot}07 & 0{\cdot}05 \end{bmatrix} \tag{5.23}$$

The ideal model matrix is given by $A_m = A + BK$. The feedforward gains are computed to ensure steady-state tracking of the pilot's command using the command generator tracker developed by Broussard (1978). If the tracked variables of the aircraft defined by

$$y_t = Hx \tag{5.24}$$

the feedforward gains are computed from the matrix

$$\Omega = \begin{bmatrix} \Omega_{11} & \Omega_{12} \\ \Omega_{21} & \Omega_{22} \end{bmatrix} = \begin{bmatrix} A & B \\ H & 0 \end{bmatrix}^{-1} \tag{5.25}$$

yielding

$$Nr = B\left[\Omega_{22} - K\,\Omega_{12}\right] r \tag{5.26}$$

It is important to note that these gains are computed using the feedback matrix K which has been previously computed using eigenstructure assignment in order to produce the desired modal shaping.

For the pitchpointing/vertical translation problem under consideration, command of pitch angle and flight path angle is required. Therefore

$$y_t = H\,x = \begin{bmatrix} \theta \\ \gamma \end{bmatrix} = \begin{bmatrix} 1 & 0 & 1 & 0 & 0 \\ 1 & 0 & 0 & 0 & 0 \end{bmatrix} x \tag{5.27}$$

and

$$r = \begin{bmatrix} \theta_c \\ \gamma_c \end{bmatrix} \tag{5.28}$$

where θ_c and γ_c are the pitch and flight path angle commands at the pilot's seat, respectively. If $\gamma_c = 0$, a pitchpointing manoeuvre will result, whereas if $\theta_c = 0$ a vertical translation manoeuvre results. It should be noted that it is possible to command both pitch and flight path angle simultaneously if required using this control type. Eqns. (5.25) and (5.26) yield

$$B_m = \begin{bmatrix} 0 & 0 \\ 0 & 0 \\ 0 & 0 \\ -53{\cdot}08 & 14{\cdot}46 \\ 7{\cdot}19 & 31{\cdot}70 \end{bmatrix} \tag{5.29}$$

The required model has now been precisely defined; development of the nonlinear control scheme can now be considered. The pair $(A_m, -B)$ is taken and a corresponding hyperplane designed using quadratic minimisation with performance index given by

$$Q = (10, 5, 5, 20, 20) \tag{5.30}$$

This choise is based upon the assumptions that the deflections of the elevator and flaps of the actual system should follow the model deflections closely and that fast response to errors in desired flight path angle is required. The following hyperplane matrix C is obtained:

$$C = \begin{bmatrix} 0{\cdot}09 & -0{\cdot}06 & -0{\cdot}03 & 0 & 1 \\ -0{\cdot}7 & -0{\cdot}42 & -0{\cdot}19 & 1 & 0 \end{bmatrix} \tag{5.31}$$

The eigenvalues of the range space dynamics, linked with the design of the linear control part Le, are chosen to be -20 and -20. This ensures rapid approach to the subspace $s = 0$ whilst accommodating the bandwidths of the elevator and flaps. The control structure has the form of eqns. (5.10) and (5.16); the L, N and M matrices are given by

$$L = \begin{bmatrix} 1{\cdot}23 & 0{\cdot}19 & 2{\cdot}16 & -0{\cdot}08 & -0{\cdot}05 \\ -1{\cdot}86 & -0{\cdot}28 & -3{\cdot}08 & 0{\cdot}13 & 0{\cdot}07 \end{bmatrix}$$

$$N = \begin{bmatrix} -0{\cdot}0009 & -0{\cdot}0005 & -0{\cdot}0002 & 0{\cdot}0012 & 0{\cdot}0 \\ 0{\cdot}0001 & -0{\cdot}0001 & 0{\cdot}0 & 0{\cdot}0 & 0{\cdot}0012 \end{bmatrix} \tag{5.32}$$

$$M = \begin{bmatrix} 0{\cdot}0021 & -0{\cdot}0016 & -0{\cdot}0007 & 0{\cdot}0 & 0{\cdot}025 \\ -0{\cdot}0176 & -0{\cdot}0105 & -0{\cdot}0047 & 0{\cdot}025 & 0 \end{bmatrix}$$

This control scheme is augmented from (5.8) to form the full model-following control law of (5.10).

The nonlinear model-following control law is now implemented upon a fully nonlinear simulation of the combat aircraft model in the presence of wind turbulence. For comparison purposes, the purely linear control law generated using eigenstructure assignment in conjunction with the Broussard command generator tracker is also implemented upon the system. Fig 5·2 shows the response of the pitch angle and flight path angle during a pitchpointing manoeuvre for both controller types; a command of $\theta = 0{\cdot}02$ rad has been made at the pilot's seat and no perturbation of gamma is required. For this case there is very little difference in system performance. Fig. 5·3 shows the response of the states during a vertical translation manoeuvre corresponding to a pilot command of $\gamma = 0{\cdot}02$ rad. Here pilot command of γ is required whilst maintaining zero perturbation of θ. For this case the response of the model-following variable structure control is considerably better than the purely linear control response; the perturbation of θ is maintained very small and the γ achieved is very close to the pilot command. The model-following nonlinear approach is thus seen to provide added robustness whilst still maintaining a required modal structure.

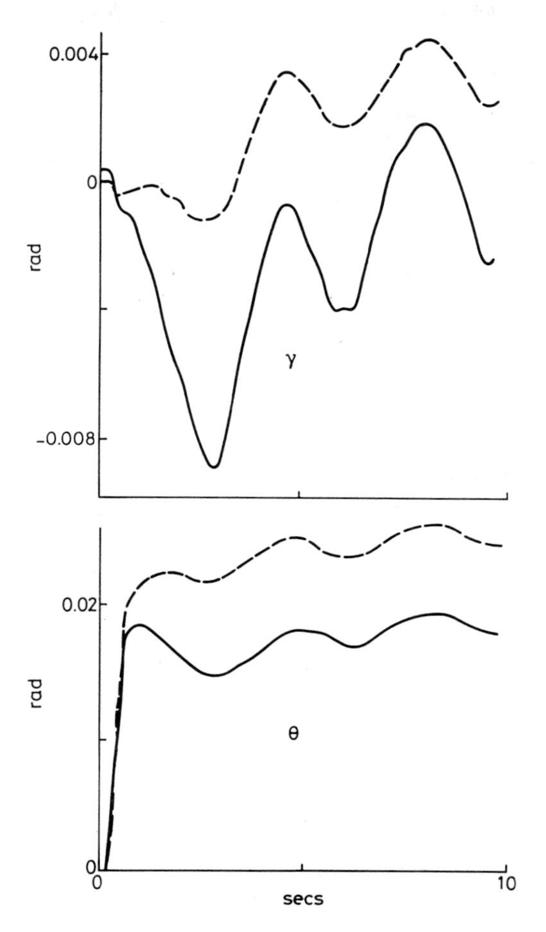

Fig. 5.2 *Pitchpointing manoeuvre in the presence of wind turbulence (linear and non-linear control schemes)*

—— linear control

---- nonlinear control

5.6 Conclusion

The technique of variable structure model-following control system design has been discussed. The design objective is to force the error between the model and the plant to zero as time tends to infinity. The initial condition of the plant and model need not be equal although the matching conditions must be satisfied for a perfect model-following. A great deal of the work in the field of variable structure flight control system design to date has concentrated upon solving the state-regulator problem (Mudge and Patton, 1988*b*; Calise and Kramer, 1984),

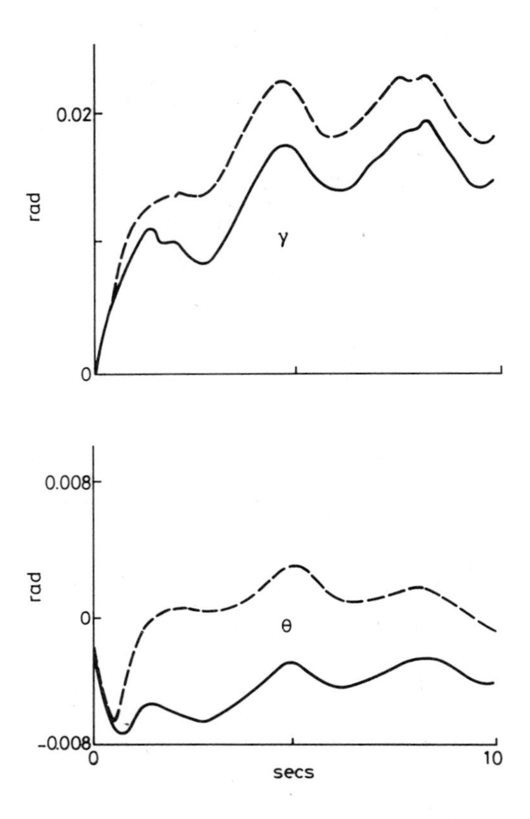

Fig. 5.3 *Vertical translation manoeuvre in the presence of wind turbulence (linear and non-linear control schemes)*

——— linear control

– – – – nonlinear control

and indeed some very robust control schemes have been designed. The examples presented here use a model-following approach to resolve the problem of pilot input; a demand at the pilot's seat may be considered a disturbance by the control system, which, at best, results in an approximation to the desired manoeuvre as the controller acts to restrain the system state to the required manifolds. In order to achieve satisfactory response to demands at the pilot's seat, the pilot demand is incorporated as a model input, and an augmented variable structure control scheme designed to force the error between the plant and model to zero. For both examples presented here, eigenstructure assignment has been used to modally shape a desired system response and so determine an ideal model. The first example considers the design of a variable structure model-following control scheme for the lateral motion stability augmentation system of an aircraft. This uses robust eigenstructure assignment to form the sliding subspace, and nonlinear simulation studies show that the plant response follows the model trajectory and the error between the two responses is rapidly eliminated. The second example considers the design of a pitchpointing/vertical translation control scheme. The example has shown that the purely linear control scheme used to define the model provides adequate performance when faced with a fully nonlinear combat aircraft simulation in the presence of wind turbulence. However, added robustness is achieved by a model-following variable structure control approach which will still modally shape the aircraft's response by the astute choice of model.

5.7 References

ANDRY, A., SHAPIRO, E. Y., and CHUNG, J. (1983): 'Eigenstructure assignment for linear systems.' *IEEE Trans.*, **AES-19,** pp. 711–729

ASLIN, P. P., PATTON R. J., and DORLING, C. (1985): 'The design of a sliding model controller for Dutch roll dampling in a non-linear aircraft system. ' Proc. IEE Conference Control '85, Cambridge, IEE Conf. 252, pp. 435–438

ASLIN, P. P. (1985): 'Aircraft simulation and robust flight control system design.' D. Phil Thesis, University of York

BALESTRINO A., DE MARIA, G., and SCIAVICCO, L. (1982): 'Hyperstable adaptive model following control of nonlinear plants,' *System and Control Lett.*, **1,** pp. 232–236

BALESTRINO, A., DE MARIA G., and ZINOBER, A. S. I. (1984): 'Nonlinear adaptive model-following control,' *Automatica,* **20,** pp. 559–568

BURTON, J. A. and ZINOBER, A. S. I. (1986): 'Continuous approximation of variable structure control,' *Int. J. System Sci.,* **17,** pp. 876–885

CALISE, A. J., and KRAMER, F. S. (1984): 'A variable structure approach to robust control of VTOL aircraft,' *J. Guidance, Control and Dynamics,* **7,** pp. 620–626

CHAN, Y. T. (1973): 'Perfect model-following with real model.' Proc. JACC, pp. 287–293

CORLESS, M., GOODALL, D. P., LEITMANN, G., and RYAN, E. P. (1985): 'Model-following controls for a class of uncertain dynamical systems.' Proc. 7th IFAC/IFORS Symp. on Identification and System Parameter Estimation, York, pp. 1895–1899

DORLING, C.M. (1985): 'The design of a variable structure control system (manual for the VASSYD CAD package).' Dept. of Applied and Computational Mathematics, University of Sheffield, 100 pp.

DORLING, C. M., and ZINOBER, A. S. I. (1985*a*): 'Hyperplane design in model-following variable structure control systems.' Proc. 7th IFAC/IFORS Symp. on Identification and System Parameter Estimation, York, pp. 1901–1905

DORLING, C. M., and ZINOBER, A. S. I. (1985*b*): 'The design and implementation of multivariable variable structure control systems.' Proc. IEE Control Conf., Cambridge, pp. 264–268

DORLING, C. M., and ZINOBER, A. S. I. (1985*c*) 'Computer-aided-design of robust multivariable structure control systems.' Proc 3rd IFAC/IFIP Symp. on CAD in Control and Engineering Systems, Copenhagen, pp. 6.32–6.38
DORLING, C. M., and ZINOBER, A. S. I. (1986): 'Two approaches to hyperplane design in multivariable variable structure control systems, *Int. J. Control,* **44,** pp. 65–82
DORLING, C. M., and ZINOBER, A. S. I. (1988): 'Robust hyperplane design in multivariable variable structure control systems, *Int. J. Control,* **48**, pp. 2043–2054
EL-GHEZAWI, O. M. E., ZINOBER, A. S. I., and BILLINGS, S. A. (1982): 'Analysis and design of variable structure systems using a geometric approach,' *Int J. Control,* **38,** pp. 657–671
ERZBERGER, H. (1968): 'Analysis and design of model-following control systems by state space techniques.' Proc. JACC, Ann Arbor, pp. 572–581
GRAYSON, L. (1985): 'The status of synthesis using Lyapunov's method,' *Automatica,* **3,** pp. 91–125
HANG, C. C., and PARK, P. C. (1973): 'Comparative studies of model reference adaptive control systems, *IEEE Trans.,* **AC-18,** pp. 419–428
ITKIS, U. (1976): 'Control systems of variable structure,' (Wiley, NY)
KAUTSKY, J., and NICHOLS, N. K. (1983): 'Robust eigenstructure assignment in state feedback controls.' University of Reading, Dept. of Mathematics, Numerical Analysis Report NA/2/83
LANDAU, I. D. (1979): 'Adaptive control: The model reference approach' (M. Dekker Inc., NY)
MIL-F-8785C (1980): 'Military Specification-Flying qualities of piloted aeroplanes'
MUDGE, S. K., and PATTON, R. J. (1988*a*): 'Analysis of the technique of robust eigenstructure assignment with application to aircraft control, *IEE Proceedings D,* **135,** pp. 275–281
MUDGE, S. K., and PATTON, R. J. (1988*b*) 'Enhanced assessment of robustness for an aircraft's sliding mode controller.' *AIAA J. Guidance, Control & Dynamics,* **11**, pp. 500–507
O'BRIEN, M. J., and BROUSSARD, J. R. (1978): 'Feedforward gains to track the output of a forced model.' Proc. 17th IEEE Conference on Decision and Control, pp. 1149–1155
RYAN, E. P., and CORLESS, M., (1984): 'Ultimate boundedness and asymptotic stability of a class of uncertain dynamical systems via continuous and discontinuous feedback control,' *IMA J. Math. Control Information,* **1,** pp. 223–242
SOBEL, K. M., and SHAPIRO, E. Y. (1985): 'A design methodology for pitchpointing flight control systems,' *J. Guidance, Control & Dynamics,* **8,** pp. 181–187
SPRINATHKUMAR, S. (1978): 'Eigenvalue/eigenvector assignment using output feedback,' *IEEE Trans.,* **AC-23,** pp. 79–81
UTKIN, V. I. (1977): 'Variable structure systems with sliding modes,' *IEEE Trans.,* **AC-22,** pp. 212–222
YEW, M. K., PATTON, R. J., and ZINOBER, A. S. I. (1987): 'Variable structure control systems with particular application to flight control: A feasibility study.' Research Report YEES/87, Dept. of Electronics, University of York
YOUNG, K. K.-D. (1978): 'Design of variable structure model-following control systems,' *IEEE Trans.,* **AC-23,** pp. 1079–1085
ZINOBER, A. S. I. (1984) 'Properties of adaptive discontinous model-following control systems.' 4th IMA International Conf. on Control Theory, Cambridge, pp. 337–346
ZINOBER, A. S. I. (1979): 'Controller design using the theory of variable structure systems.' *in* HARRIS, C., AND BILLINGS, S. A. (Eds.) 'Self tuning and adaptive control.' (Peter Peregrinus) pp. 204–229.
ZINOBER, A. S. I., EL-GHEZAWI, O. M. E., and BILLINGS, S. A. (1982): 'Multivariable variable-structure adaptive model-following control systems,' *Proc IEE,* **130D,** pp. 1–5

Chapter 6

Canonical formulation and general principles of variable structure controllers

L. Guzzella, H. P. Geering

Swiss Federal Institute of Technology (ETH)

6.1 Introduction

This Chapter has essentially two goals: first it introduces a new variable structure controller using a canonical system representation, orthogonal sliding hyperplanes, and a straightforward synthesis and analysis procedure. The second goal is more didactically oriented. Some basic facts of VSC will become evident from the special formulation.

It is a well known fact that the design of VSC is notably simplified by working with a canonical state space model known as block-companion form. In Section 6.2 it will be shown that the canonical formulation and the rank conditions introduced by Drazenovic (1969) are equivalent.

In Section 6.3 the new controller is presented which consists of two distinct terms. The first is linear state feedback which produces a pole–zero cancellation leaving only integrators in the transfer matrix of the known part of the system. The second term, a relay type controller, forces the initially nonzero outputs to zero and stabilises the pole–zero cancellation in the presence of parameter disturbances.

The linear part of the proposed controller can be calculated explicitly knowing only the constant part of the plant and the desired poles of the sliding system. This feature differs from known split formulations of regulators (Utkin, 1978). The problem of constructing the sliding hyperplanes given some desired poles has been investigated by various researchers (Utkin, 1979; El-Ghezawi *et al.*, 1983*a*; Dorling *et al.*, 1986). In the first paper no explicit algorithm is given.

The second and the third papers present some methods which are based on the construction of an admissible set of eigenvectors and therefore require a rather large amount of computations. With the method presented in Section 6.3 an easy and explicit pole placement is achieved; i.e. given the desired poles only the coefficients of their characteristic polynomials have to be computed.

The nonlinear part of the regulator guarantees the stability of the closed-loop system in the presence of parameter disturbances. The solution of this stability problem is easy; i.e. it is not required to know the order of succession of the reached sliding planes, as the so-called 'hierarchy of controls' method requires (Utkin, 1978). Only a maximum inequality has to be fulfilled and no Lyapunov equations (Gutman, 1979) or eigenvalues (Balestrino *et al.*, 1984) have to be computed.

In Section 6.5 some interesting remarks on the nature of sliding variable structure systems are listed. Section 6.6 shows that the presented regulator can be modified in a straightforward way in order to accomplish the requirements of model-following VSC as introduced by Young (1978). The numerical example in Section 6.7 demonstrates the synthesis and analysis procedure.

6.2 Plant description and transformations

This Chapter deals with a special class of linear uncertain dynamic systems which is defined by

$$\begin{aligned} \dot{x}(t) &= [A_0 + \delta A(t)]x(t) + B_0 u(t) \\ x(t) &\in \mathscr{R}^n,\ u(t) \in \mathscr{R}^m \end{aligned} \tag{6.1}$$

where A_0 and B_0 are constant and known plant matrices. The uncertainty matrix $\delta A(t)$ has to fulfil the following rank condition:

Assumption 1: $\text{rank}[\delta A(t)|B_0] = \text{rank}[B_0] = m$ for all t

This rank condition is equivalent to the decomposition:

$$\delta A(t) = B_0 R(t), \qquad R(t) \in \mathscr{R}^{mxn}$$

The elements of the uncertainty matrix $\delta A(t)$ are unknown but bounded by the following inequalities:

$$\delta a_{ij,min} \leqslant \delta a_{ij}(t) \leqslant \delta a_{ij,max} \tag{6.2}$$

Note that the parameter disturbances $\delta a_{ij}(t)$ are allowed to vary arbitrarily fast; this fact clearly distinguishes the analysed systems from those which can be handled by adaptive controllers.

Assumption 2: the pair $\{A_0, B_0\}$ is completely controllable.

It is well known (Kailath, 1980) that for linear time-invariant systems which fulfil this assumption two canonical co-ordinate transformations $T \in \mathscr{R}^n$ and $P \in \mathscr{R}^m$ exist which transform the time-invariant part of eqn. (6.1) to the block-companion canonical form:

$$A_{oc} = T^{-1} A_0 T \qquad B_{oc} = T^{-1} B_0 P$$

The explicit structure of A_{0c} and B_{0c} is displayed below:

$$A_{0c} = \begin{pmatrix} Q_1 \\ - \; - \; a_{1c} \; - \; - \\ \cdot \quad \cdot \quad \cdot \quad \cdot \quad \cdot \\ - \; - \; a_{mc} \; - \; - \end{pmatrix} \begin{matrix} \\ \cdot \leftarrow q_1 \text{th row} \rightarrow \\ \cdot \\ \leftarrow q_1 + q_2 \ldots = n\text{th row} \rightarrow \end{matrix} \qquad B_{0c} = \begin{pmatrix} 1 & & & \\ & \cdot & & \\ & & \cdot & \\ & & & \cdot \\ & & & & 1 \end{pmatrix}$$

The unspecified entries of A_{0c} and B_{0c} are zero. The entries of the matrices $Q_i \in \mathscr{R}^{(q_i - 1)xq_i}$ are almost all equal to zero; only the upper secondary-diagonal elements $[Q_i]_{j,j+1}$ ($j = 1, \ldots, q_i - 1$) are equal to one.

The controllability indices q_i denote the number of linearly independent column vectors in the controllability matrix $U = [B_0, A_0 B_0, \ldots A_0{}^{n-1} B_0]$ which are associated with the $i$the column of B_0. There are several possibilities of search procedures in the controllability matrix. Here the search is performed simply from the left to the right. Without loss of generality one can assume that the indices q_i are ordered, i.e.,

$$q_1 \geqslant q_2 \geqslant \ldots \geqslant q_m$$

This can always be attained by renumbering the states of the plant (6.1). Using the linearly independent columns found in the matrix U a square matrix S can be built in the following way:

$$S = [b_1, A_0 b_1, \ldots A_0^{q_1^{-1}} b_1, b_2, A_0 b_2, \ldots A_0^{q_m^{-1}} b_m], \qquad S \in \mathscr{R}^{nxn}$$

The columns of S are just the linearly independent columns of U but their order of has been changed such that all columns containing the ith column of B are kept together. The transformation matrix T^{-1} is built using some rows of the inverse of the matrix S:

$$S^{-1} = \begin{pmatrix} - - - \; s_1 \; - - - \\ \cdot \; \cdot \; \cdot \; \cdot \quad \cdot \; \cdot \; \cdot \\ - - - \; s_{p_1} \; - - - \\ \cdot \; \cdot \; \cdot \; \cdot \quad \cdot \; \cdot \; \cdot \\ - - - \; s_{p_m} \; - - - \end{pmatrix} \quad T^{-1} = \begin{pmatrix} - - - \; s_{p_1} \quad - - - \\ \cdot \; \cdot \; \cdot \; \cdot \quad \cdot \; \cdot \; \cdot \\ - - - \; s_{p_1} A_0^{q_1^{-1}} \; - - - \\ \cdot \; \cdot \; \cdot \; \cdot \quad \cdot \; \cdot \; \cdot \\ - - - \; s_{p_m} A_0^{q_m^{-1}} \; - - - \end{pmatrix}$$

The numbers p_i are simply the partial sums of the controllability indices q_i, i.e. $p_1 = q_1, p_2 = q_1 + q_2, \ldots, p_m = n$. The second transformation P is found using the matrix T and the matrix B_{0c} which is already known since it depends only on the controllability indices:

$$P = (B_{0c}^T T^{-1} B_0)^{-1}$$

All calculations can be implemented quite easily as a MATLAB-macro. Although the algorithm can potentially have numerical difficulties the authors have never encountered any problems applying this procedure to medium scale systems.

Lemma 1 (The proof of this lemma is given in Appendix 6.8.1)

If and only if assumption 1 is satisfied the transformed uncertain system (6.1) will also be in block-companion form; i.e. the matrix $\delta A_c(t) = T^{-1}\delta A_c(t)T$ will have the following structure (again the elements not explicitly defined are all zero):

$$\delta A_c(t) = \begin{pmatrix} - - \delta a_{1c}(t) - - \\ \cdots \qquad \cdot\,\cdot \\ \\ - - \delta a_{mc}(t) - - \end{pmatrix} \begin{matrix} \leftarrow q_1\text{th row} \\ \cdot \\ \cdot \\ \leftarrow q_1 + q_2 \ldots = n\text{th row} \end{matrix}$$

Remark: In Draženović (1969) the class of parameter variations which can be neutralised by bringing a dynamic system into the sliding mode is defined, making no assumptions on the co-ordinate system used to represent the plant. The conditions formulated by Drazenovic (1969) are (written in our notation):

$$\delta A(t)x(t) = B(CB)^{-1}C\delta A(t)x(t) \qquad \text{and} \qquad Cx(t) = 0$$

The matrix $C \in \mathcal{R}^{mxn}$ (which will be introduced in Section 6.3) defines m hyperplanes. If these hyperplanes have to be arbitrarily choosable these conditions are identical with assumption 1 which has been introduced here for a completely different purpose! This fact shows that no additional restrictions are imposed on the admissible parameter variations using the block-companion form. Thus, there are no reasons why one should not use this canonical form. The benefit of this co-ordinate transformation will be an easy and clear formulation of both the synthesis and the analysis problem.

In the rest of the Chapter (except Section 6.6) it is assumed that the abovementioned transformations have been performed; the notation will be simplified by dropping the index 'c'.

6.3 Regulator synthesis

In this Section the regulator problem is treated; i.e. a control law $u(t) = f(x(t))$

is sought which guarantees that the initially nonzero state vector $x(t)$ tends to zero as t goes to infinity. The first part of this Section shows how to construct the sliding hyperplanes given a desired pole configuration. The second part introduces the controller.

The desired dynamics of the closed loop system are expressed by choosing $(n - m)$ poles λ_{ij} $(i = 1 \ldots m, j = q_i - 1)$. These $(n - m)$ poles λ_{ij} are ordered into m groups of $q_i - 1$ elements each. For each of the m groups its characteristic polynomial is computed:

$$(s - \lambda_{1,1})(s - \lambda_{1,2}) \ldots (s - \lambda_{1,q_1-1}) = \gamma_{1,0} + \gamma_{1,1}s + \ldots + \gamma_{1,q_1-2}s^{(q_1-2)} + s^{(q_1-1)})$$

$$\cdot$$

$$\cdot$$

$$(s - \lambda_{m,1})(s - \lambda_{m,2}) \ldots (s - \lambda_{m,q_m-1}) = \gamma_{m,0} + \gamma_{m,0}s + \ldots + \gamma_{m,q_m-2}s^{(q_m-2)} + s^{(q_m-1)}$$

The matrix $C \in \mathscr{R}^{mxn}$ defining the m hyperplanes is built with the coefficients γ_{ij} of these polynomials (note that the coefficient '1' of the highest power of s is used, too):

$$C = \begin{pmatrix} \gamma_{1,0} & \cdot & \gamma_{1,q_1-2} & 1 & 0 & . & . & & . & . & & 0 \\ & & & & & & * & & & & & \\ 0 & & . & . & & & . & . & 0 & \gamma_{m,0} & \cdot & \gamma_{m,q_m-2} & 1 \end{pmatrix}$$

The construction of the matrix C is straightforward and does not imply any stability analysis of the system (6.1).

The switching variable $s(t) \in \mathscr{R}^{\mathrm{m}}$ is defined by eqn. (6.3):

$$s(t) = Cx(t) \tag{6.3}$$

As usual the equation $s(t) = 0$ defines the m sliding hyperplanes. Owing to the method used to construct the matrix C these planes are always orthogonal to each other.

The variable structure controller can now be introduced:

$$u(t) = -\, C\, A_0 x(t) - D\, \mathrm{sgn}(s(t))|x(t)| \tag{6.4}$$

The entries of the diagonal matrix D, the role of which is to counteract the parameter disturbances, will be specified in Section 6.4. The m-dimensional vector $\mathrm{sgn}(s(t))$ has the elements $\mathrm{sgn}(s_i(t))$ $(i = 1, \ldots m)$. The norm operator $|.|$ is the usual Euclidean length.

The control (6.4) has some nice properties. The clear separation into a linear and a nonlinear part has theoretical consequences (see remark 2 in the next Section) but also some practical ones. In fact the design procedure can now be separated into two distinct steps. In addition the control (6.4) is given in a closed-form expression rather than as a collection of 'if then else' statements as often used in other papers (Utkin, 1977).

6.4 Stability conditions and equivalent system

The behaviour of the closed-loop system (6.1) and (6.2) is clearly separated into two time intervals. During the first interval the nonlinear part of (6.4) forces the state $x(t)$ to reach the sliding hyperplanes. In the second time interval the state is captured on the sliding manifold and the dynamics of the sliding system are purely determined by the chosen pole configuration; i.e., the parameter disturbances will not affect the system any longer.

Theorem 1 specifies the regulator gains d_i:

Theorem 1:
If all regulator gains satisfy the condition

$$d_i > \sup_t \{/\delta a_i(t)/\} \qquad i = 1, \ldots, m$$

the system (6.1) with regulator (6.4) will reach the sliding mode in a time interval t^*. In addition, if $t^* < \infty$ the identity $s(t) \equiv 0$ will be true for all $t > t^*$. The proof of theorem 1 is given in the Appendix 6.8.2. Note the easy formulation of the stability condition! In many applications a finite hitting time t^* is important. Therefore in Section 6.5 a sufficient condition will be given which guarantees that $t^* < \infty$.

The second theorem applies to the sliding mode with times $t > t^*$

Theorem 2:

(*a*) For all times $t > t^*$ the dynamics of the sliding system are determined by

$$\dot{x}(t) = (I - B_0 C)A_0 x(t) = A_{eq} x(t)$$

ie. the influence of the parameter disturbances is completely eliminated.

(*b*) The matrix A_{eq} has the $(n - m)$ poles used to construct the matrix C plus m poles equal to zero. The latter have no influence on the sliding system since the corresponding eigenvectors are not excited.

Part (*a*) of theorem 2 is a direct consequence of the known 'equivalent control principle' (Utkin, 1978). The proof of part (*b*) will be outlined in remark 3 in the next Section.

6.5 Some remarks

Remark 1: The matrix $\Pi = (I - B_0 C)$ is a projection into the sliding manifold; i.e. the relation $\Pi^2 = \Pi$ holds. This matrix Π is the equivalent of the projector introduced by (El-Ghezawi *et al.*, 1983*a*). This equivalence is verified by noting that, owing to the canonical co-ordinate system and to the special choice of the matrix C, the relation $CB_0 = I_m$ is structurally guaranteed.

Remark 2: Assuming no parameter disturbances and using only the linear part of the controller (6.4), the switching vector $s(t)$ can be interpreted as the output of an ideal dynamic system:

$$\dot{x}(t) = \Pi A_0 x(t) + B_0 \hat{u}(t) \qquad y(t) = s(t) = Cx(t)$$

The transfer function matrix of this system can be written as

$$G(\lambda) = C\,(\lambda I - \Pi A_0)^{-1} B_0$$

The orthogonality of the sliding hyperplanes guarantees that $C\Pi = 0$. By using the matrix identity $(I - M)^{-1} = I + M(I - M)^{-1}$ (M is an arbitrary square matrix of the same dimension as the identity I) the transfer matrix can be simplified:

$$\begin{aligned} G(\lambda) &= C[\lambda^{-1} I + \lambda^{-2}\Pi A_0 (I - \lambda^{-1}\,\Pi A_0)^{-1}]B_0 \\ &= \lambda^{-1} CB_0 + \lambda^{-2} C\Pi A_0 (I - \lambda^{-1}\Pi A_0)^{-1}\, B_0 \\ &= \lambda^{-1} I_m \end{aligned}$$

Thus the transfer matrix of the ideal system is given by m decoupled integrators. This nice result allows an easy interpretation of the basic mechanisms of the *VSC*. In fact, by defining an output $s(t) = Cx(t)$, $n - m$ zeroes are introduced. With the linear part of the regulator (6.4) a pole–zero cancellation is achieved leaving only m poles at zero. The nonlinear part of (6.4) has two purposes. The first is eliminating the initial value of these integrators (i.e. bringing the system into the sliding mode). The second purpose is stabilising the pole–zero cancellation in the presence of parameter disturbances (i.e. keeping the system in the sliding mode).

For the matrix C chosen by any *VSC* design techniques the reasoning of this remark are also valid. The only slight difference concerns the transfer matrix $G(\lambda)$, which in that case will have the more general form

$$G(\lambda) = \lambda^{-1} CB_0$$

Remark 3: The matrix A_{eq} has a block-diagonal structure. This important fact is a direct consequence of the orthogonality of the m sliding hyperplanes. In addition, each block of A_{eq} is in companion form and its nontrivial row is almost identical to the negative characteristic coefficients of the ith polynomial used to construct the matrix C.

$$A_i = \begin{pmatrix} 0 & 1 & & 0 & . & 0 \\ . & . & & . & . & . \\ . & . & & . & . & . \\ 0 & . & & . & 0 & 1 \\ 0 & -\gamma_{i,0} & & . & . & -\gamma_{i,q_i-2} \end{pmatrix} \qquad A_i \in \mathcal{R}^{q_i x q_i}$$

The leftmost zero coefficient in the nontrivial row of A_i can be split off, yielding one pole at zero in each block. The rest of the eigenvalues of A_i are identical to the ith group of the desired poles. Therefore the matrix A_{eq} has the pole configuration stated in theorem 2.

In El-Ghezawi *et al.* (1983*b*) the inverse problem is investigated; i.e. the eigenvalues of the matrix A_{eq} are used in order to find the zeroes of the original system.

It is shown elsewhere in the book that the m zero poles of A_{eq} do not affect the system behaviour in the sliding mode since the corresponding eigenvectors must be unexcited in order to yield an identically vanishing output $s(t)$. (The m eigenvectors belonging to the m zero poles do not lie in the null of C.)

Remark 4: Since in most applications a finite hitting time must be guaranteed, a sufficient condition is required.

Theorem 3:
If all regulator gains d_i satisfy the condition

$$d_i > d_{i0} + /c_i/ + \alpha_i \qquad \alpha_i > 0 \; i = 1, \ldots, m$$

the hitting time will be finite (d_{i0} is given by theorem 1, exchanging the '>' symbol by the '=' symbol; $/c_i/$ is the norm of the ith row of the matrix C defined in Section 6.3).

The proof of Theorem 3 is given in Appendix 6.9.3

6.6 Reference model tracking control

The model reference approach to the tracking problem has been introduced in variable structure systems by Young (1978). In this Section, it will be shown that for this problem the treatment in the canonical state space is equivalent to some rank conditions known as 'perfect model matching conditions' (Erzberger, 1968) which always have to be fulfilled. The controller (6.4) proposed in Section 6.3 can easily be modified in order to handle the tracking problem.

In this Section the distinction between the original and the canonical co-ordinates is important. Therefore the index '$_c$' will be used again.

The reference model, given in the original co-ordinate system, has the following form:

$$\begin{aligned} \dot{z}(t) &= F_0 z(t) + G_0 w(t) \\ z(t) &\in \mathscr{R}^n, \qquad w(t) \in \mathscr{R}^P \end{aligned} \tag{6.5}$$

The prefect model matching conditions are (Erzberger, 1968):

$$\text{rank}[A_0 - F_0 | B_0] = \text{rank}[G_0 | B_0] = \text{rank}[B_0] = m$$

Again, these rank conditions are equivalent to particular decompositions, which here are:

$$F_0 = A_0 + B_0 Q_F, \qquad Q_F \in \mathscr{R}^{mxn}$$

$$G_0 = B_0 Q_G, \qquad Q_G \in \mathscr{R}^{mxp}$$

Applying the transformation $T_{Z_c}(t) = z(t)$, as introduced in Section 6.2, the following matrix F_{0c} results which is obviously in block-companion form:

$$F_{0c} = A_{0c} + T^{-1}B_0Q_FT = A_{0c} + B_{0c}P^{-1}Q_FT = A_{0c} + B_{0c}Q_{Fc}$$

The reference model tracking problem can be interpreted as a regulator problem for the error system $e_c(t) = x_c(t) - z_c(t)$, the dynamics of which are given by the following equation:

$$\dot{e}_c(t) = A_{0c}e_c(t) + \delta A_c(t)x_c(t) + B_{0c}[u_c(t) - Q_{Fc}z_c(t) - P^{-1}Q_Gw(t)] \quad (6.6)$$

The switching variable is now defined by

$$s_c(t) = C_ce_c(t) \quad (6.7)$$

The matrix C_c has the same form as in Section 6.3. Note, however, that in this case the matrix C_c determines the dynamics of the error transients rather than the desired dynamics of the state vector $x_c(t)$ which is supposed to track the state $z_c(t)$ of the reference model.
The tracking controller is derived from eqns. (6.4) and (6.6):

$$u_c(t) = -CA_{0c}e_c(t) + Q_{Fc}z_c(t) + P^{-1}Q_Gw(t) - D_c\mathrm{sgn}(s_c(t))/x_c(t)/ \quad (6.8)$$

With this controller and equation (6.7) the derivative of $s_c(t)$ can be expressed as follows:

$$\dot{s}_c(t) = C\delta A_c(t)x_c(t) - D_c\mathrm{sgn}(s_c(t))/x_c(t)/ \quad (6.9)$$

This equation, which is identical to eqn. 6.14 in Appendix 6.8.2, shows that the tracking problem is equivalent to a regulator problem for the error. Therefore, all the theorems and remarks for the regulator problem can easily be adapted to the tracking problem; i.e. the stability conditions defining the *VSC* gains D_c are given by theorem 1. Comparing this stability condition with other approaches to the reference-model tracking problem, which use hyperstability concepts (Balestrino *et al.*, 1984), the simplicity of the formulation presented here becomes evident.

6.7 A numerical example

This Section shows an example of a model following tracking controller for a plant having two input signals and five state variables. The complete analysis and synthesis procedure is shown with explicit numerical values (intermediate results have been placed in the Appendix 6.8.4). The plant is defined by the following matrices:

$$A_0 = \begin{pmatrix} 0{\cdot}00 & 1{\cdot}00 & 0{\cdot}00 & 0{\cdot}00 & 0{\cdot}00 \\ -2{\cdot}10 & -0{\cdot}80 & -1{\cdot}80 & 0{\cdot}00 & 0{\cdot}90 \\ 0{\cdot}00 & 0{\cdot}00 & -5{\cdot}30 & 1{\cdot}00 & 0{\cdot}00 \\ 0{\cdot}00 & 0{\cdot}00 & 0{\cdot}00 & 0{\cdot}00 & 1{\cdot}00 \\ 0{\cdot}00 & 1{\cdot}00 & 0{\cdot}20 & -2{\cdot}90 & -0{\cdot}57 \end{pmatrix}$$

$$B_0 = \begin{pmatrix} 0{\cdot}00 & 0{\cdot}00 \\ 0{\cdot}00 & 2{\cdot}70 \\ 0{\cdot}00 & 0{\cdot}00 \\ 0{\cdot}00 & 0{\cdot}00 \\ -2{\cdot}30 & 0{\cdot}00 \end{pmatrix}$$

The controllability indices of this system are $q_1 = 3$ and $q_2 = 2$. The transformations to block-companion form are displayed below:

$$T^{-1} = \begin{pmatrix} 0{\cdot}0000 & 0{\cdot}0000 & -0{\cdot}4348 & 0{\cdot}0000 & 0{\cdot}0000 \\ 0{\cdot}0000 & 0{\cdot}0000 & 2{\cdot}3043 & -0{\cdot}4348 & 0{\cdot}0000 \\ 0{\cdot}0000 & 0{\cdot}0000 & -12{\cdot}2130 & 2.3043 & 0{\cdot}4338 \\ 0{\cdot}3704 & 0{\cdot}0000 & -0{\cdot}3333 & 0{\cdot}0000 & 0{\cdot}0000 \\ 0{\cdot}0000 & 0{\cdot}3704 & 1{\cdot}7667 & -0{\cdot}3333 & 0{\cdot}0000 \end{pmatrix} P = I_2$$

The parameter disturbances $\delta A(t)$ have to satisfy assumption 1 of Section 6.2; i.e. they have to be in the range space of B_0. Therefore only elements in the third and fifth row can be nonzero. In the simulation the following two disturbances were used:

$$\delta a_{22}(t) = 1{\cdot}58{*}\sin(100{*}t) \qquad \delta a_{55}(t) = -0{\cdot}37{*}\sin(100{*}t)$$

The interesting variables are assumed to be $x_1(t)$ and $x_4(t)$. The reference model is separated into two independent blocks and has reasonable time constants and unity gains for the two relevant states:

$$F_0 = \begin{pmatrix} 0{\cdot}00 & 1{\cdot}00 & 0{\cdot}00 & 0{\cdot}00 & 0{\cdot}00 \\ -2{\cdot}00 & -2{\cdot}00 & 0{\cdot}00 & 0{\cdot}00 & 0{\cdot}00 \\ 0{\cdot}00 & 0{\cdot}00 & -5{\cdot}30 & 1{\cdot}00 & 0{\cdot}00 \\ 0{\cdot}00 & 0{\cdot}00 & 0{\cdot}00 & 0{\cdot}00 & 1{\cdot}00 \\ 0{\cdot}00 & 0{\cdot}00 & 0{\cdot}00 & -2{\cdot}00 & -2{\cdot}00 \end{pmatrix}$$

$$g_0 = \begin{pmatrix} 0{\cdot}00 \\ 2{\cdot}00 \\ 0{\cdot}00 \\ 0{\cdot}00 \\ 2{\cdot}00 \end{pmatrix}$$

These matrices satisfy the perfect model matching conditions and th resulting decompositions are:

$$Q_F = \begin{pmatrix} 0{\cdot}0000 & 0{\cdot}4348 & 0{\cdot}0870 & -0{\cdot}3913 & 0{\cdot}6217 \\ 0{\cdot}0370 & -0{\cdot}4444 & 0{\cdot}6667 & 0{\cdot}0000 & -0{\cdot}3333 \end{pmatrix}$$

$$q_g = \begin{pmatrix} -0{\cdot}8698 \\ 0{\cdot}7407 \end{pmatrix}$$

The error dynamics are chosen five times faster than the relevant time constants of the reference model:

$$C_c = \begin{pmatrix} 25{\cdot}0 & 10{\cdot}0 & 1{\cdot}0 & 0{\cdot}0 & 0{\cdot}0 \\ 0{\cdot}0 & 0{\cdot}0 & 0{\cdot}00 & 5{\cdot}0 & 1{\cdot}0 \end{pmatrix}$$

The following definitions introduce the matrices C and D in the original co-ordinate system:

$$C = PC_cT^{-1} = \begin{pmatrix} 0{\cdot}0000 & 0{\cdot}0000 & -0{\cdot}0391 & -2{\cdot}0435 & -0{\cdot}4348 \\ 1{\cdot}8519 & 0{\cdot}3704 & 0{\cdot}1000 & -0{\cdot}3333 & 0{\cdot}0000 \end{pmatrix}$$

$$D = PD_c = \begin{pmatrix} d_1 & 0 \\ 0 & d_2 \end{pmatrix}$$

The controller (6.8) can also be expressed in the original co-ordinates:

$$u(t) = Q_Fz(t) + q_gw(t) - CA_0e(t) - D\text{sgn}(Ce(t))/T^{-1}x(t)/ \quad (6.10)$$

The nonlinear part of (6.10) can be specified with the stability condition of theorem 1 (δA_c is specified in the Appendix 6.9.4). For the assumed disturbances the following gains result:

$$d_1 = 2{\cdot}0; \qquad d_2 = 2{\cdot}0$$

Now the controller is completely specified.
In the numerical simulation the following initial conditions and reference input were used:

$$x_1(0) = 1{\cdot}0; x_2(0) = 0{\cdot}8; x_3(0) = -0{\cdot}4; x_4(0) = 0{\cdot}5; x_5(0) = 0{\cdot}7$$

$$z(0) = 0; \; w(t) = (2{\cdot}0 \text{ if } t < 4{\cdot}0 \text{ else } -2{\cdot}0)$$

Figs. 6.1 and 6.2 display the behaviour of the closed-loop system. Fig. 6.1 shows the switching variables and the error (note that the time scale are different). The hitting time is about 0·27 s. For times $t < t^*$ the sinusoidal parameter disturbances are visible. As soon as the sliding mode is reached the error signal performs the desired robust motion.

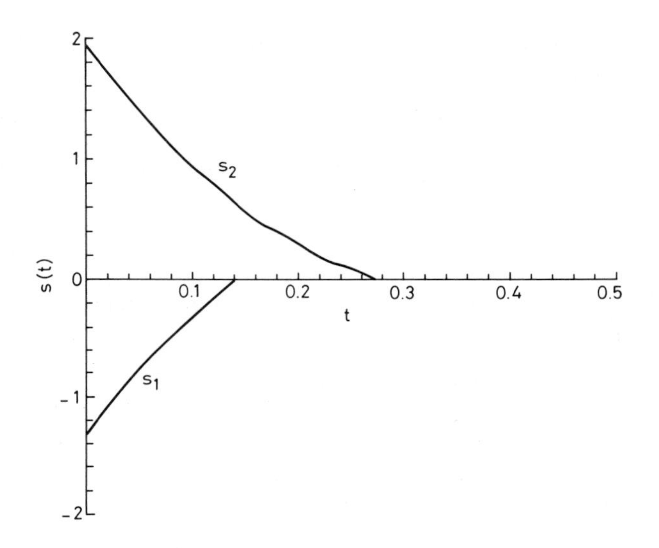

Fig. 6.1A *Switching variables*

Fig. 6.2 shows the relevant states $x_1(t)$ and $x_4(t)$. After the short error transient shown in Fig. 6.1 the state $x(t)$ perfectly tracks the reference signal $z(t)$ in spite of the continuing parameter disturbances.

6.8 Conclusions

This chapter has shown how the design of regulators and model following controllers with variable structure can be notably simplified. This simplification is achieved by using a canonical transformation, a split formulation of the regulator, and orthogonal switching hyperplanes. In addition the basic mechanism of the parameter robustness has been clarified.

The authors are convinced that these easy synthesis and analysis procedures will promote the use of variable structure regulators in practice.

6.9 References

BALESTRINO, A., DE MARIA, G., AND ZINOBER, A. S. I., (1984): 'Nonlinear adaptive model-following control', *Automatica,* **20,** pp. 559–568

DRAŽENOVIĆ, B., (1969): 'The invariance condition in variable structure systems', *Automatica,* **5,** pp. 287–295

DORLING, C.M., and ZINOBER, A. S. I., (1986): 'Two approaches to hyperplane design in multivariable variable structure control systems', *Int. J. Control,* **44,** pp. 65–69

EL-GHEZAWI, O. M. E., ZINOBER, A. S. I., and BILLINGS, S., (1983*a*): 'Analysis and design of variable structure systems using a geometric approach', *Int. J. Control,* **38,** pp. 657–671

EL-GHEZAWI, O. M. E., BILLINGS, S., and ZINOBER, A. S. I., (1983*b*): 'Variable structure systems and systems zeros', *IEE Proc.,* **130,** pp.1–5

EMELYANOV, S. V., (1969): 'Regelungssysteme mit variabler Struktur' (M: Oldenbourg Verlag, Munich) (in German)

ERZBERGER, H., (1968): 'Analysis and design of model-following control systems by state space techniques'. Proc. Joint Automat. Contr. Conf., pp. 572–581

GUTMAN, S., (1979): 'Uncertain dynamical systems: A Lyapunov min-max approach', *IEEE Trans.,***AC-24,**437–443

ITKIS, U., (1976): 'Control systems of variable structure' (Wiley, NY)

KAILATH, T., (1980): 'Linear systems' (Prentice-Hall, Englewood Cliffs, N.J)

LASSALLE, J., and LEFSCHETZ, S., (1961): 'Stability by Lyapunov's direct method with applications' (Academic Press, NY)

UTKIN, V., (1977): 'Variable structure systems with sliding modes: A survey', *IEEE Trans.,* **AC-22,**pp. 212–222

UTKIN, V., (1978): 'Variable structure systems' (MIR, Moscow)

UTKIN, V., and YANG, K., (1979): 'Methods for constructing discontinuity planes in multidimensional variable structure systems', *Automat & Remote control,* **10,** pp. 1466–1470

YOUNG, K. K., (1978): 'Design of variable structure model-following control systems', *IEEE Trans.,***AC-23,**pp. 1079–1085

6.10 Appendix

6.10.1 Proof of lemma 1

Sufficient part: All matrices $\delta A(t)$ satisfying assumption 2 can be split into two factors:

$$\delta A(t) = B_0 R(t) \qquad R(t) \in \mathscr{R}^{mxn}, \text{ arbitrary}$$

Therefore, the matrix $\delta A_c(t)$ will be

$$\begin{aligned}\delta A_c(t) &= T^{-1} B_0 R(t) T \\ &= B_{0c} P^{-1} R(t) T = B_{0c} R^*(t) \qquad R^*(t) \in \mathscr{R}^{mxn}, \text{ arbitrary}\end{aligned}$$

Owing to the multiplication with the matrix B_{0c} the structure of $\delta A_c(t)$ is as stated in lemma 1.

Necessary part: The proof can simply be reversed since all matrices $\delta A_c(t)$ which do not destroy the block-companion form can be split into two factors:

$$\delta A_c(t) = B_{0c} R^*(t)$$

This completes the proof of lemma 1.

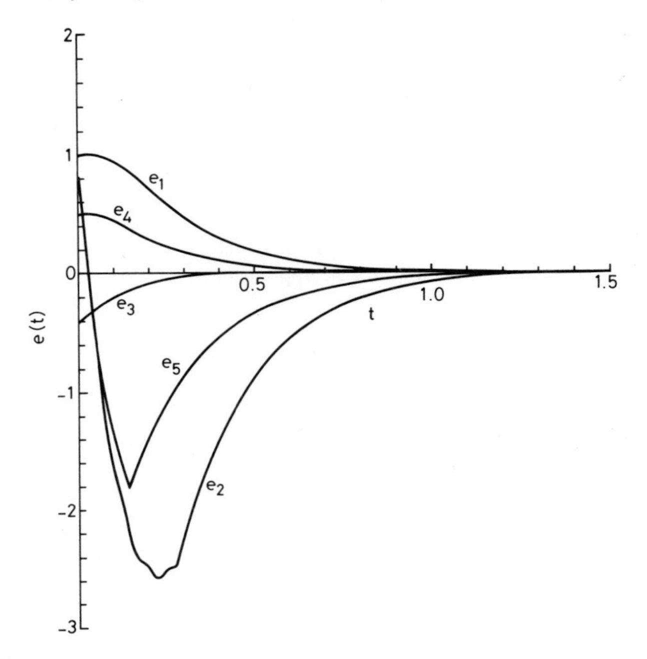

Fig. 6.1*B* *Error vector*

6.10.2 Proof of theorem 1

This proof is based on a Lyapunov method (Lasalle *et al.*, 1967). The chosen Lyapunov function, given by eqn. (6.11), is positive everywhere outside the sliding manifold:

$$v(s) = \sum_{i=1}^{m} |s_i(t)| > 0 \; \forall \; s(t) \neq 0 \tag{6.11}$$

The derivative of $v(s)$ is given by eqn. (6.12):

$$\dot{v}(s) = \sum_{i=1}^{m} \text{sgn}(s_i(t)) \; \dot{s}_i(t) \tag{6.12}$$

A sufficient stability condition is given by the following m inequalities:

$$\text{sgn}(s_i(t)) \; \dot{s}_i(t) < 0 \qquad \forall \; i = 1, \ldots m \tag{6.13}$$

Note that with this condition the sliding property is guaranteed on *each* of the m hyperplanes. The derivative of $s_i(t)$ can be calculated from eqns. (6.1), (6.3) and (6.4) (see also remarks 1 and 2 in Section 5):

$$\dot{s}(t) = [C\Pi A_0 + C\delta A(t)]x(t) - CB_0 D\text{sgn}(s(t))/x(t)/$$

$$\Rightarrow \dot{s}(t) = C\delta A(t)x(t) - D\text{sgn}(s(t))/x(t)/ \tag{6.14}$$

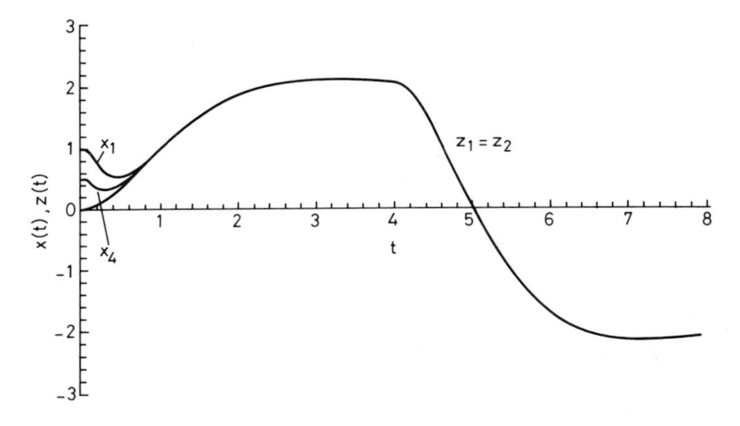

Fig. 6.2 *Tracking behaviour*

It is easy to see that owing to the special structure of the matrices $\delta A(t)$ and C, the following m relations hold:

$$i\text{th row of } (C\delta A(t)) \quad = \quad \delta a_i(t) \qquad i \;=\; 1, \ldots, m$$

Therefore, the derivative of $s_i(t)$ can be written as:

$$\dot{s}_i(t) \quad = \quad \delta a_i(t)x(t) \;-\; \mathrm{sgn}(s_i(t))d_i/x(t)/ \tag{6.15}$$

The stability condition (6.13) is guaranteed if the regulator gains d_i are all chosen as stated by theorem 1:

$$d_i \;>\; \sup_t\{/\delta a_i(t)/\} \qquad i \;=\; 1, \ldots, m$$

This condition guarantees that $s_i(t)$ and $\dot{s}_i(t)$ always have different signs. Hence, starting from an arbitrary initial point in the state space, the sliding mode will be reached on a first sliding plane. Because the corresponding component of the Lyapunov function has a minimum at that plane, the state vector will never leave that plane any more. As soon as the state vector has reached an additional sliding plane it will be captured on that plane, too. This procedure will continue until all components of $s(t)$ have become zero because (6.13) holds for all $i = 1, \ldots, m$ (note that the order of succession is irrelevant).

This completes the proof of theorem 1.

6.10.3 Proof of theorem 3

This theorem is proved by contradiction. If the hitting time t^* is infinite at least one component of $s(t)$ never becomes negative. Without loss of generality one can assume that this component $s_i(t)$ has a positive initial value and therefore has to be non-negative for all $t \leqslant \infty$. If the regulator gains are chosen as in theorem 3 the derivative of $s_i(t)$ has to fulfil the following inequality (cf. eqn. (6.15)):

$$\dot{s}_i(t) \leqslant -/c_i/\ /x(t)/ - \alpha_i/x(t)/ \tag{6.16}$$

The lower bound of $/c_i//x(t)/$ can be estimated:

$$/c_i/\ /x(t)/ \geqslant s_i(t)$$

Since $s_i(t) \neq 0$ implies that $|x(t)| \neq 0$ $\dot{s}_i(t)$ has to strictly fulfil the following inequality (remember $\alpha_i > 0$):

$$\dot{s}_i(t) < -s_i(t) \Rightarrow s_i(t) < e^{-t}s_i(0) \tag{6.17}$$

Taking the limit $t^* = \infty$ in eqn. (6.17) the contradiction $s_i(\infty) < 0$ results. Therefore the hitting time has to be finite.

This completes the proof of theorem 3.

6.10.4 Intermediate results of the example in Section 6.7

First six columns of the controllability matrix:

$$U = \begin{pmatrix} 0{\cdot}0000 & 0{\cdot}0000 & 0{\cdot}0000 & 2{\cdot}7000 & -2{\cdot}07000 & -2{\cdot}1600 & . \\ 0{\cdot}0000 & 2{\cdot}7000 & -2{\cdot}0700 & -2{\cdot}1600 & 2{\cdot}8359 & -1{\cdot}5120 & . \\ 0{\cdot}0000 & 0{\cdot}0000 & 0{\cdot}0000 & 0{\cdot}0000 & -2{\cdot}3000 & 0{\cdot}0000 & . \\ 0{\cdot}0000 & 0{\cdot}0000 & -2{\cdot}3000 & 0{\cdot}0000 & 1{\cdot}3110 & 2{\cdot}7000 & . \\ -2{\cdot}3000 & 0{\cdot}0000 & 1{\cdot}3110 & 2{\cdot}7000 & 2{\cdot}8527 & -3{\cdot}6990 & . \end{pmatrix}$$

Matrix S:

$$S = \begin{pmatrix} 0{\cdot}0000 & 0{\cdot}0000 & -2{\cdot}0700 & 0{\cdot}0000 & 2{\cdot}7000 \\ 0{\cdot}0000 & -2{\cdot}0700 & 2{\cdot}8359 & 2{\cdot}7000 & -2{\cdot}1600 \\ 0{\cdot}0000 & 0{\cdot}0000 & -2{\cdot}3000 & 0{\cdot}0000 & 0{\cdot}0000 \\ 0{\cdot}0000 & -2{\cdot}3000 & 1{\cdot}3110 & 0{\cdot}0000 & 0{\cdot}0000 \\ -2{\cdot}3000 & 1{\cdot}3110 & 3{\cdot}8527 & 0{\cdot}0000 & 2{\cdot}7000 \end{pmatrix}$$

Matrix S^{-1}:

$$S^{-1} = \begin{pmatrix} 0{\cdot}4348 & 0{\cdot}0000 & -1{\cdot}2609 & -0{\cdot}2478 & -0{\cdot}4348 \\ 0{\cdot}0000 & 0{\cdot}0000 & -0{\cdot}2478 & -0{\cdot}4348 & 0{\cdot}0000 \\ 0{\cdot}0000 & 0{\cdot}0000 & -0{\cdot}4348 & 0{\cdot}0000 & 0{\cdot}0000 \\ 0{\cdot}2963 & 0{\cdot}3704 & 0{\cdot}0000 & -0{\cdot}3333 & 0{\cdot}0000 \\ 0{\cdot}3704 & 0{\cdot}0000 & -0{\cdot}3333 & 0{\cdot}0000 & 0{\cdot}0000 \end{pmatrix}$$

Matrix δA_c:

$$\delta A_c = \begin{pmatrix} 0 & 0 & 0 & 0 & 0 \\ 0 & 0 & 0 & 0 & 0 \\ 0 & -1{\cdot}21*\sin(100*t) & -0{\cdot}37*\sin(100*t) & 0 & 0 \\ 0 & 0 & 0 & 0 & 0 \\ 0 & -1{\cdot}21*\sin(100*t) & 0 & 0 & 1{\cdot}58*\sin(100*t) \end{pmatrix}$$

Matrix CA_0:

$$CA_0 = \begin{pmatrix} 0{\cdot}0000 & -0{\cdot}4348 & 0{\cdot}1204 & 1{\cdot}2217 & -1{\cdot}7975 \\ -0{\cdot}7778 & 1{\cdot}5556 & -1{\cdot}1967 & 0{\cdot}1000 & 0{\cdot}0000 \end{pmatrix}$$

Chapter 7

Variable structure controllers for robots

L. Guzzella, H. P. Geering

Swiss Federal Institute of Technology (ETH)

7.1 Introduction

This Chapter shows how a broad class of uncertain mechanical systems (including robots) can be forced to behave like a chosen and therefore perfectly known reference model. This model can be used to calculate off-line some particular trajectories (e.g. time-optimal solutions) or to design some model based control schemes (computed torque (Craig, 1986)).

The presented results are well suited for applications where the robot has to repeatedly perform certain tasks, e.g. in assembly operations with very high precision requirements. As a first step the trajectories of the task are optimised using the reference model. The second step consists in finding a nonlinear on-line controller which guarantees that the robot performs exactly the same motions as predicted by the mathematical model (of course, a good model/plant matching requires smaller on-line corrections than a bad one).

The problem of controlling a nonlinear uncertain system using a reference model and variable structure controller (VSC) has been tackled by several authors (e.g. Young, 1978; Balestrino *et al.*, 1983; Slotine, 1985; Hiroi *et al.*, 1986). These papers use linear reference models and the control schemes and the stability proofs turn out to be rather complicated.

In this Chapter a nonlinear reference model and a simple regulator with variable structure control is used. The advantages of using a nonlinear reference model are twofold. Espeically in high speed applications the neglected nonlinear terms would require very high on-line effort for perfect linearisation and in time-optimal schemes the nonlinear couplings can be exploited in order to reduce the transfer times (see example).

The stability proofs are simplified by assuming a zero model-plant state error at start-up. This assumption is not very restrictive since most robots perform some kind of reference mark localisation at the beginning.

The complete design and analysis procedure is shown in an example which includes a numerical simulation of the time-optimal motion of a robot in an assembly task. The rest of the paper is organised as follows:

- Section 2 introduces the considered system, the reference model, and a simple transformation.
- Section 3 presents the structure of the controller.
- Section 4 analyses the behaviour of the closed-loop system in the sliding mode.
- Section 5 gives and proves a theorem which states sufficient conditions for the regulator gains in order to guarantee a persistent sliding mode.
- Section 6 finally resumes the complete design and analysis cycle for a simple but realistic two degree of freedom robot.

7.2 System description

A holonomial finite-dimensional mechanical system can always be described by the second order vector differential equation (7.1) provided every degree of freedom has its own actuator:

$$\begin{aligned} M(y(t))\,\ddot{y}(t) &= f(y(t), \dot{y}(t)) + u(t) \\ y(t), u(t) \in \mathscr{R}^p, \qquad M(y(t)) &= M^T(y(t)) > 0 \end{aligned} \tag{7.1}$$

The vector $y(t)$ represents the position co-ordinates and the vector $u(t)$ contains all the torques and forces. The matrix $M(y(t))$ is assumed to be a mass matrix and therefore symmetric and positive-definite. Note that all robots are included in this class of systems (Paul, 1981). In this Chapter it is assumed that all positions and velocities are measured. This holds for most robots.

The next step is transforming this equation into a first order one. Introducing the state vector $x(t)$:

$$\begin{aligned} x_i(t) &= y_j(t) \qquad i = 1, 3, \ldots 2p-1 \qquad j = 1, 2, \ldots p \\ x_i(t) &= \dot{y}_j(t) \qquad i = 2, 4, \ldots 2p \qquad j = 1, 2, \ldots p \end{aligned} \tag{7.2}$$

Eqn. (7.1) can be written as a first order vector differential equation:

$$\begin{aligned} \dot{x}(t) &= Qx(t) + BM(x(t))^{-1}(f(x(t)) + u(t)) \\ & x(t) \in \mathscr{R}^n \qquad (n = 2p) \end{aligned} \tag{7.3}$$

In eqn. (7.3) two matrices have been introduced consisting of zeroes and ones only ('structural matrices'). The explicit form of these matrices is shown below:

$$Q = \begin{pmatrix} Q_0 & 0 & . & . & 0 \\ 0 & Q_0 & 0 & . & 0 \\ . & . & . & . & . \\ 0 & . & . & Q_0 & 0 \\ 0 & . & . & 0 & Q_0 \end{pmatrix} \quad \text{with } Q_0 = \begin{pmatrix} 0 & 1 \\ 0 & 0 \end{pmatrix} \quad Q \in \mathscr{R}^{nxn}$$

$$B = \begin{pmatrix} 0 & . & . & . & 0 \\ 1 & 0 & . & . & 0 \\ . & . & . & . & . \\ 0 & . & . & . & 0 \\ 0 & . & . & 0 & 1 \end{pmatrix} \quad B \in \mathscr{R}^{nxp}$$

Its very important to clarify the consequences of this structure, i.e., its connections to the 'invariance condition' introduced by Drazenovic (1969). In fact, the nonlinear coupling $f(x(t))$ in eqn. (7.3) structurally lies only in the range space of B. Therefore, the influence of any present model/plant mismatch can be neutralised by an appropriate variable structure controller. This is the key property which distinguishes the class of nonlinear systems (7.1) from general nonlinear systems.

Because of modelling errors and uncertain or time-varying parameters it is assumed that eqn. (7.3) can only be approximated by

$$\dot{z}(t) = Qz(t) + N(z(t))^{-1}(g(z(t)) + w(t)) \qquad z(t) \in \mathscr{R}^n \qquad w(t) \in \mathscr{R}^p \tag{7.5}$$

where N and g are the approximations for M and f, respectively. Note that the reference model (7.5) and the plant (7.3) necessarily have the same structure which is only determined by the matrices Q and B. This fact corresponds to the 'perfect model matching' conditions in the linear case (Erzberger, 1968). The mass matrix N and the vector-valued function g need not be equal to their counterparts M and f, respectively. Of course, the model/plant matching will require less control action if the model (7.5) is close to the plant (7.3).

The synthesis of the nominal trajectories $z(t)$, which is not in the scope of this Chapter, is based on the model (7.5). Therefore, usually a close model/plant matching is highly desirable. Once a desired trajectory and its corresponding reference control $w(t)$ have been found the VSC task can be formulated.

The tracking problem considered is stated as follows:

> Given a plant (7.3), a reference model (7.5), and a reference input $w(t)$ find a control $u(t)$ which guarantees that the state $x(t)$ perfectly tracks the reference state $z(t)$.

7.3 Control system structure

The complete structure of the closed-loop system is shown in Fig. 7.1. In order to be able to define $u_{VSC}(t)$ some auxiliary signals have to be introduced. The error $e(t)$ is defined by eqn. (7.6):

$$e(t) \quad = \quad x(t) \; - \; z(t) \tag{7.6}$$

As usual when dealing with VSC an important step is the definition of the switching vector $s(t)$:

$$s(t) \quad = \quad Ce(t) \tag{7.7}$$

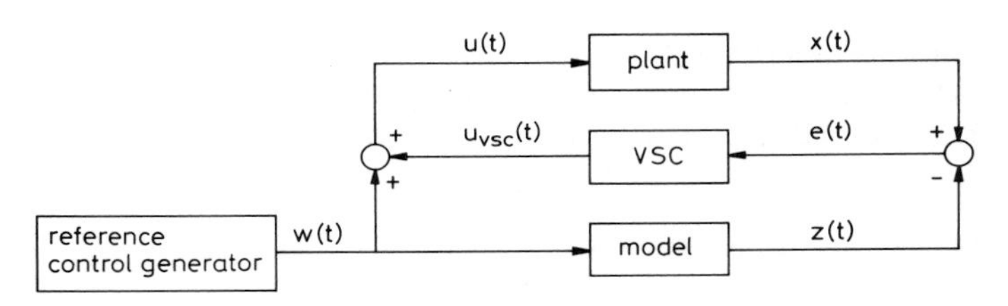

Fig. 7.1 *General structure*

As can be seen from eqn. (7.7), in a model-reference approach, instead of the state $x(t)$ the model-plant state-error $e(t)$ has to be forced into the sliding mode. The desired error dynamics in the sliding mode are expressed by choosing p real (and, of course, stable) eigenvalues λ_i which are used to construct the matrix C:

$$C = \begin{pmatrix} -\lambda_1 & 1 & 0 & . & . & . & . & 0 \\ 0 & 0 & -\lambda_2 & 1 & 0 & . & . & 0 \\ . & . & . & . & . & . & . & . \\ 0 & . & . & . & . & 0 & -\lambda_p & 1 \end{pmatrix}$$

This matrix C has a special form, vis. its rows are all orthogonal to each other. This special choice will have some nice consequences. A very important one is the fact that $CB = I_p$. (Other ones are shown in the next Section.)

If the switching variable $s(t)$ vanishes identically during a finite time interval the closed-loop system is in the 'sliding model', i.e. the first time with $s(t) = 0$, is denoted by t^*.

Now the VSC can be introduced:

$$u(t) \quad = \quad w(t) \; - \; d_x \operatorname{sgn}(s(t))|x(t)| \; - \; d_w \operatorname{sgn}(s(t))|w(t)| \tag{7.8}$$

The vector $\operatorname{sgn}(s(t))$ is simply formed by tacking the signum values of the corresponding elements of $s(t)$. The norm operator $|.|$ is the usual Euclidean length. The two unknown scalars d_x and d_w will be determined in Section 7.5.

Note the simple form of the VSC (7.8)! Only the norm calculation requires a certain amount of computation which could be reduced by using a simpler norm operator. The reference control $w(t)$ can be taken from a storage device or can be generated on-line depending on the complexity of the synthesis procedure used for that purpose.

Instead of using only two scalar gains as shown in (7.8) other VSC using $2p$ gains are realisable, too. The resulting stability conditions can be slightly less conservative, but the derivation of the stability conditions for that case is very cumbersome.

7.4 Sliding system

This Section analyses the behaviour of the system in the sliding mode, i.e., for times $t > t^*$ and therefore with $s(t)$ identically vanishing. Using the 'equivalent control' principle (Utkin, 1977) the equivalent control vector $u_{eq}(t)$ can be evaluated:

$$\begin{aligned}\dot{s}(t) &= C\dot{e}(t) = C\,\{Qe(t) + B\,[M(x)^{-1}\,(f(x) + u(t)) \\ &\quad - N(z)^{-1}\,(g(z) + w(t))]\} \equiv 0\end{aligned} \tag{7.9}$$

Using the above-mentioned simplification $CB = I_p$ this equation can be solved for the interesting variable $u_{eq}(t)$:

$$\begin{aligned}\Rightarrow u_{eq}(t) &= -M(x)\,\{CQe(t) + M(x)^{-1}f(x) \\ &\quad - N(z)^{-1}\,[g(z) + w(t)]\}\end{aligned} \tag{7.10}$$

Using this equivalent control vector and eqns. (7.3) and (7.5) the equivalent error dynamics can be expressed:

$$\begin{aligned}\dot{e}(t) &= Qe(t) + B[M(x)^{-1}\,(f(x) + u_{eq}(t)) \\ &\quad - N(z)^{-1}\,(g(z) + w(t))] \Rightarrow \dot{e}(t) \\ &= (I - BC)\,Qe\,(t) = \Pi Qe(t)\end{aligned} \tag{7.11}$$

This nice result is quite surprising, viz. starting with two nonlinear systems (7.3) and (7.5) and applying a nonlinear control (7.8) the resulting error dynamics are linear and time-invariant (in the sliding mode).

The matrix Π is a projection (El-Ghezawi *et al.*, 1983) and the matrix ΠQ has the following block-diagonal structure:

$$\Pi Q = \begin{pmatrix} \Lambda_1 & 0 & . & . & 0 \\ 0 & \Lambda_2 & 0 & . & 0 \\ . & . & . & . & . \\ . & . & . & . & . \\ 0 & . & . & 0 & \Lambda_p \end{pmatrix} \qquad \Lambda_i = \begin{pmatrix} 0 & 1 \\ 0 & \lambda_i \end{pmatrix}$$

Therefore, the error vector is governed by a linear differential equation with p eigenvalues λ_i (which where used to construct the matrix C and therefore can be chosen to be stable) and p eigenvalues equal to zero. In addition, the error dynamics have been decoupled into p independent second order blocks. This separation is a consequence of the special structure of the matrix C.

It has already been shown elsewhere in this book that the poles lying on the origin cannot influence the behaviour of the sliding system, since at $t = t^*$ the corresponding eigenvectors are not excited. Hence, in the sliding mode the effective order of the error dynamics (apparently) is halved.

The dynamics of the remaining p error states are supposed to be chosen to be stable. Therefore, if a time t^{**} exists with $e(t^{**}) = 0$ then the error vanishes for all times $t > t^{**}$, provided the closed-loop system remains in the sliding mode.

7.5 Stability conditions

This Section will give sufficient conditions for the VSC gains which guarantee that, if the system has a zero error (and therefore is in sliding mode) at a given time t^{**}, the sliding mode will never be left for all times $t > t^{**}$. Of course, in this case the error remains zero, too.

As stated in the Section 7.1 the robot is assumed to perform some kind of model/plant state matching at start-up; i.e. it is assumed that $t^{**} = t_0$. Since the state vector $x(t)$ is assumed to be completely measured the zero initial error can always be attained. In an assembly operation this has to be performed only once before starting a new task so that this assumption is reasonable. The main result is summarised in the following theorem:

Theorem:
If a time t^{**} with $e(t^{**}) = 0$ exists and if the regulator gains d_x and d_w fulfil the condition (7.12) then the error vector $e(t)$ will remain zero for all times $t > t^{**}$.

$$\begin{aligned} d_x &> \max_{z(t)} \left(\frac{|f(z(t)) - M(z(t))N(z(t))^{-1}g(z(t))|}{|z(t)|} \right) \\ d_w &> \max_{z(t)} (\sigma_{\max}(I - M(z(t))N(z(t))^{-1})) \end{aligned} \tag{7.12}$$

The norm operator $|.|$ used in the first condition is simply the Euclidean length. The norm operator σ_{max} used in the second condition is the largest singular value of its argument $(I - M(z(t))N(z(t))^{-1})$.

Note that the matrices and vector-valued functions in (7.12) all have the same argument, viz. the reference state $z(t)$. Therefore, the calculation of the sufficient regulator can be performed off-line simply by simulation of the reference model with the desired reference input $w(t)$.

Proof: The proof proceeds along the following lines: assume the initial error is zero; find sufficient conditions for the VSC gains assuring the asymptotic stability of the sliding mode (Lyapunov approach); in this condition the error dynamics are governed by the linear and homogeneous differential equation (7.11), the asymptotic stability of which can be granted by the very construction of the matrix C via pole placement.

The explicit proof is based on the following Lyapunov function:

$$v(t) = s^T(t) M(x(t)) s(t) \tag{7.13}$$

Since the mass matrix $M(x(t))$ is symmetric and positive-definite for all states $x(t)$, the function $v(t)$ is a valid candidate for a Lyapunov function. It remains to show that its first time derivative is negative-definite.

Since it is assumed that the error is zero only the neighbourhood of $e(t) = 0$ is studied where $s(t)$ approaches zero. The derivative of $v(t)$ is given by

$$\dot{v}(t) = 2s^T(t) M(x(t)) \dot{s}(t) + s^T(t) \dot{M}(x(t)) s(t) \Rightarrow \lim_{e \to 0} \dot{v}(t)$$
$$= 2s^T(t) M(x(t)) \dot{s}(t) \tag{7.14}$$

The second term of $\dot{v}(t)$ can be neglected since it is quadratic in $s(t)$. The time derivative of the switching variable $s(t)$ is given by eqn (7.9). In the neighbourhood of $e(t) = 0$ this equation becomes (using $CB = I_p$):

$$\Rightarrow \lim_{e \to 0} \dot{s}(t) = M(x)^{-1} (f(x) + u(t)) - N(z)^{-1} (g(z) + w(t)) \tag{7.15}$$

Combining the equations (7.8), (7.15) and (7.14) the time derivative of the Lyapunov function (7.13) can be expressed as follows:

$$\begin{aligned}\lim_{e \to 0} \dot{v}(t) = \; & s^T(t)\{f(x) - M(x) N(z)^{-1} g(z) \\ & + (I - M(x) N(z)^{-1}) w(t) - d_x \operatorname{sign}(s)|x(t)| \\ & - d_w \operatorname{sign}(s)|w(t)|\}\end{aligned} \tag{7.16}$$

Since the error is assumed to be zero, the state $x(t)$ can now be substituted by $z(t)$. Using the triangle inequality an upper bound for $\dot{v}(t)$ is found:

$$\begin{aligned}\lim_{e \to 0} \dot{v}(t) \leqslant \; & |s(t)| \, |f(x) - M(x) N(z)^{-1} g(z)| - d_x s^T(t) \operatorname{sign}(s)|x(t)| \\ & + |s(t)| \, |(I - M(x) N(z)^{-1}) w(t))| - d_w s^T(t) \operatorname{sign}(s)|w(t)|\end{aligned} \tag{7.17}$$

The relation $s^T(t) \operatorname{sign}(s) \geqslant |s(t)|$ is always true. Moreover, an upper bound of the term $|(I - M(x) N(z)^{-1} w(t))|$ can be found using the singular values of the matrix $(I - M(x) N(z)^{-1})$.

The final form of the upper bound of (7.14) is:

$$\begin{aligned}\lim_{e \to 0} \dot{v}(t) \leqslant \; & |s(t)| \, (|f(x) - M(x) N(z)^{-1} g(z)| - d_x |x(t)|) \\ & + |s(t)| \, (\sigma_{max}(I - M(x) N(z)^{-1}) |w(t)| - d_w |w(t)|) < 0\end{aligned} \tag{7.18}$$

The conditions (7.12) are immediately recognised from eqn. (7.18). But since these conditions guarantee that the Lyapunov function (7.14) has a negative-definite time derivative the error dynamics have to remain in the sliding mode. If this is true the error is governed by eqn. (7.11) and therefore remains zero for all times.

This completes the proof.

7.6 Example

In Geering *et al.* (1986) the minimum principle of Pontryagin was used to find some time-optimal trajectories for several robots with constrained inputs. In the following example the ideas developed above are applied to the cylindrical robot analyzed in that paper (cf. Fig. 7.2).

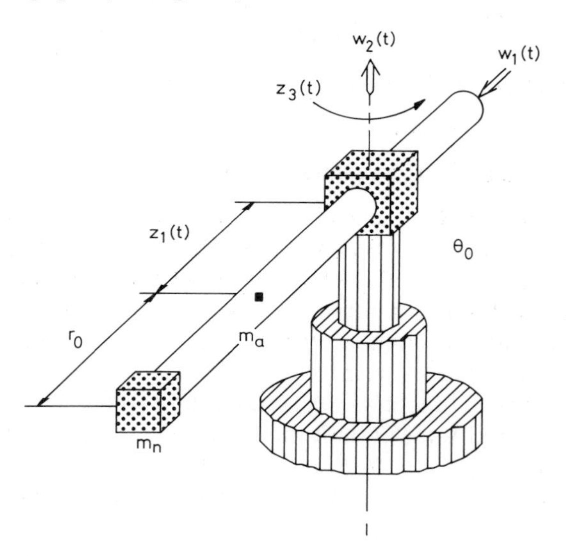

Fig. 7.2 *Sketch of the robot*

The robot has two degrees of freedom, viz. the length of the arm $z_1(t)$ and the angle of rotation $z_3(t)$ (the vertical co-ordinate, being decoupled, will be disregarded). Both degrees of freedom have their own actuator (exerting the force $w_1(t)$ and the torque $w_2(t)$, respectively). The mass of the arm is represented by m_a and the payload by m_n. The constant r_0 indicates the distance between the gripper and the centre of gravity of the arm. Finally, the constant Θ_0 represents the mass moment of inertia of the arm (without payload) with respect to its centre of gravity S and the constant Θ_t represents the mass moment of inertia of the rotating centre part.

The dynamic equations of this robot are easily found by using the method of

Lagrange. The reference model (7.5) of this robot used to calculate the time-optimal control $w(t)$ is defined by:

$$\dot{z}(t) = Qz(t) + BN(z(t))^{-1}(g(z(t)) + w(t)) \qquad z(t) \in \mathscr{R}^4,\ w(t) \in \mathscr{R}^2$$

$$Q = \begin{pmatrix} 0 & 1 & 0 & 0 \\ 0 & 0 & 0 & 0 \\ 0 & 0 & 0 & 1 \\ 0 & 0 & 0 & 0 \end{pmatrix} \qquad B = \begin{pmatrix} 0 & 0 \\ 1 & 0 \\ 0 & 0 \\ 0 & 1 \end{pmatrix}$$

$$N(z(t)) = \begin{pmatrix} m_a + m_n & 0 \\ 0 & \Theta(z(t)) \end{pmatrix}$$

$$g(z(t)) = \begin{pmatrix} [m_a z_1(t) + m_n(r_0 + z_1(t))]z_4^2(t) \\ -2[m_a z_1(t) + m_n(r_0 + z_1(t))]z_2(t)z_4(t) \end{pmatrix}$$

$$\text{with } \Theta(z(t)) = \Theta_0 + z_1^2 m_a + (r_0 + z_1(t))^2 m_n$$

The explicit numerical values are:

$$m_a = 3{\cdot}7\,\text{kg} \qquad m_n = 4{\cdot}6\,\text{kg} \qquad r_0 = 0{\cdot}37\,\text{m} \qquad \Theta_0 = 0{\cdot}37\,\text{kg}\,\text{m}^2$$

This model was used in Geering *et al.* (1986) in order to find some time-optimal trajectories for bounded control signals. Using Pontryagin's minimum principle one can show that the optimal controls have to be discontinuous ('bang-bang' control). Therefore, both actuators have to start with either positive or negative maximum values and switch at a given time(s) to the corresponding maximal reversing forces and torques, respectively.

The resulting control functions will combine both degrees of freedom in order to perform the desired transfer with minimal inertia. As an example the pick-and-place trajectory $z_1(t_0) = 0{\cdot}15$, $z_1(t_f) = 0{\cdot}15$, $z_3(t_0) = 0$ and $z_3(t_f) = \pi/2$ (start and stop at standstill) has the following sequence of inputs:

$$\begin{array}{lll} 0 \leqslant t < 0{\cdot}382 & w_1(t) = w_{1,min} & w_2(t) = w_{2,max} \\ 0{\cdot}382 \leqslant t \leqslant 0{\cdot}582 & w_1(t) = w_{1,max} & w_2(t) = w_{2,max} \\ 0{\cdot}582 \leqslant t \leqslant 0{\cdot}782 & w_1(t) = w_{1,max} & w_2(t) = w_{2,min} \\ 0{\cdot}782 \leqslant t \leqslant 1{\cdot}164 & w_1(t) = w_{1,min} & w_2(t) = w_{2,min} \end{array}$$

with the explicit values $w_{1,min} = -15\,\text{N}$, $w_{1,max} = +15\,\text{N}$, $w_{2,min} = -5\,\text{Nm}$ and $w_{2,max} = +5\,\text{Nm}$.

The determination of the explicit switching points is done numerically and requires a rather large amount of computations. Once these off-line computations have been done the on-line control is generated by some memory-device which holds the correct switching points.

Of course, the optimal switching times depend on the model used in the computations; i.e. if the control is applied to a different robot the resulting trajectory will be neither time-optimal nor will it satisfy the desired requirements of the final position. In assembly operations the second effect is not acceptable; therefore one has to counteract the model/plant mismatches by some on-line device.

The VSC introduced here is well suited for such applications; i.e. it guarantees that, regardless of the unavoidable parameter errors, the real robot will precisely follow the refrence state. Of course one has to pay a price for a perfect tracking; i.e. a VSC will require a part of the actuator power for its tasks!

In the following study it is assumed that the real robot has to perform the assembly task with payloads of uncertain (or time-varying) weight $m_n + \delta m_n$. In this case the resulting stability conditions for the regulator gains (7.11) are determined by:

$$f(z(t)) - M(z(t))\,N(z(t))^{-1}g(z(t))$$

$$= \begin{pmatrix} \delta m_n\left(\dfrac{m_a r}{m_a + m_n}\right) z_4^2 \\ 2\delta m_n\left(\dfrac{r_0 + z_1}{\Theta(t)}(m_a z_1 + m_n(r_0 + z_1)) - 1\right)(r_0 + z_1) z_2 z_4 \end{pmatrix}$$

$$I - M(z(t))\,N(z(t))^{-1} = \begin{pmatrix} \dfrac{-\delta m_n}{m_a + m_n} & 0 \\ 0 & \dfrac{-\delta m_n (r_0 + z_1)^2}{\Theta(t)} \end{pmatrix}$$

The calculation of the norm of $f(z(t)) - M(z(t))N(z(t))^{-1}g(z(t))$ is obvious. Since the matrix $I - M(z(t))N((t))^{-1}$ is diagonal its singular values are just the absolute value of its diagonal elements.

By simulation of the reference model these values can be calculated off-line. Assuming a maximum extra payload δm_n of 2 kg the following values for the VSC gains are found (those used in the simulation shown in Fig. 7.3 are indicated between parentheses):

$$d_w > 0{\cdot}318\ldots(0{\cdot}32) \qquad d_x > 1{\cdot}10\ldots(1{\cdot}11) \tag{7.18}$$

Finally the matrix C is defined which is used in the calculation of the switching variable $s(t)$:

$$C = \begin{pmatrix} 5 & 1 & 0 & 0 \\ 0 & 0 & 5 & 1 \end{pmatrix}$$

As shown by C, the error dynamics of both movements are chosen to be equal with a time constant of 0·2 s.

Fig. 7.3 shows the behaviour of the closed-loop system in the case of the above regulator gains. If the regulator gains are chosen smaller than prescribed by the stability conditions (7.18) the robot cannot follow the reference model, as shown in Fig. 7.4.

A last remark concerns the practical feasibility of the proposed VSC. Many people believe that VSCs have some very attractive properties but that a VSC

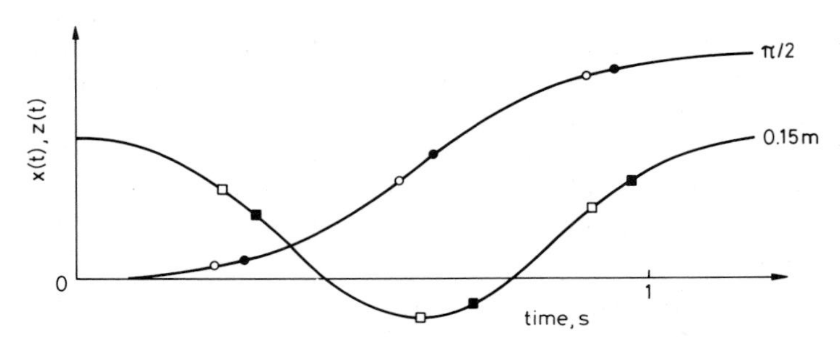

Fig. 7.3 *Sufficient VSC-gains*
□ = $x_1(t)$; ○ = $x_3(t)$; ■ = $z_1(t)$; ● = $z_3(t)$

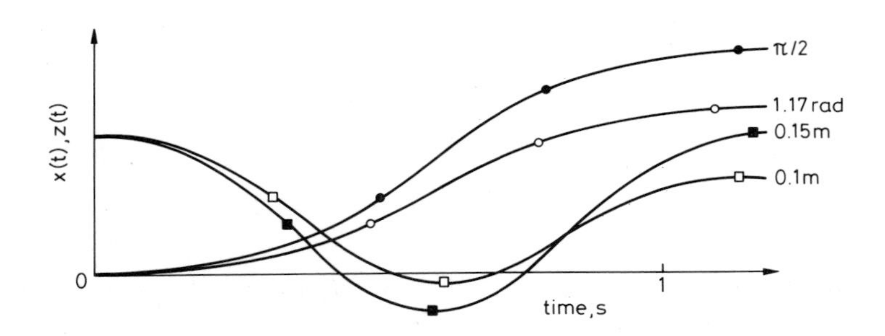

Fig. 7.4 *Insufficient VSC-gains*
□ = $x_1(t)$; ○ = $x_3(t)$; ■ = $z_1(t)$; ● = $z_3(t)$

cannot be used because of its chattering behaviour which may excite the elastic modes of the mechanical structure. A possible solution to this problem consists in smoothing the sgn function in order to yield a continuous 'switching' law (Slotine, 1985); but in this case the perfect sliding properties are lost.

We believe that the correct solution to this problem is the design of adapted hardware; i.e. very fast switching and computing devices are needed. Today's power MOS-FETs can switch considerable currents at rates of up to several hundred kilohertz. The prospect of even faster and more powerful devices is rather positive.

The problem of computing speed is also solvable because the reference model (7.5) and the controller (7.8) can be implemented using different hardware. The reference model has not to be very fast since it represents a rigid mechanical system. The controller (7.8), which has to be fast, performs rather simple operations and could easily be implemented using a digital signal processor. Judging from experience in other applications with digital signal processors, sampling rates of several tens of kilohertz should be achievable. Cearly, such a high-frequency excitation of a mechanical structure cannot produce harmful chattering motions!

7.7 References

BALESTRINO, A., DE MARIA, G. and SCIAVICCO, L., (1983): 'An adaptive model-following control for robotic manipulators', *J. Dynamic Systems, Measurement and Control*, **105**, pp. 143–151

CRAIG, J. J., (1986): 'Introduction to robotics', (Addison-Wesley, Reading, Mass.)

EL-GHEZAWI, O. M. E., ZINOBER, A. S. I. and BILLINGS, S. A., (1983), 'Analysis and design of variable structure systems using a geometric approach', *Int. J. Control*, **38**, pp. 657–671

ERZBERGER, H., (1968): 'Analysis and design of model-following control systems by state space techniques', Proc. Joint Automatic Control Conf., pp. 572–581

GEERING, H. P., GUZZELLA, L., HEPNER, S. A. R. and ONDER, C. H., (1986): 'Time optimal motions of robots in assembly tasks', *IEEE Trans.* **AC-31**, pp. 512–518

HIROI, M., HASAYUKI, M., HASHIMOTO, Y., ABE, Y. and DOTE, Y. (1986): 'Microprocessor-based decoupled control of manipulators using modified model-following method with sliding mode', *IEEE Trans.* **IE-33**, pp. 110–113

PAUL, R. P. (1981): 'Robot manipulators', (MIT Press, Cambridge, Mass.)

SLOTINE, J., (1985): 'The robust control of robot manipulators', *Int. J. Robotics Research*, **4**, pp. 49–64

UTKIN, V. I. (1977): 'Variable structure systems with sliding modes: A survey', *IEEE Trans.* **AC-22**, pp. 212–222

YOUNG, K. K., (1978): 'Controller design for a manipulator using theory of variable structure systems', *IEEE Trans. Systems, Man and Cyb.*, **8**, pp. 101–109

Chapter 8

Applications of output feedback in variable structure control

B. A. White

Royal Military College of Science

8.1 Introduction

Most of the techniques for the design of Variable Structure Control Systems (VSCS) for both scalar and multivariable systems that is presented in the literature assumes that either the state vector is directly measurable or that an observer is used to reconstruct the state (Utkin, 1977; Itkis, 1976; El-Ghezawi, Zinober, Billings, 1983; Zinober, El-Ghezawi, Billings, 1982; Young, 1978). In systems which do not have the full state vector available or that have a set of output measurements $\boldsymbol{y}$, the usual technique for VSCS design of equivalent control cannot be applied. This chapter will look at the design of VSCS for the case of output feedback. State feedback will first be treated as a special case of the more general case of output feedback. This will be followed by a section extending the ideas to the output feedback case. Two applications are then presented: the first applying output feedback to a roll autopilot and the second using the outputs in a state estimator to generate states for a VSCS to control the vibrational modes of a flexible system. The VSCS takes two different forms in each case: the roll autopilot study switches on a fixed subspace given by the switching operator $\boldsymbol{S}$ and has a third order range space dynamics. The vibration suppression example has two subspaces that represent bounds on the damping ratio of the vibrational modes and switches on a segmented set of subspaces.

8.2 State feedback

The design of VSCS in this chapter assumes, as in previous chapters, that the state space description of a system is described by

$$\dot{x} = Ax + Bu$$
$$y = Cx \tag{8.1}$$

where $x \in \mathscr{R}^n$, $u \in \mathscr{R}^m$, $y \in \mathscr{R}^l$. The system triple (A, B, C) is assumed to be controllable and observable. For the system controlled by state feedback ($C = I_n$, $y \in \mathscr{R}^n$) the input u is of the form

$$u = -(K + \Delta K)x \tag{8.2}$$

such that the variable structure switching function $z \in \mathscr{R}^m$ is maintained as close as possible to zero where

$$z = Sx = \begin{pmatrix} s_1^t \\ s_2^t \\ \cdot \\ \cdot \\ \cdot \\ s_m^t \end{pmatrix} x \tag{8.3}$$

where $S \in \mathscr{R}^{m \times n}$. The dynamic description of the system with feedback of the form in eqn. (8.2) is

$$\dot{x} = A_c x - B\Delta K x \tag{8.4}$$

where the closed loop matrix A_c is given by

$$A_c = A - BK \tag{8.5}$$

The dynamic behaviour of the switching function z is then given by

$$\dot{s} = SA_c x - SB\Delta K x \tag{8.6}$$

8.2.1 Range space dynamic equations

The fixed gain component of eqn. (8.2) can be designed using the equivalent control principle such that $n - m$ of the eigenvectors of A_c in eqn. (8.4) are placed in the kernel of the operator S. This can be normally achieved by pole placement or eigenstructure assignment techniques (Davison, 1970; Patton, Wilcox, Winter, 1986) to get a satisfactory set of eigenvalues for A_c with an associated operator S. The system can also be designed using model following techniques (Zinober, El-Ghezawi, Billings, 1982) that produce a fixed gain and S operator by optimal control techniques designed to minimise the error between the system performance in eqn. (8.2) and a model describing the desired system behaviour. The remaining m eigenvalues are normally set to zero to yield

$$\dot{z} = -SB\Delta K x \tag{8.7}$$

These cases have been presented in previous chapters. It is not necessary, however, to force all the remaining m eigenvalues to zero. For the multivariable case consider the following theorem:

Theorem 1: The dynamic behaviour of the range space z is described by the spectrum of $\boldsymbol{SBQ}$, if the fixed gain $\boldsymbol{K}$ is chosen as $(\boldsymbol{SB})^{-1}\boldsymbol{SA} - \boldsymbol{QS}$, where $\boldsymbol{Q}$ is an arbitrary square matrix $\boldsymbol{Q} \in \mathscr{R}^{m \times m}$

The proof of this theorem can be shown by substitution. From eqn. (8.6) we have:

$$\dot{z} = \boldsymbol{S}(\boldsymbol{A} - \boldsymbol{BK})\boldsymbol{x} - \boldsymbol{SB}\Delta\boldsymbol{Kx} \tag{8.8}$$

Substituting for $\boldsymbol{K}$ yields

$$\boldsymbol{S}(\boldsymbol{A} - \boldsymbol{BK})\boldsymbol{x} = \boldsymbol{SA} - \boldsymbol{SB}[(\boldsymbol{SB})^{-1}\boldsymbol{SA} - \boldsymbol{QS}] \tag{8.9}$$

$$= (\boldsymbol{SBQ})\boldsymbol{S} \tag{8.10}$$

Hence eqn. (8.8) can be written as

$$\dot{z} = (\boldsymbol{SBQ})z - \boldsymbol{SB}\Delta\boldsymbol{Kx} \tag{8.11}$$

The dynamic behaviour of the range space of $\boldsymbol{S}$ is governed by the spectrum of $(\boldsymbol{SBQ})$ and the theorem is proved. A corollary of Theorem 1 is:

Corollary 1: The spectrum of $\boldsymbol{SBQ}$ can be arbitrarily set to any Λ by:

$$\boldsymbol{Q} = (\boldsymbol{SB})^{-1}\Lambda \tag{8.12}$$

provided $(\boldsymbol{SB})^{-1}$ is full rank.

The dynamic behaviour of the range space is then

$$\dot{z} = \Lambda z - \boldsymbol{SB}\Delta\boldsymbol{Kx} \tag{8.13}$$

The stability of eqn. (8.13) can therefore be guaranteed by a suitable choice of $\boldsymbol{Q}$ from eqn. (8.12) and if the switched gain matrix $\Delta\boldsymbol{K} = 0$, the resulting feedback reduces to pole placement (Munro, Vardulakis, 1973). The case of equivalent control can be treated as a special case of corollary 1, where

$$\Lambda = 0, \Rightarrow \boldsymbol{Q} = 0 \tag{8.14}$$

so that all the range space eigenvalues are placed at the origin in the s-plane.

8.2.2 Range space dynamic stability

The problem of reachability (White, Silson, 1984) can be addressed by considering the stability of eqn. (8.11). If the switched gain matrix $\Delta\boldsymbol{K}$ is used for active range space control, then performance for the m uncoupled dynamic systems described by eqn. (8.13) can be enhanced by considering the following theorem:

Theorem 2: The stability of the range space of $\boldsymbol{S}$ can be ensured by switched feedback of the form

$$\Delta\boldsymbol{Kx} = (\boldsymbol{SB})^{-1}\boldsymbol{F}(\boldsymbol{x})z \tag{8.15}$$

where

$$\boldsymbol{F}(\boldsymbol{x}) = \text{diag}\,\{f_i(\boldsymbol{x})\} \tag{8.16}$$

where $f_i(\boldsymbol{x})$ are functions satisfying

$$f_i(\boldsymbol{x}) \geqslant \lambda_i \tag{8.17}$$

Proof of the theorem uses Lyapunov on the resulting uncoupled dynamic first order systems. Substituting eqn. (8.15) into eqn. (8.11) yields

$$\dot{\boldsymbol{z}} = \Lambda \boldsymbol{z} - \boldsymbol{F}(\boldsymbol{x})\boldsymbol{z} \tag{8.18}$$

or in component form, using eqn. (8.17)

$$\dot{z}_i = \lambda_i z_i - f_i(\boldsymbol{x}) z_i, \; i = 1, \ldots, m \tag{8.19}$$

Stability of eqn. (8.19) can be investigated by a simple Lyapunov equation of the form

$$V_i = z_i^2 \tag{8.20}$$

so that

$$\begin{aligned} \dot{V} &= 2z_i \dot{z}_i \\ &= 2z_i[\lambda_i z_i - f_i(\boldsymbol{x}) z_i] \\ &= 2[\lambda_i - f_i(\boldsymbol{x})]\, z_i^2 \end{aligned} \tag{8.21}$$

$\dot{V}$ is negative provided

$$\lambda_i - f_i(\boldsymbol{x}) < 0 \tag{8.22}$$

or

$$f_i(\boldsymbol{x}) > \lambda_i \tag{8.23}$$

and the theorem is proved.

For the case of equivalent control design, eqn. (8.23) reduces to

$$f_i(\boldsymbol{x}) > 0 \tag{8.24}$$

The choice of $f_i(\boldsymbol{x})$ is made by considering each row of $\Delta \boldsymbol{K}\boldsymbol{x}$:

$$\Delta \boldsymbol{K}\boldsymbol{x} = \begin{pmatrix} \Delta \boldsymbol{k}_1^t \\ \Delta \boldsymbol{k}_2^t \\ \cdot \\ \cdot \\ \cdot \\ \Delta \boldsymbol{k}_m^t \end{pmatrix} \boldsymbol{x} \tag{8.25}$$

Each row is then treated as a switched function, where

$$\Delta \boldsymbol{k}_i^t = \boldsymbol{\Phi}_i^t \,\text{sgn}\,(z_i) \tag{8.26}$$

and $\boldsymbol{\Phi}_i^t$ is a vector given by

$$\boldsymbol{\Phi}_i = \begin{pmatrix} \phi_{i1} \operatorname{sgn}(x_1) \\ \phi_{i2} \operatorname{sgn}(x_2) \\ \cdot \\ \cdot \\ \cdot \\ \phi_{im} \operatorname{sgn}(x_m) \end{pmatrix} \tag{8.27}$$

Now

$$\operatorname{sgn}(z_i) = \frac{z_i}{|z_i|} \tag{8.28}$$

Substituting for sgn (z_i) in eqn. (8.25) yields

$$\Delta \boldsymbol{k}_i^t \boldsymbol{x} = \frac{\boldsymbol{\Phi}_i^t \boldsymbol{x}}{|z_i|} \tag{8.29}$$

Substituting for $|z_i|$ from eqn. (8.3) gives

$$\Delta \boldsymbol{k}_i^t \boldsymbol{x} = \frac{\boldsymbol{\Phi}_i^t \boldsymbol{x}}{|\boldsymbol{s}_i^t \boldsymbol{x}|} z_i$$

$$= f_i(\boldsymbol{x}) z_i \tag{8.30}$$

Hence the functions $f_i(\boldsymbol{x})$ are given by

$$f_i(\boldsymbol{x}) = \frac{\boldsymbol{\Phi}_i^t \boldsymbol{x}}{|\boldsymbol{s}_i^t \boldsymbol{x}|} \tag{8.31}$$

From Theorem 2 stability is assured for any λ_i by

$$\frac{\boldsymbol{\Phi}_i^t \boldsymbol{x}}{|\boldsymbol{s}_i^t \boldsymbol{x}|} > \lambda_i \tag{8.32}$$

or

$$\boldsymbol{\Phi}_i^t \boldsymbol{x} > |\boldsymbol{s}_i^t \boldsymbol{x}| \lambda_i \tag{8.33}$$

Using the inequality

$$\left| \sum_{j=1}^{m} s_{ij} x_j \right| < \sum_{j=1}^{m} |s_{ij}| |x_j| \tag{8.34}$$

and the fact that

$$\boldsymbol{\Phi}_i^t \boldsymbol{x} = \sum_{j=1}^{m} \phi_{ij} \operatorname{sgn}(x_i) x_i \tag{8.35}$$

$$= \sum_{j=1}^{m} \phi_{ij}|x_i| \tag{8.36}$$

eqn. (8.33) can be satisfied by satisfying

$$\sum_{j=1}^{m} \phi_{ij}|x_i| > \lambda_i \sum_{j=1}^{m} |s_{ij}||x_i| \tag{8.37}$$

The solution to eqn. (8.37) for all state vectors $\boldsymbol{x}$ is given by

$$\phi_{ij} > \lambda_i|s_{ij}| \tag{8.38}$$

As the subsystems are first order, the Lyapunov function directly represents the energy in the modes of the system (the square of the eigenvalues along the eigenvector) and so the more negative the derivative, the faster the decay in the range space. Hence a choice of ϕ_{ij} should be as large as possible for this system. Practical considerations will, however, restrict the magnitude of ϕ_{ij} owing to unmodelled dynamics and actuator limitations.

Other choices of the function $\boldsymbol{F}(\boldsymbol{x})$ are possible that do not have the relay or bang-bang action described in eqn. (8.26). A softer control for the range space dynamics is the continuous switching control defined by

$$\Delta\boldsymbol{k}_i^t = (\alpha_i + \beta_i|z_i|)\,\boldsymbol{s}_i^t \tag{8.39}$$

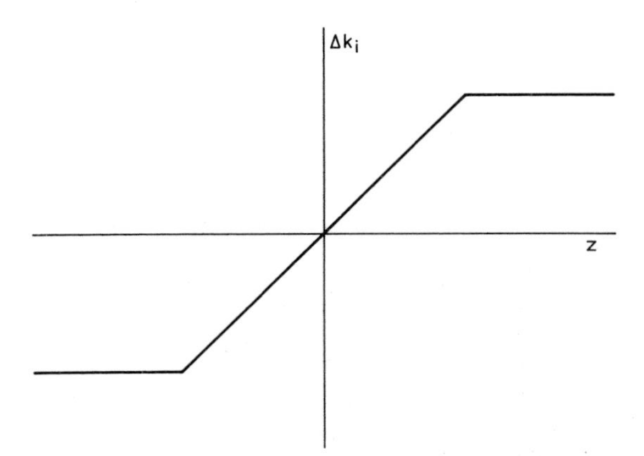

Fig. 8.1 *Soft range space controller characteristic*

where $\alpha_i > 0$ and $\beta_i > 0$ are arbitrary variables that can be chosen to shape the dynamic performance. This form of control is shown in Fig. 8.1. For this form of switched control component the functions $f_i(\boldsymbol{x})$ are given by

$$f_i(\boldsymbol{x}) = (\alpha_i + \beta_i|\boldsymbol{x}^t\boldsymbol{s}_i|) \tag{8.40}$$

From eqn. (8.40) it is clear that each $f_i(\boldsymbol{x}) > 0$ for all $\boldsymbol{x}$ not in the null space of the operator $\boldsymbol{s}^t$. This will guarantee stability everywhere outside of the null space of S for $\lambda_i \leqslant 0$. It can be shown (White, 1983) that, for the case $\beta = 0$, the control design reduces to a pole shifting algorithm using the right handed eigenvector set and conssitutes linear state feedback.

8.3 Output feedback

If the available outputs from the system is not the state vector x and a state estimator is not used, then the design of the VSC system becomes more restrictive (White, 1985). The feedback control of eqn. (8.2) becomes

$$u = -(K + \Delta K)y \tag{8.41}$$

with the switching function x given by

$$\begin{aligned} z &= Sy \\ &= SCx \end{aligned} \tag{8.42}$$

The closed loop system description is given by

$$\dot{x} = A_c x - B\Delta K y \tag{8.43}$$

where

$$A_c = A - BKC \tag{8.44}$$

For the output feedback case the range space dynamics are described by

$$\dot{z} = SCA_c x - SCB\Delta K y \tag{8.45}$$

The application of Theorem 1 to the output feedback case is not generally possible as the fixed gain K must now be chosen as

$$K = (SCB)^{-1}\, SCA + QSC \tag{8.46}$$

The selection of S to make SCB full rank is not always possible and the case of rank deficiency in SCB must be addressed. The effect of output feedback is mainly to restrict the number of closed loop, fixed gain eigenvectors that can be placed into the null space of the operator S. The operator S is itself restricted in that it operates on the output space and not the state space.

8.3.1 Range space dynamics

The dynamic behaviour of the range space of S is governed by the choice of S and the fixed gain matrix K. As the freedom of choice is now restricted for both S and K, the simultaneous solution stated in Theorem 1 cannot be used. So consider a set of equations described in eqn. (8.43) where K has been chosen to meet some specification using some feedback algorithm (Hanmandlu, Shantaram, 1986; Seraji, 1978; Fallside, 1977). The dynamics of the range space of S are given by

$$\dot{x}_s = A_s x_s - B_s \Delta K y$$
$$z = C_s x_s \qquad (8.47)$$

The form of these equations is governed by the following theorem:

Theorem 3: The dynamic equations for the range space of S are given by eqns. (8.47), where

$$A_s = UAV$$
$$= \Lambda_s$$
$$B_s = UB$$
$$C_s = SCW \qquad (8.48)$$

where the matrices U and W are formed from the left and right eigenvector set of A_c (V and Z, respectively) such that

$$U = M^{-1} T_s V \qquad (8.49)$$

$$W = X T_s^t M \qquad (8.50)$$

The $T_s \in \mathscr{R}^{m \times n}$ matrix is given by

$$T_s = (I_m 0) \qquad (8.51)$$

and the S operator is given by the solution of

$$SC = N T_s V \qquad (8.52)$$

with M and N being arbitrary matrices such that

$$\text{Rank}\,(M) = m$$
$$C_s = NM \qquad (8.53)$$

and the couple (A_s, C_s) is observable.

The proof of the theorem is obtained by considering a projection of the state space onto a subspace by the following partition:

$$x = Z T_s M x_s + Z T_n M x_n \qquad (8.54)$$

where

$$T_s = (I_m\ 0) \qquad (8.55)$$

and

$$T_n = (0\ I_{n-m}) \qquad (8.56)$$

such that

$$I_n = \begin{pmatrix} T_s \\ T_n \end{pmatrix} \qquad (8.57)$$

The matrix $\boldsymbol{M} \in \mathscr{R}^{m \times m}$ is an arbitrary matrix of rank m and $\boldsymbol{T}_s \in \mathscr{R}^{m \times n}$ is a matrix which selects m of the fixed gain, closed loop, right hand eigenvectors from $\boldsymbol{Z}$. The choice of $\boldsymbol{T}_s$ is general as the eigenvectors can be ordered arbitrarily, assuming complex conjugate eigenvalues and their associated eigenvectors are treated as a compound pair, so that both must be included or excluded from the selection.

The dynamics of the component $\boldsymbol{x}_s$ can be obtained by the reverse transformation

$$\begin{aligned} \boldsymbol{x}_s &= \boldsymbol{M}^{-1}\boldsymbol{T}_s\boldsymbol{V}\boldsymbol{x} \\ &= \boldsymbol{M}^{-1}\boldsymbol{T}_s\boldsymbol{V}\boldsymbol{Z}\boldsymbol{T}_s\boldsymbol{M}\boldsymbol{x}_s + \boldsymbol{Z}\boldsymbol{T}_n\boldsymbol{M}\boldsymbol{x}_n \\ &= \boldsymbol{x}_s \end{aligned} \tag{8.58}$$

Performing this projection on eqns. (8.43) yields

$$\begin{aligned} \dot{\boldsymbol{x}}_s &= \boldsymbol{MTVA}_c\boldsymbol{x} - \boldsymbol{MTVB}\Delta\boldsymbol{Ky} \\ &= \boldsymbol{MTV}\Lambda\boldsymbol{x} - \boldsymbol{B}_s\Delta\boldsymbol{Ky} \end{aligned} \tag{8.59}$$

where Λ_c is the spectral matrix of $\boldsymbol{A}_c$. Using eqn. (8.58) then yields

$$\begin{aligned} \dot{\boldsymbol{x}} &= \boldsymbol{MT}\Lambda_c\boldsymbol{T}^t\boldsymbol{M}^{-1}\boldsymbol{x}_s - \boldsymbol{B}_s\Delta\boldsymbol{Ky} \\ &= \boldsymbol{M}\Lambda_s\boldsymbol{M}^{-1}\boldsymbol{x}_s - \boldsymbol{B}_s\Delta\boldsymbol{Ky} \end{aligned} \tag{8.60}$$

where Λ_s is a spectral matrix containing m of the eigenvalues of $\boldsymbol{A}_c$. The range space vector z is then obtained from

$$\begin{aligned} z &= \boldsymbol{SCx} \\ &= \boldsymbol{SCZT}^t\boldsymbol{M}^{-1}\boldsymbol{x}_s \end{aligned} \tag{8.61}$$

If $\boldsymbol{S}$ is chosen so that $\boldsymbol{SC}$ has the same form as the transformation in eqn. (54) i.e.

$$\boldsymbol{SC} = \boldsymbol{NTV} \tag{8.62}$$

then

$$\begin{aligned} \boldsymbol{SCZT}^t\boldsymbol{M}^{-1}\boldsymbol{x}_s &= \boldsymbol{NTT}^t\boldsymbol{M}^{-1}\boldsymbol{x}_s \\ &= \boldsymbol{NM}^{-1}\boldsymbol{x}_s \\ &= \boldsymbol{C}_s\boldsymbol{x}_s \end{aligned} \tag{8.63}$$

The matrix $\boldsymbol{C}_s$ can then be chosen arbitrarily provided that the couple $(\boldsymbol{A}_s, \boldsymbol{C}_s)$ is observable.

8.3.2 Range space stability

The range space dynamic equations of eqn. (8.47) will have dynamic characteristics determined by the choice of the switched gain matrix $\Delta\boldsymbol{Ky}$. The switched gain component can take many forms in the literature, but the form presented here will be a switched form [White, 1987]. The main assumption will be that the dynamics of the null space and range space of $\boldsymbol{S}$ can be treated independently as the range space dynamics are much faster than the null space dynamics. The switched gain component $\Delta \boldsymbol{K}\boldsymbol{y}$ will be realised by a relay type action operating on the range space outputs z, such that

$$\Delta \boldsymbol{K}\boldsymbol{y} = \{\Delta \boldsymbol{k}_i^t \boldsymbol{y}\}, \quad i = 1, \ldots, m \tag{8.64}$$

where

$$\Delta \boldsymbol{k}_i^t \boldsymbol{y} = \operatorname{sgn}\,[\boldsymbol{q}_i^t z]\, U \tag{8.65}$$

where U is the magnitude of the relay action and $\boldsymbol{q}_i^t$ is a design parameter vector. Hence

$$\Delta \boldsymbol{K}\boldsymbol{y} = \mathrm{SGN}\,[\boldsymbol{Q}z]\, U \tag{8.66}$$

where

$$\begin{aligned} \mathrm{SGN}\,[\boldsymbol{Q}z] &= \{\operatorname{sgn}\,[\boldsymbol{q}_i^t z]\} \\ &= \mathrm{SGN}\,[z] \end{aligned} \tag{8.67}$$

For variable structure control systems, the state space trajectory will be forced into $\ker(\boldsymbol{S})$ by maintaining z at zero. If any limit cycle conditions occur then the size and frequency must be established to assess if the excursion around $\ker(\boldsymbol{S})$ arising from the limit-cycle condition will be significant. Overall stability must also be ensured to contain the state vector $\boldsymbol{x}$ in $\ker(\boldsymbol{S})$. To investigate stability and limit-cycling, a segment of the state trajectory with both initial and terminal states in $\ker(\boldsymbol{S})$ is considered.

The equations given in eqn. (8.47) describe the motion of the state vector in the range space of the operator $\boldsymbol{S}$. Consider the partitioning of the vector $\boldsymbol{x}$, into the sum of two vectors

$$\boldsymbol{x}_s = \boldsymbol{P}_k \boldsymbol{g}_k + \boldsymbol{P}_r \boldsymbol{g}_r \tag{8.68}$$

where $\boldsymbol{P}_k$ and $\boldsymbol{P}_r$ are an orthonormal basis which span $\ker(\hat{z})$ and range $(\hat{z})$ respectively, and $\boldsymbol{g}_k$ and $\boldsymbol{g}_r$ are the projections of $\boldsymbol{x}_s$ onto $\boldsymbol{P}_k$ and $\boldsymbol{P}_r$. The transition from $\ker(\hat{z})$ at time $t = 0$ to $\ker(\hat{z})$ at time $t = T$ is given by

$$\boldsymbol{x}_s[T] = \boldsymbol{E}_s(T)\boldsymbol{x}_s[0] - \boldsymbol{F}_s(T)\Delta \boldsymbol{K}\boldsymbol{y} \tag{8.69}$$

with $\boldsymbol{g}_r[0] = \boldsymbol{g}_r[T] = 0$. The discrete equations associated with this transition are

$$\begin{aligned} \boldsymbol{x}_s[T] &= \boldsymbol{E}_s(T)\boldsymbol{x}_s[0] - \boldsymbol{f}_s(T)\Delta K y \\ &= \boldsymbol{E}_s(T)\boldsymbol{x}_s[0] - \boldsymbol{f}_s(T)\operatorname{sgn}\,[\boldsymbol{q}^t z]\, U \end{aligned} \tag{8.70}$$

There exists the possibility that several disjoint limit-cycle conditions could exist, each oscillating in a disjoint subspace in the state space. Assuming that all disjoint spaces have either been eliminated by a suitable choice of $\boldsymbol{Q}$ or by treating the disjoint subspaces separately, then the following assumptions will be deemed to hold:

- There exists the possibility of one limit cycle condition.
- The state trajectory will pass through $\ker(z_i)$ cyclically with one and only one transition through each $\ker(z_i)$

The first assumption will hold if there are no disjoint subspaces and the second is assumed to occur for linear systems with relay control action, where the kernel spaces all include the origin and the relay action is symmetrical.

The switching will occur on transition throughout the kernal space of any of the elements of $\boldsymbol{z}$, so consider the transition between two adjacent kernel spaces, $\ker(z_i)$ and $\ker(z_j)$:

$$\boldsymbol{P}_{kj}\boldsymbol{g}_{kj}[T] + \boldsymbol{P}_{rj}\boldsymbol{g}_{rj}[T] = \boldsymbol{E}_s(T)\,\boldsymbol{P}_{ki}\boldsymbol{g}_{ki}[0] - \boldsymbol{f}_s(T)\,\mathrm{SGN}\,[\hat{\boldsymbol{z}}]\,U \tag{8.71}$$

As the ordering of the kernel spaces in $\ker(\boldsymbol{z})$ is arbitrary, the sequence of transitions will be taken in order of $\ker(z_i)$. Consider a state $\boldsymbol{x}_s$ which is such that

$$\begin{aligned} z_i &= 0, \qquad i = 1 \\ z_i &> 0, \qquad i = 2, \ldots, m \end{aligned} \tag{8.72}$$

The transition from $\ker(z_1)$ to $\ker(z_2)$ is obtained from eqn. (8.71)) as

$$\boldsymbol{P}_{k2}\boldsymbol{g}_{k2}[T_1] + \boldsymbol{P}_{r2}\boldsymbol{g}_{r2}[T_1] = \boldsymbol{E}_s(T_1)\,\boldsymbol{P}_{k1}\boldsymbol{g}_{k1}[0] - \boldsymbol{f}_{z1}(T_1)\,U \tag{8.73}$$

where

$$\boldsymbol{f}_{z1}(T_1) = \boldsymbol{f}_s(T_1)\,\mathrm{SGN}\,(\hat{\boldsymbol{Z}}) \tag{8.74}$$

Solving for U yields

$$U = [\boldsymbol{P}_{r2}^t\boldsymbol{f}_{z1}(T_1)]^{-1}\,\boldsymbol{P}_{r2}^t\boldsymbol{E}_s(T_1)\,\boldsymbol{P}_{k1}\boldsymbol{g}_{k1}[0] \tag{8.75}$$

with:

$$\boldsymbol{g}_{k2}[T_1] = \boldsymbol{P}_{k2}^t[\boldsymbol{I} - \boldsymbol{f}_{z1}(T_1)\,\{\boldsymbol{P}_{r2}^t\boldsymbol{f}_{z1}(T_1)\}^{-1}\boldsymbol{P}_{r2}^t]\,\boldsymbol{E}_s(T_1)\,\boldsymbol{P}_{k1}\boldsymbol{g}_{k1}[0] \tag{8.76}$$

For an m input system the following theorem holds.

Theorem 4: For the system described by eqn. (8.47), stability will be assured if the m eigenvalue problem

$$\boldsymbol{P}_{k1}^t\left[\sum_{i=1}^{m}\boldsymbol{E}_s(T_i)\,\boldsymbol{P}_{k1} - \boldsymbol{h}_m V_j\right], \qquad j = 1, \ldots, m \tag{8.77}$$

yields a set of eigenvalues that lie in the unit circle in the z domain where

$$V_j = [\boldsymbol{P}_{r(j+1)}^t\boldsymbol{h}_j]^{-1}\,\boldsymbol{P}_{r(j+1)}^t\prod_{i=1}^{j}\boldsymbol{E}(T_i)\boldsymbol{P}_{k1} \tag{8.78}$$

and

$$\boldsymbol{h}_j = \sum_{i=1}^{j} \boldsymbol{E}(T_{j+1-i})\, \boldsymbol{f}_s\,(\boldsymbol{T}_i) \tag{8.79}$$

The proof follows by examining a set of transitions from $\ker(z_1)$ to $\ker(z_2)$ described in eqns. (8.76) and (8.75); the transition from $\ker(z_1)$ to $\ker(z_3)$ can be obtained as

$$\boldsymbol{P}_{k3}\boldsymbol{g}_{k3}\,[T_1 + T_2] + \boldsymbol{P}_{r3}\boldsymbol{g}_{r3}[T_1 + T_2] = \boldsymbol{E}_s(T_2)\,\boldsymbol{x}_s[T_1] - \boldsymbol{f}_{z2}(T_2)\,U \tag{8.80}$$

where, as before

$$\boldsymbol{f}_{z2}(T_2) = \boldsymbol{f}_s(T_2)\ \mathrm{SGN}\ (\hat{\boldsymbol{z}}) \tag{8.81}$$

(Note that $\mathrm{SGN}(\hat{\boldsymbol{z}})$ will have a sign change on the transition through $\ker(z_1)$). Substituting for $\boldsymbol{x}_{z1}[T_1]$ yields

$$\begin{aligned}\boldsymbol{P}_{k3}\boldsymbol{g}_{k3}[T_1 + T_2] + \boldsymbol{P}_{r3}\boldsymbol{g}_{r3}[T_1 + T_2] &= \boldsymbol{E}_s(T_2)\,\boldsymbol{E}_s(T_1)\,\boldsymbol{P}_{k1}\boldsymbol{g}_{k1}[0] \\ &\quad - [\boldsymbol{E}_s(T_2)\,\boldsymbol{f}_{z1}(T_1) + \boldsymbol{f}_{z2}(T_2)]\,U\end{aligned} \tag{8.82}$$

which can be solved for $\boldsymbol{g}_{r3}[T_1 + T_2] = 0$ to yield

$$U = [\boldsymbol{P}_{r3}^t(\boldsymbol{E}_s(T_2)\,\boldsymbol{f}_{z1}(T_1) + \boldsymbol{f}_{z2}(T_2)]^{-1}\,\boldsymbol{P}_{r3}^t\boldsymbol{E}_s(T_2)\,\boldsymbol{E}_s(T_1)\,P_{k1}\boldsymbol{g}_{k1}[0] \tag{8.83}$$

The jth transition will then given rise to the following equations:

$$U_j = V_j\boldsymbol{g}_{k1}[0] \tag{8.84}$$

where

$$\boldsymbol{V}_j = [\boldsymbol{P}_{r(j+1)}^t\boldsymbol{f}_j]^{-1}\,\boldsymbol{P}_{r(j+1)}^t \prod_{i=1}^{j} \boldsymbol{E}_s(T_i)\,\boldsymbol{P}_{k1} \tag{8.85}$$

and

$$\boldsymbol{g}_{kj}\left[\sum_{i=1}^{j} T_i\right] = \boldsymbol{P}_{kj}^t\left(\sum_{i=1}^{j} \boldsymbol{E}_s(T_i)\,\boldsymbol{P}_{k1}\boldsymbol{g}_{k1} - \boldsymbol{h}_jU\right) \tag{8.86}$$

where

$$\boldsymbol{h}_j = \sum_{i=1}^{j} \boldsymbol{E}(T_{j+1-i})\,\boldsymbol{f}_s(T_i) \tag{8.87}$$

For each kernel space transition $\boldsymbol{g}_{rj}[\sum_{i=1}^{j} T_j] = 0$; hence m equations using each U_j will yield

$$\boldsymbol{g}_{km}\left[\sum_{i=1}^{m} T_i\right] = \boldsymbol{P}_{km}^t\left(\sum_{i=1}^{m} \boldsymbol{E}_s(T_i)\,\boldsymbol{P}_{k1}\boldsymbol{g}_{k1}[0] - \boldsymbol{h}_mU_j\right),\ j = 1, \ldots, m \tag{8.88}$$

where $\boldsymbol{f}_m$ is given by eqn. (8.87) and U_j by eqn. (8.84).

After m transitions, the state vector is back in $\ker(z_1)$ and eqn. (8.88) can be written

$$\boldsymbol{g}_{k1}\left[\sum_{i=1}^{m} T_i\right] = \boldsymbol{P}^t_{k1}\left(\sum_{i=1}^{m} \boldsymbol{E}_s(T_i)\,\boldsymbol{P}_{k1}\boldsymbol{g}_{k1}[0] - \boldsymbol{h}_m U_j\right), \quad j = 1, \ldots, m \tag{8.89}$$

and the matrices in eqn. (8.89) can be examined for stability in the same way as the single-input case. Hence stability is assured if the eigenvalues of

$$\boldsymbol{P}^t_{k1}\left(\sum_{i=1}^{m} \boldsymbol{E}_s(T_i)\,\boldsymbol{P}_{k1} - \boldsymbol{h}_m V_j\right), \qquad j = 1, \ldots, m \tag{8.90}$$

lie within the unit circle in the z domain and the theorem is proved.

Corollary 2: A limit cycle will exist if all but one of the eigenvalues of the eqns. (8.90) lie within the unit circle in the z domain, with the remaining eigenvalue lying in the circle at $z = -1$.

8.4 Lateral motion autopilot for a remotely piloted vehicle

Considerable attention has been focused on the problem of flight control of aircraft over many years. As aircraft performance and design have improved, so have the demands placed on the flight control system, and modern military aircraft rely heavily on active control to keep them stable and perform adequately. The design of flight control systems is not easy as airframes are highly nonlinear and have a complex dynamic description.

Many techniques have been applied to aircraft control systems to overcome the interaction and nonlinear aerodynamic effects as well as to compensate for wide variations in parameters due to speed, altitude and incidence variations. Some efforts have been put into designing nonlinear controllers (Gill, 1979; Gill, McLean, 1979; McLean, 1983) which will compensate for parameter variations and nonlinearities. This section will look at the application of VSCS to the lateral motion control of a Machan aircraft. The Machan system considered in this chapter is an unmanned aircraft and the problem considered is that of designing a roll and sideslip autopilot.

8.4.1 Aircraft dynamic model

The equations of motion of the airframe are normally split into two dynamic systems describing the longitudinal motions relating to forward velocity and pitching excursions and that of the lateral motions relating to roll, yaw and sideslip. The airframe under study is described in earlier papers (Aslin, Patton, Dorling, 1985; White, Patton, Mudge, Aslin, 1987) and is restricted to the study of lateral motion. A stick-fixed linearisation of the nonlinear dynamic airframe system yields the following linear model:

$$A = \begin{pmatrix} 0 & 1 & 0 & 0 & 0 & 0 \\ 0 & a_{22} & a_{23} & a_{24} & a_{25} & 0 \\ 0 & 0 & a_{33} & 0 & 0 & 0 \\ a_{41} & 0 & 0 & a_{44} & a_{45} & a_{46} \\ 0 & 0 & 0 & a_{54} & a_{55} & a_{56} \\ 0 & 0 & 0 & 0 & 0 & a_{66} \end{pmatrix}, \quad B = \begin{pmatrix} 0 & 0 \\ 0 & 0 \\ b_{31} & 0 \\ 0 & 0 \\ 0 & 0 \\ 0 & b_{62} \end{pmatrix} \tag{8.91}$$

The elements $\{a_{ij}\}$ are functions of the airframe geometry and aerodynamic derivatives and of states such as airframe speed and incidence angles. The state vector for the lateral motion comprises the following variables:

$$\boldsymbol{x} = \begin{cases} \phi & = \text{roll angle, rad} \\ p & = \text{roll rate, rad/s} \\ \zeta & = \text{aileron angle, rad} \\ v & = \text{sideslip velocity, m/s} \\ r & = \text{yaw rate, rad/s} \\ \tau & = \text{rudder angle, rad} \end{cases} \tag{8.92}$$

The control vector comprises:

$$\boldsymbol{u} = \begin{cases} \zeta_d & = \text{aileron angle demand, rad} \\ \tau_d & = \text{rudder angle demand, rad} \end{cases} \tag{8.93}$$

and the output vector is given by

$$\boldsymbol{y} = \begin{cases} \phi & = \text{roll angle, rad} \\ p & = \text{roll rate, rad/s} \\ v & = \text{sideslip velocity, m/s} \\ r & = \text{yaw rate, rad/s} \end{cases} \tag{8.94}$$

For the stick fixed condition with level flight at 33 m/s, the A and B matrices are given in Appendix 8.6.1, together with the variations in the linearisation for two pitch angles up to 0·25 rad and two roll angles up to 0·25 rad.

In the design of autopilots it has been established for some time that certain modes are associated with particular dynamic subsystems. Thus the Dutch roll mode is associated with the yaw subsystem states v and r, while the spiral mode dominates the roll response. These subsystem modes have established acceptable responses associated with them. So, for this study, the Dutch roll mode will be specified to have a damping ratio of at least 0·5 and a natural frequency of at least 2·0 rad/s and the spiral mode having a time constant as fast as possible.

8.4.2 Output VSCS control

The application of output feedback VSCS to the lateral autopilot requires the use of an output feedback algorithm. Reduction to upper Hessenberg form of the state equations matrices $\boldsymbol{A}$ and $\boldsymbol{B}$ showed that both the rudder and aileron input would control the modes of the system and no convenient split into roll

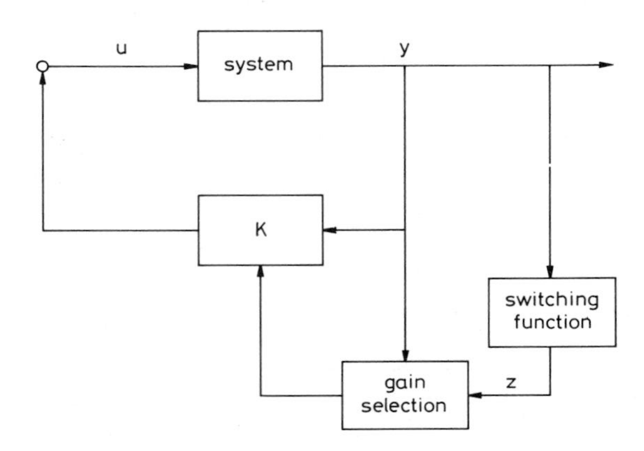

Fig. 8.2 *VSCS controller block diagram*

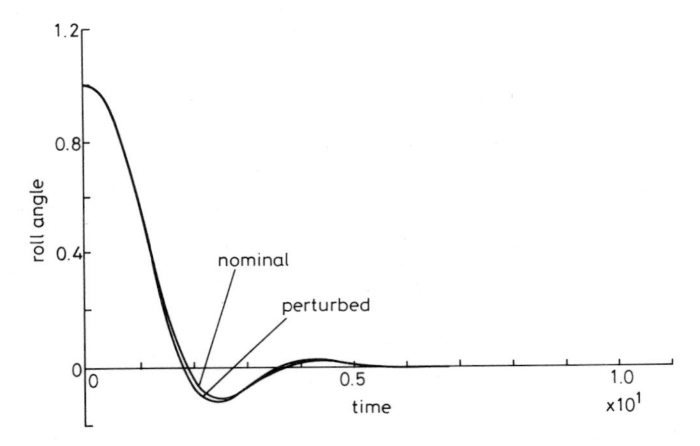

Fig. 8.3 *Roll angle response*

and sideslip control via aileron and rudder control, respectively, was possible. As a result, an output feedback algorithm using unity rank feedback [Hanmandlu, Shantaram, 1987] was used to place the closed loop fixed gain eigenvalues. The resulting fixed gain matrix is

$$\boldsymbol{K} = \begin{pmatrix} -0{\cdot}0336 & -0{\cdot}00481 & -0{\cdot}0194 & -0{\cdot}022 \\ -0{\cdot}0168 & -0{\cdot}00241 & -0{\cdot}0097 & -0{\cdot}11 \end{pmatrix} \tag{8.95}$$

The resulting poles of the fixed gain system are

$$\Lambda_c = \text{diag}\{-1 \pm j1{\cdot}735 \quad -3{\cdot}29 \quad -3{\cdot}64 \quad -6{\cdot}23 \quad -8{\cdot}26\} \tag{8.96}$$

The pole positions have been selected to give a dominant Dutch roll mode with a natural frequency of 2 rad/s and a damping ratio of 0·5, with the remaining

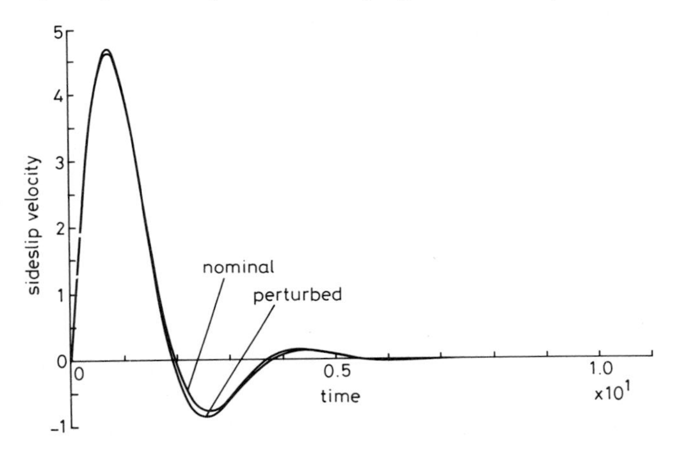

Fig. 8.4 *Sideslip velocity response*

poles as fast as possible. With this choice the range space dynamics can be found $\boldsymbol{M} = \boldsymbol{I}$ as

$$\boldsymbol{x}_s = \begin{pmatrix} -3{\cdot}64 & 0 & 0 \\ 0 & -6{\cdot}23 & 0 \\ 0 & 0 & -8{\cdot}26 \end{pmatrix} + \begin{pmatrix} -654 & -4392 \\ 56{\cdot}4 & 1055 \\ 49{\cdot}4 & -344 \end{pmatrix} \Delta \boldsymbol{K}\boldsymbol{y}$$

$$z = (0{\cdot}02 \quad 0{\cdot}124 \quad 0{\cdot}124)\,\boldsymbol{x}_s \tag{8.97}$$

Using Theorem 4, the switched gain range space feedback takes the form:

$$\Delta \boldsymbol{K}\boldsymbol{y} = \begin{pmatrix} 0{\cdot}01 \text{ SGN } (x_1 z) \\ 0{\cdot}01 \text{ SGN } (x_4 z) \end{pmatrix} \tag{8.98}$$

The resulting VSCS is shown in Fig. 8.2 in block diagram form. The time history for the roll angle and the sideslip velocity are shown in Figs. 8.3 ad 8.4, respectively. These confirm the design for the Dutch roll modes and the spiral modes for roll and sideslip. The range space dynamics are shown in Fig. 8.5 which illustrate the faster third order dynamics associated with the range space. The VSCS controller response for a different initial roll angle of $\phi = 0{\cdot}25$ rad is also shown in the figures. The response is essentially unchanged from the straight and level flight condition and shows the invariance of the VSCS design.

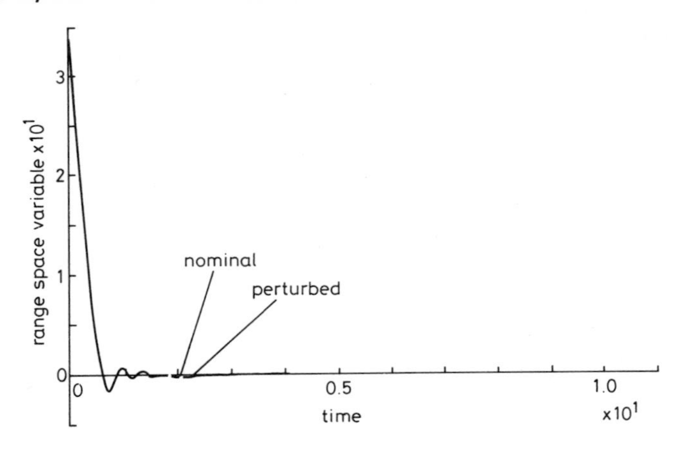

Fig. 8.5 *Range space variable response*

8.5 Flexible structure control

In recent years many research workers have tackled the problem of designing control systems for use on flexible structures (Gupta, Lyons, Aubun, Marguiles, 1981; West-Vukovich, Davison, Hughes, 1984; Balas, 1978; Dodds, Ferguson, 1981). Applications have been aimed at the use of such techniques on space structures which are inherently flexible owing to the need to minimise weight of the structure and because of the tendency to go for extended structures rather

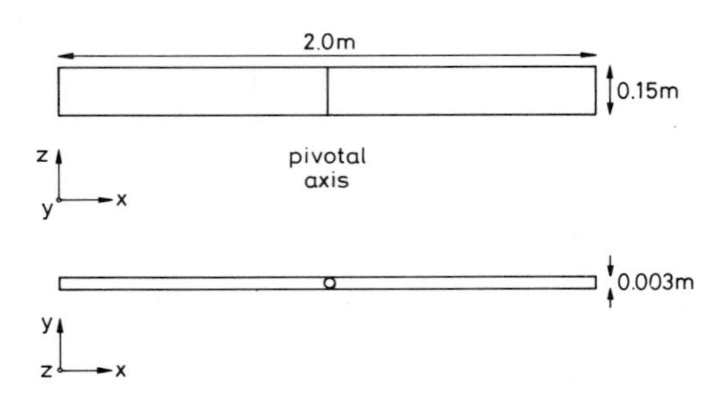

Fig. 8.6 *Experimental flexible beam*

than the current compact centre-bodies type (Holloway, 1983).

Because of the flexible nature of these structures, dynamic models used to describe their behaviour, or form the basis of control system synthesis, tend to have a very large order (typically from 6th order to 200th order) (Salehi, 1985). In many cases, the mode frequencies and shapes for the models are not well known and result in system uncertainty. The damping for each mode is the most uncertain system parameter and can vary considerably.

In order to examine the problems associated with this type of structure, an experimental rig has been constructed by the author consisting of a centrally mounted aluminium beam with a series of distributed masses along its length. (Fig. 8.6). The point masses enable some adjustment of the modal frequencies and allow the mode shapes of the beam to be changed to study modal frequency bunching, asymmetrical modal shapes and the effect of significant point masses on the dynamic behaviour. The beam has several actuators and sensors but the problem addressed in the Chapter will be confined to a torque actuator at the beam centre and an angular displacement and velocity sensor also at the beam centre.

The control problem is that of controlling a model based on the fixed mode and the first flexible mode. A variable structure control system will be developed to control the fixed mode and contain the reaction of the flexible mode. This control system will be simulated and the effect of parameter uncertainty and unmodelled dynamics investigated. A new form of VSCS will be developed to broaden the switching subspace into a switching sector bounded by a set of subspaces. This technique will enable the damping ratio of the flexible modes to be maintained within a specified range, rather than being specified exactly by the switching subspace.

8.5.1 Beam model

The model for the beam is developed initially in finite element form as

$$\boldsymbol{M}\dot{\boldsymbol{d}} + \boldsymbol{K}\boldsymbol{d} = \boldsymbol{f}, \qquad \boldsymbol{d}, \boldsymbol{f} \in \mathscr{R}^{n/2} \tag{8.99}$$

where $\boldsymbol{d}$ is a vector of translational and rotational displacements at the finite element nodes of the structure, $\boldsymbol{f}$ is a vector of forces and torques applied at the nodes, and $\boldsymbol{M}$ and $\boldsymbol{K}$ are generalised mass and stiffness matrices describing the properties of the structure between the nodes. Using the well known modal transformation of Zienkiewicz (Zienkiewicz, 1983) given by

$$\boldsymbol{d} = \boldsymbol{\Phi}\boldsymbol{q}, \qquad \boldsymbol{q} \in \mathscr{R}^{n/2} \tag{8.100}$$

where $\boldsymbol{q}$ is a vector of modal co-ordinates and $\boldsymbol{\Phi}$ is the modal shape matrix, normalised such that

$$\boldsymbol{\Phi}^t\boldsymbol{M}\boldsymbol{\Phi} = \boldsymbol{I} \tag{8.101}$$

Eqn. (8.101) can then be transformed into modal form:

$$\ddot{\boldsymbol{q}} + \boldsymbol{W}\boldsymbol{q} = \boldsymbol{\Phi}^t\boldsymbol{f} \tag{8.102}$$

where $\boldsymbol{W}$ is the modal frequency matrix, given by

$$\boldsymbol{W} = \boldsymbol{\Phi}^t\boldsymbol{K}\boldsymbol{\Phi} = \text{diag}\,\{w_i\} \tag{8.103}$$

Note that these equations are normally developed assuming zero structural damping. The modal equations can then be written in the standard state space form by assigning states as

$$x^t = \{q_i, \dot{q}_i\}, \quad x \in \mathscr{R}^n \tag{8.104}$$

to yield the state equations

$$\dot{x} = Ax + Bu, \quad u \in \mathscr{R}^m \tag{8.105}$$

$$y = Cx, \quad y \in \mathscr{R}^l \tag{8.106}$$

where

$$A = \text{diag}\left\{\begin{pmatrix} 0 & 1 \\ -\omega_i^2 & -2\zeta\omega_i \end{pmatrix}\right\}, \quad i = 1, \ldots, \frac{n}{2} \tag{8.107}$$

$$C = \left\{\begin{pmatrix} c_{i1} & 0 \\ 0 & c_{i2} \end{pmatrix}\right\}, \quad i = 1, \ldots, \frac{n}{2} \tag{8.108}$$

and the $\boldsymbol{B}$ matrix is zero everywhere except in rows $2i$. For this study a fourth order model and an eighth order model are developed for the beam with a central torque actuator and an angular displacement and velocity transducer. The state matrices are given in Appendix 8.6.2 which include an extra integrator in the system description (cf. Section 8.52). The fourth order model is used to develop the control system strategies and the eighth order model is used to assess the effect of unmodelled frequency modes on the control system design.

8.5.2 Variable structure control system design

As an alternative to direct output feedback, the designs will incorporate an observer that is derived from a steady state Kalman filter. The observer is designed to reconstruct the states from the system measurements based on the undisturbed fourth order model of the system. In subsequent simulations with the disturbed system and including the unmodelled dynamics, the observer is not redesigned. The basic control system is shown in Figure 8.7. The variable

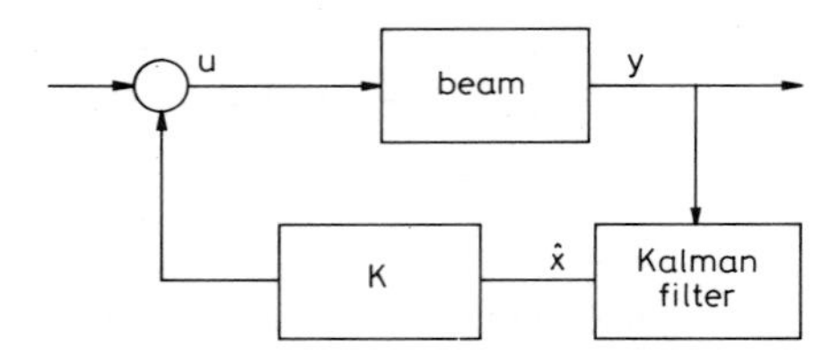

Fig. 8.7 *Control system configuration*

structure control system design is based around the definition of an invariant subspace that specifies the dynamic behaviour of the closed loop system. In order to define a suitable subspace that will describe the behaviour of the fixed modes and the flexible modes in a suitable form, an integrator is inserted in the forward path of the system (Fig. 8.7). The integrator also provides a smoothing action on the conventional VSCS driving function which is normally discontinuous.

For the nominal design of the variable structure control system, a desired closed loop behaviour is chosen to have a dominant fixed mode response of $\omega_n = 1{\cdot}5$ rad/s with a damping ratio of $\zeta = 0{\cdot}7$ and the first flexible mode to have $\omega_n = 33{\cdot}39$ rad/s with a damping ratio of $\zeta = 0{\cdot}018$ with the 'integrator' pole at $-5{\cdot}0$. The flexible mode is designed to retain the natural frequency of the free or open loop response but increase the damping ratio to provide some small amount of damping. For these closed loop eigenvalues the fixed gain matrix $\boldsymbol{K}$ is given by

$$\boldsymbol{K} = (-9{\cdot}79 \quad -11{\cdot}1 \quad 8{\cdot}13 \quad 318 \quad -1{\cdot}19) \tag{8.109}$$

with switching matrix $\boldsymbol{S}$ given by the left eigenvector of the fixed gain closed loop system $\boldsymbol{A} - \boldsymbol{BK}$ to yield

$$\boldsymbol{S} = (1 \quad 0{\cdot}9353 \quad -0{\cdot}511 \quad 0{\cdot}2503 \quad 0{\cdot}1472) \tag{8.110}$$

with the function in eqn. (8.31) defined as

$$f_i = \sum_{j=1}^{n} \phi_{ij} \frac{|x_i|}{|z_i|} \tag{8.111}$$

The resulting system was simulated and a disturbance on the fixed mode velocity was injected. The resulting dynamic behaviour is shown in Fig. 8.8 for the fixed

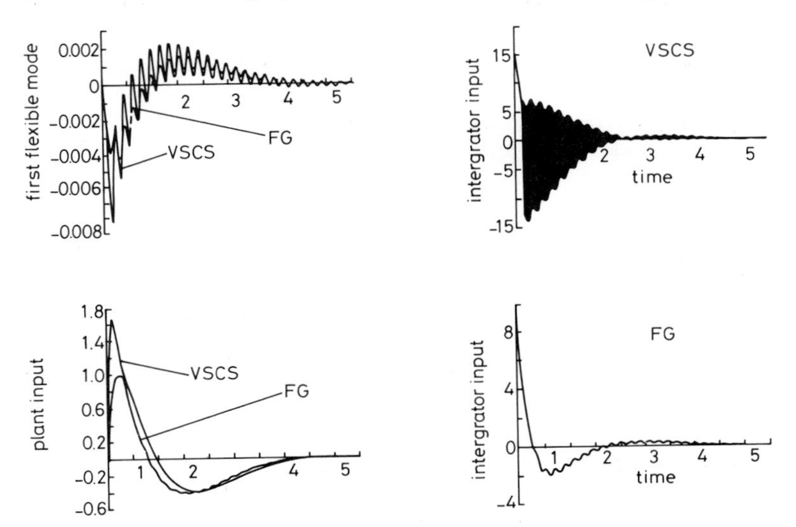

Fig. 8.8 *VSCS and fixed gain responses*

gain behaviour and for the VSCS responses with $\alpha_{1,1} = 5$. The figures shows the fixed mode response and the first flexible mode response. The figure also shows the switched gain characteristic of the VSCS system and the system input (the output of the forward path integrator). Although the VSCS control is highly discontinuous, it does not show up significantly on the system input. It is sufficient, however, to disturb the flexible mode and the performance is degraded compared with the fixed gain control system.

8.5.3 Improving the VSCS design

Although the elimination of the discontinuous behaviour of the control signal in a VSCS is well known (White, 1983; White, 1985; White, 1986) by the inclusion of a 'soft' segment close to $\ker(\boldsymbol{S})$, an alternative approach was adopted. The damping ratio of the flexible mode is the crucial parameter to control as the inherent damping ratio is so low. Rather than impose a soft sector around $\ker(\boldsymbol{S})$ it was proposed to consider two switching subspaces representing an upper and lower limit on the damping ratio.

Consider the switching function given by

$$z = \boldsymbol{S}\boldsymbol{x} \tag{8.112}$$

$$= \boldsymbol{S}_1\boldsymbol{x} + \boldsymbol{S}_2\boldsymbol{x} \tag{8.113}$$

$$= z_1 + z_2 \tag{8.114}$$

such that

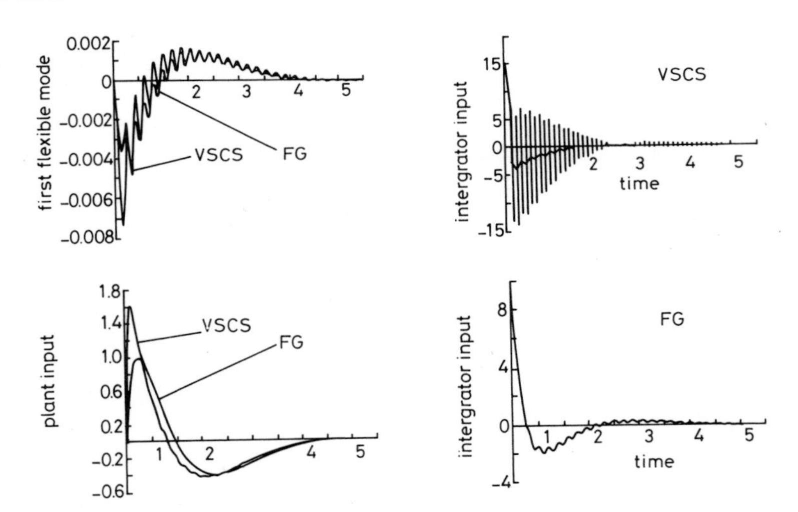

Fig. 8.9 *Segmented VSCS and fixed gain responses*

$$z_1 z_2 < 0, \quad \forall\, \boldsymbol{x} \text{ such that } \{\boldsymbol{x} : \boldsymbol{S}\boldsymbol{x} = 0\} \tag{8.115}$$

and:

$$z > 0 \;\forall\; z_1 > 0,\; z_2 > 0 \tag{8.116}$$

$$\boldsymbol{x} < 0 \;\forall\; z_1 < 0,\; z_2 < 0 \tag{8.117}$$

The subspace defined by $\ker(\boldsymbol{S})$ is then trapped between the two subspaces defined by $\ker(\boldsymbol{S}_1)$ and $\ker(\boldsymbol{S}_2)$ and it can be shown that the two subspaces can be defined to bracket a system parameter such as the flexible mode damping ratio in such a manner. The stability of the range space dynamic behaviour described by z in eqn. (8.47) will still be stable in the conventional VSCS manner if Λ_S has a stable spectrum and $SB\Delta Kx$ is chosen as before but with

$$\phi_{ij} = 0, \forall z_1 z_2 < 0 \tag{8.118}$$

$$< 0, \forall z_1 z_2 > 0 \tag{8.119}$$

The effect of this segmentation by subspaces is to produce no switching when the state vector is close to $\ker(\boldsymbol{S})$ and contained between the two subspaces that bracket the desired switching subspace $\ker(\boldsymbol{S})$. If the state trajectory crosses one of the enclosing subspaces, then switching occurs to drive the state vector back

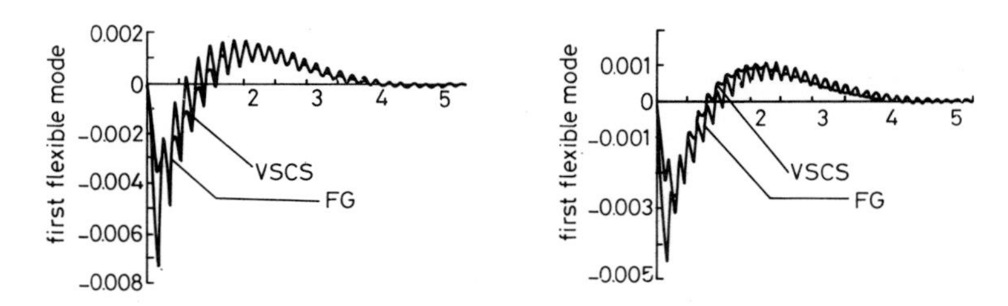

Fig. 8.10 *Perturbed responses*

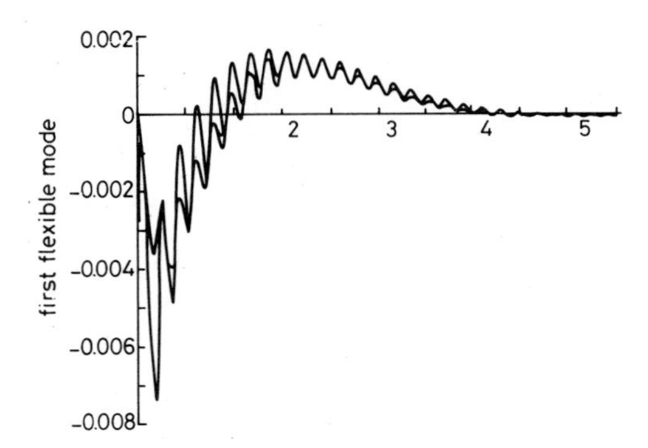

Fig. 8.11 *Unmodelled dynamic responses*

into the containing segment. A corollary of this definition is that the switching subspace and the two spanning subspaces all intersect in an $n - p$th order subspace, where p is the number of disturbed parameters. Applying the technique to the first flexible mode of the beam, a damping ratio range can be established such that

$$0{\cdot}1 > \zeta > 0{\cdot}015 \tag{8.120}$$

which yields the corresponding subspaces defined by

$$\boldsymbol{S}_1 = (1 \quad 0{\cdot}9352 \quad -0{\cdot}5111 \quad 0{\cdot}1747 \quad 0{\cdot}1187) \tag{8.121}$$

and

$$\boldsymbol{S}_2 = (0{\cdot}5232 \quad 0{\cdot}4919 \quad -0{\cdot}2673 \quad 1 \quad 0{\cdot}4851) \tag{8.122}$$

The simulations were rerun with identical parameter and gain values to the conventional VSCS and the results shown in Fig. 8.9. The Figure shows that the segmented switching region has reduced the switched behaviour of the VSCS control considerably. The damping ratio of the first mode is now greater than the fixed gain control.

To investigate the robustness of the design, the natural frequency of the systems fixed mode and first flexible mode were perturbed by 20%. The resulting responses are shown in Fig. 8.10. This Figure shows that the segmented VSCS is still producing more damping than the equivalent fixed gain system. If the next two modes of the beam are included in the simulation, the responses are as shown in Fig. 8.11. Again, the VSCS performs better than the fixed gain system.

8.6 References

ASLIN, P.P., PATTON, R.J., DORLING, C.M., (1985), The design of a sliding mode controller for dutch roll damping in a non-linear aircraft system? *IEE Control* '85, pp. 252, 435

BALAS, M.J., (1978), Direct output feedback of large flexible space structures', *SIAM J. Control and Optimisation,* **16**, pp. 450

DAVISON, E.J., (1970) 'On pole assignment in linear systems with incomplete state feedback,' *IEEE trans.*, **AC-15**, pp. 348

DODDS, S.J., FERGUSON, (1981) 'A predicted signed switching time, high precision satellite attitude control law,' *Int. J. Control*, **40**, pp. 795

EL-GHEZAWI, O.M.E., ZINOBER, A.S.I., BILLINGS, S.A., (1983) 'Analysis and design of variable structure systems using a geometrical approach,' *Int. J. Control*, **88**, pp. 657

FALLSIDE, F. (ed) (1977) '*Control System Design using Pole-Zero Assignment*' (Academic Press)

GILL, F.R., 11979, *RAE tech. report* 'Non-linear pitch rate to elevator control laws for combat aircraft', TR 79075

GILL, F.R., McLEAN, D., (1979) (Computer study of a self adaptive and longitudinal control study employing adjustments to a command input filter,' *RAE tech. report*, TR 79093

GUPTA, N.K., LYONS, M.G., AUBUN, J.N., MARGUILES, G., (1981) 'Modelling control and system identification methods for flexible structures,' *AGARD*, 350

HANMANDLU, M. SHANTARAM, V., (1986) 'Eigenvalue assignment by unity rank output feedback,' *Int. J. Control,* **36**, pp. 473

HOLLOWAY, P.F., (1983) 'Space station technology,' *British Interplanatary Society,* **36**, pp. 409

ITKIS, U. '*Control Systems of Variable Structure*,' (John Wiley)

McLEAN, D. (1983) 'Globally stable non-linear flight control systems', *IEE Proc. Pt. D,* **130**

MUNRO, N., VARDULAKIS, A.I. (1973) 'Pole shifting using output feed-back,' *Int. J. Control*, **18**, pp. 1267

SALEHI, S.V., (1985), 'Application of adaptive observers to the control of flexible spacecraft,' 10th IFAC Symposium on Automatic Control in Space

SESAK J.R., CORADETTI T., (1979) 'Decentralised control of large space structures via forced singular perturbations,' *AIAA* 17th Aerospace Sciences Meeting, pp. 79–0195

UTKIN, V.I., (1977) 'Variable Structure systems with sliding modes,' *IEEE Trans.*, **AC-22**, pp. 212

WEST-VUKOVICH, G.S., DAVISON, E.J., HUGHES, P.C., (1984) 'Decentralised control of large flexible space structures,' *IEEE Trans.*, **AC-29**, pp. 866

WHITE, B.A., (1983) 'Reduced order switching functions in variable structure control systems, *IEE Proc. Pt. D*, **130**, pp. 33

WHITE, B.A., (1985) 'Range space dynamics of scalar variable structure control systems,' *IEE Proc. Pt. D*, **132**, pp. 35

WHITE, B.A., (1985) 'Some problems in output feedback in variable structure control systems,' *Proceedings of the 7th IFAC Symposium on Identification and system parameter estimation*, York

WHITE, B.A., (1987) 'Stability of relay variable structure control systems,' IEEE Conference on Control and Applications, Jerusalem

WHITE, B.A., SILSON, P.M. (1984) 'Reachability in variable structure control systems,' *IEE Proc. Pt. D*, **131**, pp. 85

ZIENKIEWICZ, O.C., (1983) 'The Finite Element method', (Mc-Graw Hill)

ZINOBER, A.S.I., EL-GHEZAWI, O.M.E., BILLINGS, S.A., (1982) 'Multi-variable variables structure adaptive model following systems', *IEE Proc. Pt. D*, **129**, p.6

8.7 Appendices

8.7.1 Variation in the A matrix with roll and pitch angle

All conditions are taken at an airspeed of 33 m/s

The nominal system for level flight is

$$A = \begin{pmatrix} 0 & 1 & 0 & 0 & 0 & 0 \\ 0 & -7{\cdot}867 & -24{\cdot}4 & -0{\cdot}00939 & 3{\cdot}58 & 0 \\ 0 & 0 & -5{\cdot}0 & 0 & 0 & 0 \\ 9{\cdot}8 & 0 & 0 & -0{\cdot}255 & -30{\cdot}35 & -4{\cdot}69 \\ 0 & 0 & 0 & 0{\cdot}3366 & -0{\cdot}304 & -8{\cdot}02 \\ 0 & 0 & 0 & 0 & 0 & -10{\cdot}0 \end{pmatrix} \tag{8.123}$$

$$B = \begin{pmatrix} 0 & 0 \\ 0 & 0 \\ 10 & 0 \\ 0 & 0 \\ 0 & 0 \\ 0 & 20 \end{pmatrix} \tag{8.124}$$

$\phi = 0{\cdot}25\,\text{rad}$

$$A = \begin{pmatrix} 0{\cdot}0079 & 1 & 0 & 0 & -0{\cdot}0025 & 0 \\ 0 & -7{\cdot}76 & -23{\cdot}76 & -0{\cdot}122 & 3{\cdot}94 & 0 \\ 0 & 0 & -5{\cdot}0 & 0 & 0 & 0 \\ 9{\cdot}8 & -0{\cdot}0018 & 0 & -0{\cdot}267 & -29{\cdot}08 & -4{\cdot}5 \\ 0 & -0{\cdot}215 & 0 & 0{\cdot}332 & -0{\cdot}288 & -7{\cdot}812 \\ 0 & 0 & 0 & 0 & 0 & -10{\cdot}0 \end{pmatrix} \tag{8.125}$$

8.7.2 *A, B and C matrices for the flexible beam model*

5th order model

$$\boldsymbol{A} = \begin{pmatrix} 0 & 1 & 0 & 0 & 0 \\ 0 & 0 & -1{\cdot}148 & 0 & 0 \\ 0 & 0 & 0 & 0 & 0 \\ 0 & 0 & 0 & 0 & 1 \\ 0 & 0 & -3{\cdot}582 & -1115 & -0{\cdot}167 \end{pmatrix} \tag{8.126}$$

$$\boldsymbol{B} = \begin{pmatrix} 0 \\ 0 \\ 1 \\ 0 \\ 0 \end{pmatrix} \tag{8.127}$$

9th order model

$$A = \begin{pmatrix} 0 & 1 & 0 & 0 & 0 & 1 & 0 & 0 & 0 \\ 0 & 0 & -1{\cdot}148 & 0 & 0 & 0 & 0 & 0 & 0 \\ 0 & 0 & 0 & 0 & 0 & 0 & 0 & 0 & 0 \\ 0 & 0 & 0 & 0 & 1 & 0 & 0 & 0 & 0 \\ 0 & 0 & -3{\cdot}582 & -115 & -0{\cdot}167 & 0 & 0 & 0 & 0 \\ 0 & 0 & 0 & 0 & 0 & 0 & 1 & 0 & 0 \\ 0 & 0 & -6{\cdot}93 & 0 & 0 & -11600 & -0{\cdot}593 & 0 & 0 \\ 0 & 0 & 0 & 0 & 0 & 0 & 0 & 0 & 1 \\ 0 & 0 & 9{\cdot}2 & 0 & 0 & 0 & 0 & -55000 & -1{\cdot}2 \end{pmatrix} \tag{8.128}$$

$$B = \begin{pmatrix} 0 \\ 0 \\ 1 \\ 0 \\ 0 \\ 0 \\ 0 \\ 0 \\ 0 \end{pmatrix} \tag{8.129}$$

Chapter 9

The hyperstability approach to VSCS design

A. Balestrino and M. Innocenti

University of Pisa

9.1 Introduction

Adaptive model-following control is an efficient control technique which includes a reference model, to provide a behaviour objective for the controlled plant, and an adaptation mechanism to synthesise a plant input signal so that the deviations between the plant state and the model state are minimised (Landau, 1979). In this way we overcome the difficulties inherent in the choice of a performance index for a multivariable system, and we are also able to obviate both the high degree of uncertainty concerning plant parameters and their time variability.

Assuming that the plant state variables are available for measurement, under some mild conditions a large class of nonlinear plants can be controlled using adaptive model-following control with nonlinear high gain subsystems. Such a systematic and effective design procedure is denoted as the variable structure control (VSC) approach. It allows a complete specification of the transient behaviour by means of sliding modes (Utkin, 1974). The existence and stability of the sliding modes can be established using a Lyapunov method. The global stability may also be assured by systematic use of hyperstability theory (Popov, 1973), usually by means of discontinuous control laws (Balestrino, De Maria and Sciavicco, 1982).

In this Chapter a thorough description of the hyperstability approach to variable structure control systems (HVSCS) is given. In the next Section some required preliminaries on hyperstability theory are put foward and the design objectives are posed in Section 9.3. A preliminary single-input/single-output example of a robotic DC drive is given in Section 9.4, while a multivariable example in Section 9.5 shows how to design a nonlinear observer for a class of nonlinear plants.

In Section 9.6 a systematic approach to the design of HVSCS is set out including a bank of integrators on the outputs. In Section 9.7 the integrator outputs are neglected in the feedback realising the sliding modes, and the synthesis follows along the same lines of Balestrino, De Maria and Siciliano (1985).

The results of Section 9.6 are applied to the control synthesis of a robot manipulator system in Section 9.8. It is possible to decouple the system dynamics, following a second order linear time-invariant model for each degree of freedom and overcoming the parametric uncertainties and the environmental disturbances. In Section 9.9 the simplified design technique of Section 9.7 is applied to a flight control problem. Once again it is demonstrated that the hyperstability approach to VSCS can lead to a successfully robust design in the case of highly nonlinear plant. Some final comments are given in Section 9.10.

9.2 Preliminaries: Hyperstability

There are many paths leading to VSCS design, but the hyperstability approach is one of the most straightforward and practical methods.

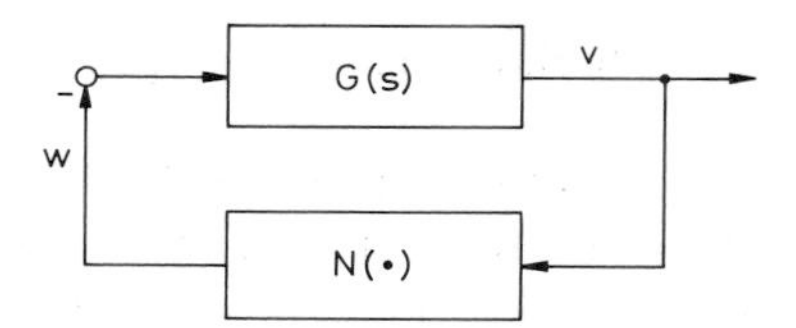

Fig. 9.1 *A basic feedback system*

In order to introduce this approach let us consider the dynamic system in Fig. 9.1, where $G(s)$ is an $m \times m$ transfer function matrix with a minimal realisation (A, B, C, D), i.e.

$$\dot{x}(t) = Ax(t) + Bu(t) \qquad x \in R^n, \quad u \in R^m \tag{9.2}$$

$$y(t) = Cx(t) + Du(t) \qquad y \in R^m,$$

$$G(s) = C(sI - A)^{-1} B + D \tag{9.2}$$

(A, B) is completely controllable and (A, C) completely observable. Moreover the block N is generally a nonlinear time-variant subsystem:

$$w = f\,(y\,(0,\,t),\,t) \tag{9.3}$$

where f denotes a functional dependence between w and the values of $y(\tau)$, $0 \leqslant \tau \leqslant t$.

Definition 1: A transfer function matrix $G(s)$ is strictly positive real (SPR) if:

(i) Poles of $G(s)$ lie in the half plane $\text{Re}(s) < 0$
(ii) $G(j\omega) + G^T(-j\omega)$ is positive definite hermitian for all real ω.

Here we will not attempt to give a formal definition of passivity (Willems, 1972) but limit ourselves to the following statement.

Definition 2: The nonlinear time-varying block N (9.3) is passive if it satisfies the following integral inequality for all $T > 0$:

$$\int_0^T w^T y dt \geqslant -c_0 - c_1 \|y_T\|^2 - c_2 \|w_T\|^2 \tag{9.4}$$

where g_T is the standard Euclidean norm in the extended space (Praly, 1986) and the constants c_i, $i = 0, 1, 2$, are nonnegative.

Definition 3: A system is hyperstable if there are two constants $k_1 \geqslant 0$, $k_2 \geqslant 0$, such that any state solution of the system satisfies the inequality

$$\|x(t)\| < k_1 \|x(0)\| + k_2 \qquad \forall t \geqslant 0; \tag{9.5}$$

the system is asymptotically hyperstable if it is hyperstable and moreover

$$\lim_{t \to \infty} x(t) = 0$$

We can now state the hyperstability theorem (Popov, 1973).

Theorem 1: With reference to Fig. 9.1 the feedback system is asymptotically hyperstable if and only if the linear block is characterised by a strictly positive real transfer function matrix $G(s)$ (9.2) and the nonlinear time-varying block N (9.3) is passive.

Comment: We explicitly note that by connecting hyperstable blocks in parallel or in feedback we get a hyperstable block; i.e. the hyperstability is preserved. Of course this property is not valid for series connection.

In order to verify the passivity property of a nonlinear block N (9.3) we must verify the inequality (9.4); note that for many nonlinearities, e.g. for memoryless nonlinearities, this proof is straightforward.

Usually it is not an easy task to prove that a given transfer function matrix $G(s)$ is SPR using definition 1, whenever the dimension of G is greater than one. More powerful testing procedures are reported in Anderson and Vongpanitlerd (1973), where the main tool is the positive real lemma.

Theorem 2: (The positive real lemma): Let $G(.)$ be an $m \times m$ matrix of real rational functions of a complex variables, with $G(\infty) < \infty$. Let (A, B, C, D) (9.1) be a minimal realisation of $G(s)$ (9.2). Then $G(s)$ is positive real if and only if there exist real matrices P, L and W with P positive definite symmetric such that

$$
\begin{aligned}
A^T P + PA &= -LL^T \\
PB &= C^T - LW \\
W^T W &= D + D^T
\end{aligned} \tag{9.6}
$$

(The number of rows of W and columns of L are unspecified, while the other dimensions of P, L and W are automatically fixed).

9.3 Design Objectives

In developing the design of a multivariable control system four objectives are considered:

(*a*) Strong stability characteristics
(*b*) Simple control laws
(*c*) High speed of response
(*d*) Systematic design method.

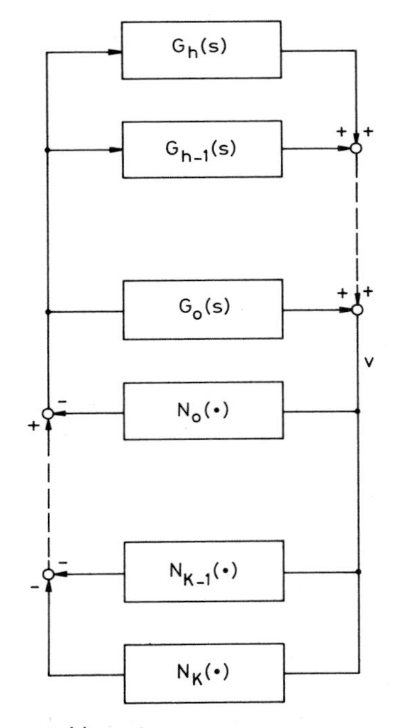

Fig. 9.2 *A generalised hyperstable system*

Let us briefly comment upon these points.

(*a*) In order to ensure strong stability characteristics for the global system the hyperstability approach is appropriate. Indeed, with reference to Fig. 9.2, we can state that, if all linear time-invariant blocks are characterised by transfer functions $G_i(s)$ strictly positive real and all nonlinear time-variant blocks N_i are passive, then the global system is asymptotically hyperstable. Note that some blocks N_h could be linear and time-invariant and the hyperstability will be preserved if the corresponding transfer function matrices are SPR. Therefore the stability of the multivariable control system is assured whenever we are able to reformulate the design problem as in Fig. 9.2 with all blocks hyperstable.

We can also disregard some blocks, e.g. blocks taking into account the uncertain parts of the plant, provided that they are hyperstable; otherwise we must establish the robustness of an uncertain dynamical system by means of appropriate algorithms (Chen, 1987; Yedavalli and Liang, 1987; Corless and Leitmann, 1981).

(*b*) If in Fig. 9.2 a block, say N_0, is representative of our choice for the control laws, then we can verify that simple control laws such as proportional, integral, proportional–integral, relay, saturation etc., may be used because they all lead to a passive block N_0.

(*c*) In order to obtain high speed of response we must use control laws with high gain for the linear blocks and with the highest admissible levels for the non-linearities such as relay or saturation. The realisation of this objective is one of the main motivations for using variable structure control (VSC) theory. Indeed by using high gain and relay type nonlinearities we get a control discontinuous on a number of switching hypersurfaces (usually hyperplanes) (Utkin, 1974). On the intersection of these hypersurfaces the system dynamics are in the sliding mode and the system becomes less sensitive to parameter variations and noise disturbances.

(*d*) A systematic design method may be developed by specifying the design objectives by means of a reference model. Model-following control and adaptive model-following control (AMFC) are well established design techniques (Landau, 1979).

By combining model-following control and variable structure control theory we obtain variable structure model-following control (Utkin, 1974). In this Chapter we shall solve the control design problem by using both the hyperstability approach and VSC theory, i.e. by means of the hyperstable variable structure control (HVSC) approach.

9.4 A SISO example

In this Section the HVSC approach is developed and clarified with reference to a single-input/single-output (SISO) example. Let us consider a DC drive as in Fig. 9.3 (Buhler, 1979).

Assuming a proper choice of PI parameters in the inner current loop we obtain the equivalent block diagram depicted by solid lines in Fig. 9.4, where the transfer function $k_i/(1 + sT_i)$ is representative of the inner current loop, T is a load torque disturbance and T_m depends on mechanical inertia. Note that if the DC drive moves a robotic joint then the variations of T_m may be very large.

With $T = 0$, and without the nonlinear block N, the transfer function between the output speed y and the input voltage r is

$$Y(s)/(R(s) = g(1 + sT_r)/(g + sgT_r + s^2 + T_i s^3) \tag{9.7}$$

where

$$g = kk_i/T_m \tag{9.8}$$

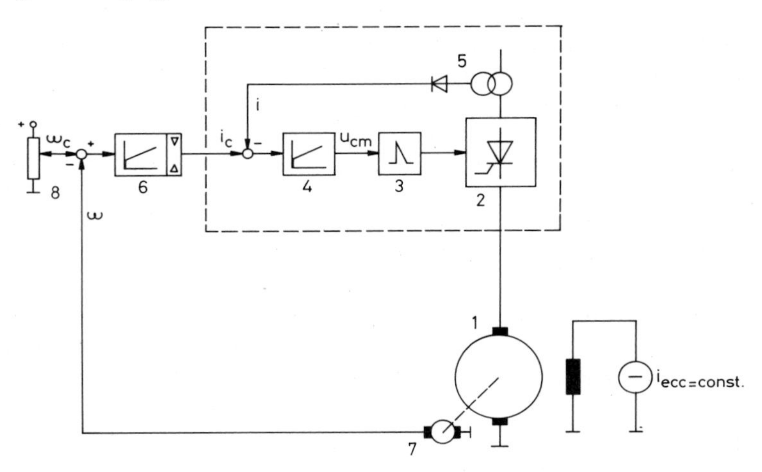

Fig. 9.3 *A DC drive*

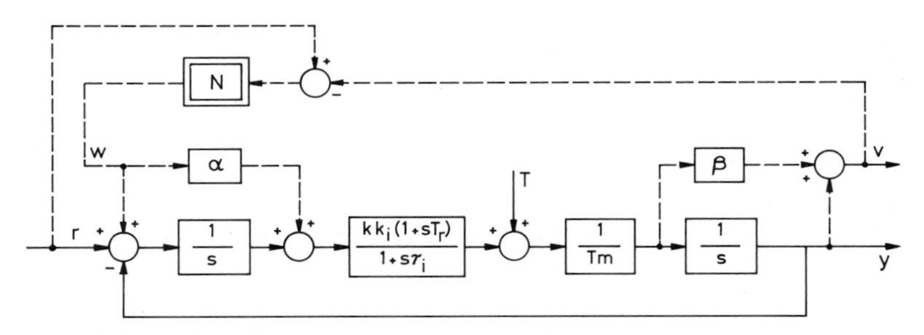

Fig. 9.4 *DC drive block diagram*

This tranfer function is not positive real. In order to realise strict positive realness we can introduce an additional block denoted by α and represented by dotted lines in Fig. 9.4. Now the transfer function between the output v and the input w, with $\beta = 0$, is given by

$$V(s)/W(s) = g(1 + sT_r)(1 + \alpha s)/(g + sgT_r + s^2 + s^3 T_i) \tag{9.9}$$

The positive real property is assured in view of definition 1 if

$$g + \omega^2(gT_r^2 - 1) + \omega^4(\alpha T_r - \alpha T_i - T_r T_i) > 0, \forall\omega \tag{9.10}$$

A sufficient condition is given by

$$gT_r^2 > 1 \tag{9.11a}$$

$$\alpha T_r/(\alpha + T_r) > T_i \tag{9.11b}$$

The condition (9.11*a*) must be tested by taking into account (9.8) and the range of variation of the parameter T_m. Therefore we have

$$kk_iT_r^2 > \max(T_m) \tag{9.12}$$

i.e. we must guarantee a sufficiently high gain in the open loop. The condition (9.11*b*) assures a sufficient phase margin.

Note that usually T_i is very small so that engineers often assume $T_i = 0$ in (9.7) and they choose g and T_r accordingly to (9.11*a*) in order to make the transfer function (9.7) SPR. Now if the gain k is implemented by means of a relay, i.e. $k \to \infty$, in order to realise a high gain closed-loop performance to enlarge the bandwidth, the parasitic constant T_i just disgregarded plays a fundamental role, leading to unsatisfactory performance and also to the instability of the whole system. It clearly appears that we must take much care in modelling and particularly in using reduced order models.

Now by taking into account $T_i \neq 0$ we have shown that it is possible to satisfy the SPR condition (9.10) by means of (9.11); another equivalent possible choice could be $\alpha = 0$ and $\beta \neq 0$, but usually this choice is not practicable because the acceleration is not available, raising the question of the feasibility of using observers.

Assuming for the sake of simplicity $\beta = 0$, $\alpha = T_r > 2T_i$, and $gT_r^2 > 1$, the transfer function (9.9) is SRP. Now by choosing a passive nonlinear block N in accordance with theorem 1 we assure the hyperstability of the global system.

In particular by choosing N as a relay, i.e.

$$w = C\,\mathrm{sgn}(e) = C\,\mathrm{sgn}(v - r) \tag{9.13}$$

we have that after a finite time transient the closed-loop system forces the variable e to be zero; i.e. in terms of the Laplace transform we have

$$(1 + sT_r)^2\,(v - r) = 0 \tag{9.14}$$

This equation represents the sliding motion and clearly shows dynamics independent of the DC-motor and load parameters as long as (9.11) are satisfied. The simulation results for a particular DC drive are reported in Balestrino, Gallanti and Kalas (1988).

9.5 A MIMO example: nonlinear observers

In this Section we exemplify the HVSC approach for a particular multi-input multi-output (MIMO) problem.

Let us consider a plant described by

$$\dot{x}_p = A_p x_p + B_p (u_p + d_p) \quad (9.15a)$$

$$x_p = C_p x_p \quad (9.15b)$$

where $x_p \in R^n$ is the state vector, $u_p \in R^r$ is the input, $y_p \in R^m$ is the output, and $d_p \in R^r$ is a disturbance vector. The dynamic and input matrices are rewritten as

$$\begin{aligned} A_p &= A'_p + A''_p \\ B_p &= B'_p + B''_p \end{aligned} \quad (9.15c)$$

in order to differentiate the nominal components A'_p and B'_p and the uncertain components A''_p and B''_p, respectively. For the sake of simplicity we will assume that the nominal components are known and constant, while the nonlinear parts may be nonlinear time-variant but are bounded. The matrices B_p and C_p are of full rank, moreover the pair (A_p, C_p) is completely observable.

Now we will construct a nonlinear observer such that its state vector x_0 tracks asymptotically the plant state x_p. To this end we choose

$$\dot{x}_0 = A_0 x_0 + B'_p u_p - K_0 C_p x_p - N(y_p, x_0, t) \quad (9.16)$$

where A_0 has all its eigenvalues with negative real parts.

The error dynamics, with

$$e = x_0 - x_p \quad (9.17)$$

become

$$\dot{e} = A_0 e - N(y_p, x_0, t) + (A_0 - A_p - K_0 C_p)\, x_p - B''_p u_p - B_p d_p \quad (9.18)$$

In order to solve the observer design via the HVSC approach, we must be able to restate the problem as in Fig. 9.1. This can be done by using the positive real lemma (9.6) assuming $D = W = 0$, $A = A_0$, $C = C_p$, and obtaining the matrix P_0 as the solution of the Lyapunov equation.

$$A_0^T P_0 + P_0 A_0 = -LL^T \quad (9.19a)$$

$$B_0 = B = P_0^{-1} C_p^T \quad (9.19b)$$

Putting

$$N(y_p, x_0, t) = B_0 w(y_p, x_0) = P_0^{-1} C_p^T w(y_p, x_0), \qquad w \in R^m \quad (9.20)$$

the error dynamics become

$$\dot{e} = A_0 e + [(A_0 - A_p - K_0 C_p)\, x_p - B''_p d_p - B_p d_p] - P_0^{-1} C_p^T w(y_p, x_0) \quad (9.21)$$

Introducing the output vector

$$v = C_p e \tag{9.22}$$

we obtain the scheme of Fig. 9.5 where d_0 is an equivalent disturbance vector corresponding to the terms in the square brackets in (9.21). By comparing the diagrams in Fig. 9.1 and 9.5, assuming that d_0 is bounded and belongs to the range space of C_p, i.e.

$$d_0 = P_0^{-1} C_p^{\mathrm{T}} h(t) \tag{9.23}$$

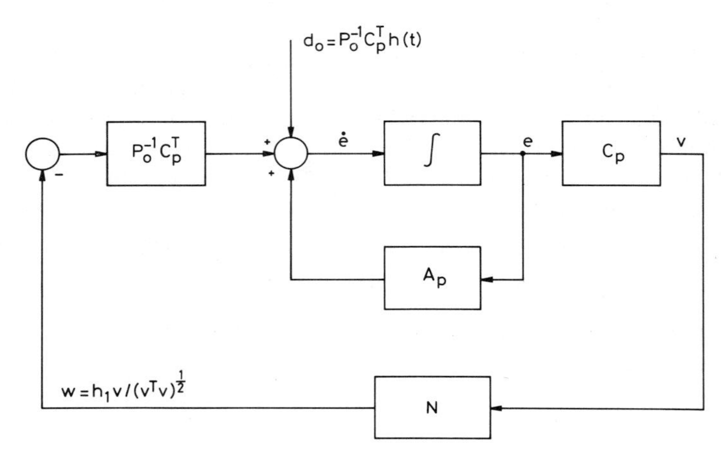

Fig. 9.5 *Observer error dynamics*

with $\|h(t)\| < h_0$, it clearly appears that w can be chosen to be passive and such that the disturbance d_0 is exactly compensated. A possible choice for w is the unit-vector control law (Balestrino, De Maria and Sciavicco, 1982)

$$w = h_1 v/(v^T v)^{1/2}, \qquad h_1 > h_0 \tag{9.24}$$

(see Walcott and Żak, 1988; Slotine, Hedrik and Misawa, 1986).

It can be easily checked that $e^T P_0 e$ is a Lyapunov function for the system (9.21) with its time derivative negative semidefinite along the system trajectories. Therefore if $e(0) = 0$, the control law (9.24) will force $e(t)$ to zero for $t > 0$. If $e(0) \neq 0$ the dynamics can be studied by rewriting the generalised error $e(t)$ as

$$e = B_0(B_0^T P_0 B_0)^{-1} v + B_1 z = P_0^{-1} C_p^T (C_p P_0 C_p^T)^{-1} v + B_1 z \tag{9.25}$$

with

$$\text{rank}\,(B_0 B_1) = n \tag{9.26a}$$

$$B_0^T P_0 B_1 = C_p B_1 = 0 \tag{9.26b}$$

$$v = C_p e, \qquad z = (B_1^T P_0 B_1)^{-1} B_1^T P_0 e \tag{9.26c}$$

The output dynamics, by using (9.21) and (9.26) are given by

$$\begin{aligned}\dot{v} &= C_p A_0 P_0^{-1} C_p^T (C_p P_0^{-1} C_p^T)^{-1} v - C_p P_0^{-1} C_p^T [(h_1 v/(v^T v)^{1/2} - h(t)] \\ &\quad + C_p A_0 B_1 z = A_v v - B_v [(h_1 v/(v^T v)^{1/2} - h(t) \\ &\quad - B_v^{-1} C_p A_0 B_1 z)] \end{aligned} \tag{9.27}$$

Note that the transfer function matrix

$$I(sI - A_v)^{-1} B_v \tag{9.28a}$$

is SPR; indeed the positive real lemma is satisfied by taking into account (9.19*a*):

$$A_v^T B_v^{-1} + B_v^{-1} A_v = -B_v C_p P_0^{-1} L L^T P_0^{-1} C_p^T B_v = -L_v L_v^T \tag{9.28b}$$

$$C_v = B_v^T B_v^{-1} = I, \qquad B_v = C_p P_0^{-1} C_p^T > 0 \tag{9.28c}$$

Therefore

$$V(v) = v^T (C_p P_0^{-1} C_p^T)^{-1} v = v^T B_v^{-1} v \tag{9.29}$$

is a suitable Lyapunov function for the output dynamics (9.27). In view of the hyperstability theorem we can prove that the system (9.27) is hyperstable if the following passivity condition is satisfied:

$$v^T (h_1 v/(v^T v)^{1/2} - h(t) - B_v^{-1} C_p A_0 B_1 z) > 0 \tag{9.30}$$

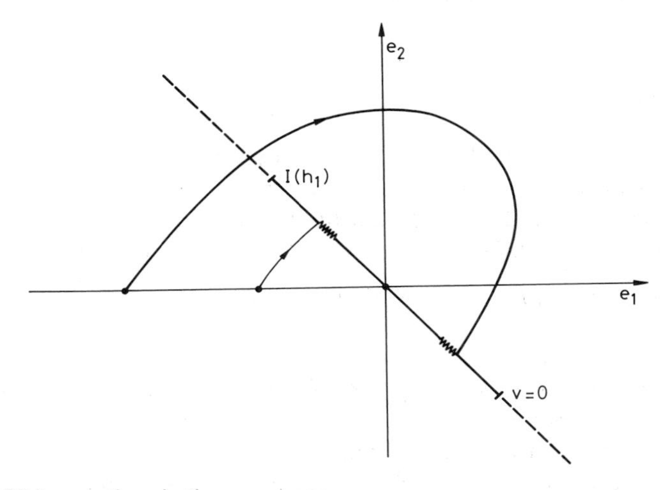

Fig. 9.6 *Sliding motions in the error space*

We know that the error dynamics (9.21) are asymptotically stable; i.e. $e(t)$ is bounded and asymptotically converging to zero so that two positive constants a and b exist such that

$$\| B_v^{-1} C_p A_0 B_1 z(t) \| < b \exp(-at) \tag{9.31}$$

Hence the passivity condition is certainly satisfied if

$$h_1 \; > \; h_0 + b \exp(-at) \; > \; \|h(t) + B_v^{-1} C_p A_0 B_1 z(t)\| \tag{9.32}$$

If $h_1 > h_0$ then after a finite time interval the system reaches the hyperplane $v = 0$. This hyperplane is invariant in a suitable neighbourhood of $e = 0$. The larger is h_1 the larger is the neighbourhood $I(h_1)$ where the invariance of $v = 0$ is guaranteed. Of course, the maximum allowable level of h_1 usually depends on physical constraints.

If the variables belong to the invariant set $I(h_1)$, then $v = \dot{v} = 0$, ad from (9.27) we can compute the equivalent control

$$w_{eq} \; = \; [h_1 v/(v^T v)^{1/2}]_s \; = \; B_v^{-1} C_p A_0 B_1 z(t) + h(t \tag{9.33}$$

From (9.33) we have that during the sliding motion the signal $h_1 v/(v^T v)^{1/2}$ chatters at a very high frequency, ideally infinite, and its low frequency component is given by (9.33) (Utkin, 1974). By means of the equivalent control we can characterise the extension of the domain $I(h_1)$ in the state space (Buhler, 1986).

Some typical trajectories in the error space are shown in Fig. 9.6.

The observer state variables x_0 may be affected by a considerable amount of ripple due to the control chattering; therefore, whenever we use this observer we must care about real chattering which occurs at a finite frequency because of the physical constraints and the high frequency inaccurate modelling.

9.6 HVSCS design

A systematic design method of the hyperstable variable structure control for a large class of uncertain systems can be developed using the model following control as the reference frame (Balestrino *et al.*, 1985).

Let us consider a nonlinear time-invariant plant satisfying the equations

$$\dot{x}_p \; = \; A_p(x_p, t)x_p + B_p R(x_p, t)u_p + d_p \tag{9.34a}$$

$$y_p \; = \; C_p x_p \tag{9.34b}$$

where $x_p \in R^n$ is the plant state, $u_p \in R^m$ is the input, $y_p \in R^m$ is the output and $d_p \in R^n$ is a disturbance vector. The $m \times m$ matrix R is of full rank; the matrices B_p and C_p are constant and of full rank.

The design objectives are specified for the sake of simplicity by a linear time-invariant model:

$$\dot{x}_m \; = \; A_m x_m + B_m u_m \tag{9.35a}$$

$$y_m \; = \; C_m x_m \tag{9.35b}$$

with $x_m \in R^n$, $u_m \in R^m$, $y_m \in R^m$, the pair (A_m, B_m) completely controllable, the pair (A_m, C_m) completely observable, and by the requirement

$$\lim_{t \to \infty} y_p(t) = y_m(t) \tag{9.36}$$

In order to satisfy this requirement we introduce a bank of m integrators

$$\dot{x}_I = B_I(y_m - y_p) = B_I C_m x_m - B_I C_p x_p \tag{9.37}$$

along with some feedfoward and feedback actions, as in Fig. 9.7, realising

$$u_p = M(K_u u_m + K_y C_m x_m - K_y C_p x_p + K_I x_I + K_p x_p) + MN(v), \tag{9.38a}$$

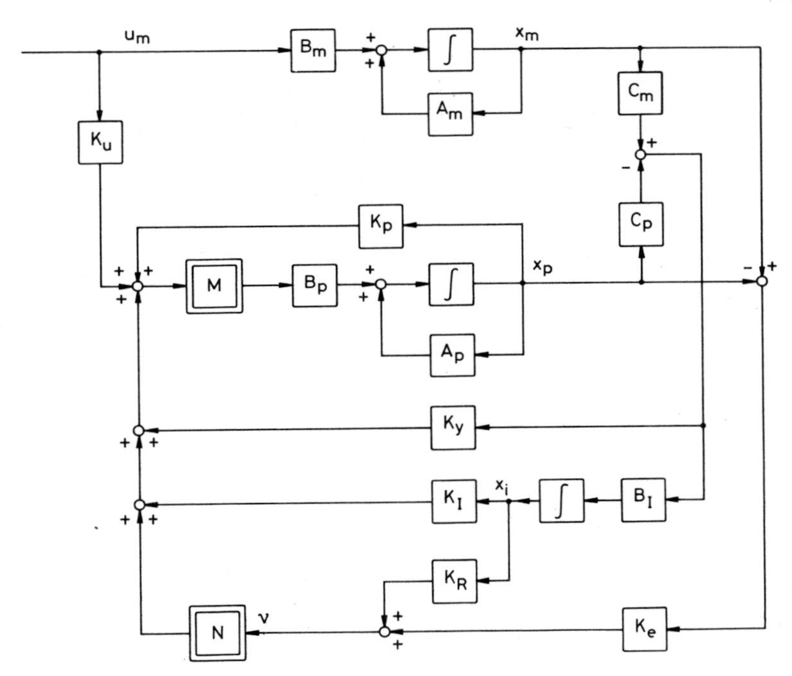

Fig. 9.7 *HVSCS block diagram*

$$v = K_e(x_m - x_p) + K_R x_I \tag{9.38b}$$

For the sake of simplicity we assume

$$\begin{aligned} C_m &= C_p = C \\ B_I C_m &= B_I C \\ K_e &= HC \end{aligned} \tag{9.39}$$

and putting

$$e = x_m - x_p \tag{9.40}$$

the error dynamics satisfy

$$\dot{e} = (A_m - B_pRMK_yC)e + (A_m - A_p + B_pRMK_p)x_p + (B_m - B_pRMK_u)u_m - B_pRMK_Ix_I - B_pRMN(v) - d_p \quad (9.41)$$

with

$$v = HCe + K_Rx_I \quad (9.42)$$

$$\dot{x}_I = B_ICe \quad (9.43)$$

We assume that the state variables x_p are available for measurement, otherwise a nonlinear observer must be used, and that the perfect model-following conditions are satisfied, i.e.

$$(I - B_mB_m^+)(B_p, A_m - A_p, d_p) = (0, 0, 0) \quad (9.44)$$

with $B^+ = (B^TB)^{-1}B^T$ the pseudo-inverse matrix of B.

The disturbance d_p is in the range space of B_m; moreover it is assumed

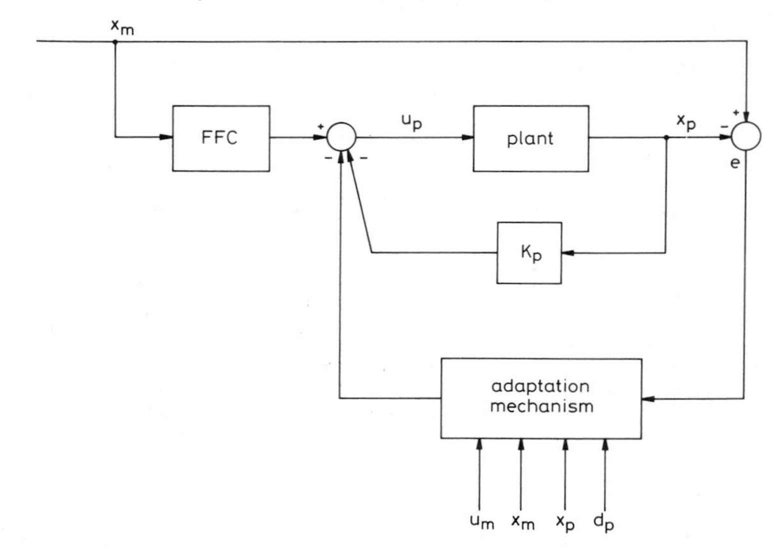

Fig. 9.8 *An inverse model following control system*

bounded. An alternative formulation of the control system in Fig. 9.7 is shown in Fig. 9.8, and both may be considered as an implementation of an adaptive model-following control (AMFC) system via signal synthesis adaptation (Landau, 1979).

Of course, the matrices K_p and K_u could also be generated by means of an adaptation mechanism in order to have

$$\begin{aligned} A_p - B_pRMK_p &= A_m \\ B_pRMK_u &= B_m \end{aligned} \quad (9.45)$$

by using parameter adaptation, but in the following the design is performed assuming no parameter adaptation. The hyperstability approach and the existence of a sliding motion on $v = 0$ are the main tools used for the design.

If the system dynamics are in the sliding mode, i.e. if $v = \dot{v} = 0$, then b using (9.41)(9.43), and denoting the variables in the sliding mode by means of a subscript s, we have

$$\frac{d}{dt}(Ce)_s = -H^{-1}K_R B_I(Ce)_s = F_s(Ce)_s \tag{9.46}$$

$$\dot{x}_{Is} = B_I Ce_s, \; x_{Is} = -K_R HCe_s \tag{9.47}$$

$$\begin{aligned} C\dot{e}_s = {} & (CA_m - CB_p RMK_y C)e_s + C(A_m - A_p + B_p RMK_p)x_p \\ & + C(B_m - B_p RMK_u)u_m + CB_p RMK_I K_R^{-1} HCe_s \\ & - Cd_p - CB_p RMN(v)_s \end{aligned} \tag{9.48}$$

By comparing (9.46) with (9.48), assuming that $CB_p RM$ is a full rank matrix, we get the equivalent control

$$\begin{aligned} w_{eq} = N(v)_s = {} & (CB_p RM)^{-1}(H^{-1}K_R B_I C + CA_m)e_s \\ & + (K_p - (CB_p RM)^{-1}C\,(A_m - A_p))x_p \\ & + ((CB_p RM)^{-1}CB_m - K_u)u_m - (CB_p RM)^{-1}Cd_p \end{aligned} \tag{9.49}$$

In order to ensure sliding motion the output of the nonlinear block $N(v)$ must be able to satisfy the above condition; therefore for a given domain of the variables x_m, x_p, u_m and d_p, we can compute what amplitude level is needed in order to sustain the sliding motion (Buhler, 1986).

Note that the output error dynamics during the sliding motion are completely at the designer's disposal; indeed we can freely choose the dynamic matrix F_s in (9.46); for instance, a possible choice is

$$F_s = -H^{-1}K_R B_I = -Q_s \tag{9.50}$$

with Q_s a positive definite symmetric matrix so that the asymptotic stability of the output error dynamics during the sliding motion is automatically assured. In view of the complete observability of the error dynamics it follows also that not only Ce but also the error $e(t)$ goes asymptotically to zero, hence assuring the asymptotic stability of the error dynamics during the sliding motion.

Having designed the output error dynamics during the sliding mode, now we must assure the existence of the sliding mode. By using (9.41)–(9.43), we get

$$\dot{v} = HC\dot{e} + K_R B_I Ce = A_v v - B_v N(v) + d_v \tag{9.51}$$

where

$$A_v = -HCB_p RMK_I K_R^{-1}$$

$$B_v = HCB_p RM$$

$$d_v = (HCA_m - HCB_pRM(K_y - K_IK_R^{-1}H)C + K_RB_IC) \times e + HC(A_m - A_p + B_pRMK_p)x_p + HC(B_m - B_pRMK_u)u_m - HCd_p$$

In order to consider the dynamic system (9.51) in terms of the hyperstability theory as in Fig. 9.1, we can choose the matrix M so that

$$R(x_p, t)M(x_p, t) = N_0 + N_1(x_p, t) \tag{9.52a}$$

with

$$\| N_0 \| \gg \| N_1(x_p, t) \| \tag{9.52b}$$

Correspondingly we split the matrices A_v and B_v into a nominal part and an uncertain part:

$$A_v = A'_v + A''_v(x_p, t) \tag{9.53a}$$

$$B_v = B'_v + B''_v(x_p, t) \tag{9.53b}$$

where A'_v and B'_v are constant matrices. Now we can rewrite the dynamic system (9.51) as follows:

$$v \quad A'_v v - B'_v N(v) + d'_v \tag{9.54a}$$

where

$$d'_v = d_v + A''_v - B''_v N(v) \tag{9.54}$$

The transfer function

$$G_v(s) = I(sI - A'_v)^{-1} B'_v \tag{9.55}$$

is SPR if and only if a positive symmetric matrix P_v exists such that

$$A'^T_v P_v + P_v A'_v = -LL^T \tag{9.56a}$$

$$B'^T_v P_v = I \tag{9.56b}$$

Hence the hyperstability will be realised if the nonlinear block

$$N(v) - (B^T_v)^{-1} d'_v \tag{9.57}$$

is passive. If the passivity of the block (9.57) is verified only on a finite domain of the variables x_m, x_p, u_m, d_p, then the existence and the stability of the sliding motion can be established only on a suitable domain including $e = 0$.

When we implement the above design procedure we have a wide choice of solutions at many intermediate steps. We have just noted this fact for the dynamic matrix F_s during the sliding motion, see (9.46) and (9.50). For the matrix M a few typical choices in (9.52) are:

(i) $M = R^{-1}(x_p, t)$

$$\begin{aligned}
&\text{(ii)} \quad M = R^T(x_p, t) \\
&\text{(iii)} \quad M = I \qquad\qquad (9.58) \\
&\text{(iv)} \quad M = \operatorname{diag}(m_i)
\end{aligned}$$

provided that (9.52*b*) is satisfied. The choice of M is often such that RM is a positive definite matrix or a Hadamard matrix. Note that the choices (i) and (ii) in (9.58) require a complete knowledge of the process dynamics and usually they involve a heavy computational burden.

From (9.56*b*) we get

$$B_v'^T = P_v^{-1} = B_v' = HCB_pN_0 > 0, \tag{9.59}$$

and from (9.56*a*)

$$P_v A_v' = -K_I K_R^{-1} \tag{9.60}$$

Therefore the positive real lemma can be satisfied by choosing

$$H = Q_H N_0^{-1}(CB_p)^{-1}, \qquad Q_H > 0 \tag{9.61a}$$

$$K_I K_R^{-1} + (K_I K_R^{-1})^T = LL^T \tag{9.61b}$$

for some positive definite matrices Q_H and LL^T. A simple choice leads to $Q_H = I$ and to a symmetric positive definite matrix $K_I K_R^{-1}$.

The passivity of the nonlinear block (9.57) can be realised if

$$v^T N(v) > v^T(B_v')^{-1}\,\mathrm{d}_v' = v^T(B_v')^{-1}[d_v + A_v''v - B_v''N(v)]$$

i.e. if

$$v^T[I + (B_v')^{-1}B_v'']N(v) > v^T(B_v^1)(d_v^{-1} + A_v''v). \tag{9.62}$$

Many choices of $N(v)$ are possible (Balestrino *et al.*, 1984); for instance, a generalised unit vector control law

$$N(v) = c_0 v/(v^T v)^{1/2} + c_1 v, \quad c_0 > 0,\ c_1 > 0, \tag{9.63}$$

leads to the sufficient condition

$$(c_0(v^T v)^{1/2} + c_1 v^T v) \min_{x_p, t} \sigma\,(I + N_0^{-1}N_1) > \max_{x_p, t} v^T(B_v')^{-1}(d_v + A_v''v) \tag{9.64}$$

where σ is the minimum singular value of the matrix.

For a given numerical range of the variables x_m, x_p, u_m, d_p we can always compute a suitable pair c_0, c_1 so that the inequality (9.64) is satisfied. Note that instead of c_0 and c_1 we can also use some positive definite matrices C_0 and C_1, eventually time-variant, on condition that the inequality (9.62) holds. Without restrictions on the allowable control amplitudes we can choose c_0 and c_1 very large but usually physical constraints prevents us from implementing such a solution. Therefore the sliding motion will be guaranteed only inside a finite domain, the extension of which depends on c_0 and c_1. The same comments apply if, with B'_v a diagonal matrix, we use relay control laws instead of the unit vector control laws.

As soon as the sliding motion is attained its invariance is assured by the nonlinear block $N(v)$. In the output of this block we recognise a low-frequency component, i.e. the equivalent control (9.49), and a very high frequency component with zero mean referred to as chattering in the technical literature.

In engineering practice chattering is uaully an undesirable phenomenon highly dependent on the unmodelled dynamics. The unwanted effects due to chattering can be reduced or eliminated by dithering (Balestrino, De Maria and Sciavicco, 1983), i.e. by adding to the signal $v(t)$ any zero mean signal $s(t)$ with a frequency outside the plant bandwidth, or by replacing the unit vector law by a high gain block (Kalman and Bertram, 1960; Emelyanov, 1985).

9.7 Design procedure in the limit case $K_R = 0$

The design procedure set out in the previous Section is directly practicable only if K_R is of full rank, i.e. if all integrators take part in forming the signal v.

Many intermediate cases arise by relaxing the full rank assumptions; here we deal with the limit case $K_R = 0$, leaving the development of the intermediate cases as an exercise for the skilled reader.

The equations (9.41)–(9.43) are rewritten as

$$\begin{aligned}\dot{e} = {} & (A_m - B_p RMK_y C)\, e + (A_m - A_p + B_p RMK_p)\, x_p \\ & + (B_m - B_p RMK_u)\, u_m) + - B_p RMK_I x_I - d_p - B_p RMN(v)\end{aligned} \tag{9.65}$$

$$\dot{x}_I = B_I Ce \tag{9.66}$$

$$v = HCe \tag{9.67}$$

Assuming again that

$$R(x_p, t)\, M(x_p, t) = N_0 + N_1(x_p, t) \tag{9.68a}$$

with

$$\| N_0 \| > \| N_1(x_p, t) \| \tag{9.68b}$$

by means of a suitable similarity transformation we can rearrange (9.65) as follows:

$$\dot{e}_1 = A_{11} e_1 + A_{12} e_2 \tag{9.69a}$$

$$\dot{e}_2 = A_{21} e_1 + A_{22} e_2 + B_2 (d_2 - G_I x_I - RMN(v)) \tag{9.69b}$$

where the matrices B_2, and A_{ij}, $i, j = 1, 2$, are constant, and d_2 is an equivalent disturbance vector taking into account all nonlinear and uncertain components not included in $RMN(v)$.

The signal v is computed as

$$v = K_1 e_1 + K_2 e_2 = K_e e \tag{9.70}$$

while (9.66) becomes

$$\dot{x}_I = G_1 e_1 + G_2 e_2 \tag{9.71}$$

Now we assume that K_2 is invertible and that the pair (A, K_2) is completely observable so that the pair (A_{11}, A_{12}) is controllable (Young, Kokotovic and Utkin, 1977).

By means of a simple co-ordinate transformation the dynamic equations become:

$$\dot{e}_1 = F_{11} e_1 + F_{12} v \tag{9.72a}$$

$$\dot{v} = F_{21} e_1 + F_{22} v + K_2 B_2 d_2 - K_2 B_2 G_I x_I - K_2 B_2 R\, MN(v) \tag{9.72b}$$

$$\dot{x}_I = (G_1 - G_2 K_2^{-1} K_1)\, e_1 + G_2 K_2^{-1} v = G_1' e_1 + G_2' v \tag{9.72c}$$

where

$$\begin{aligned} F_{11} &= A_{11} - A_{12} K_2^{-1} K_1; \qquad F_{12} = A_{12} K_2^{-1}; \\ F_{21} &= K_1 A_{11} + K_2 A_{21} - K_1 A_{12} K_2^{-1} K_1 - K_2 A_{22} K_2^{-1} K_1 \\ F_{22} &= K_1 A_{12} K_2^{-1} + K_2 A_{22} K_2^{-1} \\ G_1' &= G_1 - G_2 K_2^{-1} K_1; \qquad G_2' = G_2 K_2^{-1} \end{aligned} \tag{9.73}$$

If the sliding mode exists on $v = 0$ in a suitable neighbourhood of the zero state, then from $v = \dot{v} = 0$, denoting the variables in sliding mode by a subscript s, we get

$$\dot{e}_{1s} = F_{11} e_{1s} \tag{9.74a}$$

$$\dot{x}_{Is} = G_1' e_{1s} \tag{9.74b}$$

and from (9.72c) we can compute the equivalent control

$$w_{eq} = N(v)_s = (K_2 B_2 R M)^{-1} (F_{21} e_{1s} + K_2 B_2 d_2 - K_2 B_2 G_I x_{Is}) \tag{9.75}$$

In order to assure the asymptotic stability of (9.72a) because of the controllability of the pair (A_{11}, A_{12}), we can assign the eigenvalues of F_{11} (9.73) everywhere in the half plane with $\text{Re}(s) < 0$ by means of a suitable choice of K_1, K_2.

Choosing the block $N(v)$ as a generalised unit vector control law and separating in (9.72b) the linear time-invariant terms from the nonlinear time-variant terms we get

$$\dot{v} = F_{22}v - K_2B_2N_0(D_1v + D_v v/(v^Tv)^{1/2}) + K_2B_2d_2' \quad (9.76a)$$

with

$$d_2' = (K_2B_2)^{-1} F_{21}e_1 + d_2 - G_Ix_I - N_1(D_1v + D_v v/(v_Tv)^{1/2}) \quad (9.76b)$$

In order to establish that v asymptotically goes to zero we require that the transfer function matrix

$$(sI - F_{22})^{-1} K_2B_2 \quad (9.77)$$

be SPR, and that the following passivity condition holds:

$$v^TK_2B_2N_0(D_1v + D_v v/(v^Tv)) - v^TK_2B_2d_2' > 0 \quad (9.78)$$

The positive real lemma leads to the conditions

$$F_{22}^TP_2 + P_2F_{22} = -L_2L_2^T \quad (9.79a)$$

$$(K_2B_2)^TP_2 = I \quad (9.79b)$$

which are satisfied with $P_2 > 0$ if the matrix F_{22} is chosen so that all its eigenvalues have negative real parts, and if

$$K_2 = P_2^{-1}B_2^{-1} \quad (9.80)$$

The passivity condition (9.78) can be verified by choosing

$$D_1 = Q_1N_0^TP_2^{-1}, \quad D_v = Q_vN_0^{-1}P_2^{-1} \quad (9.81)$$

with Q_1 and Q_v suitable positive definite matrices, e.g. c_1I and c_vI, respectively, with sufficiently large positive constant c_1 and c_v. If there are some constraints on the amplitude of the control inputs then the passivity condition can be assured only on a finite domain containing the origin; therefore the sliding motion on $v = 0$ occurs only inside this domain. The comments at the end of Section 9.6 are still in order here. Note that the control law (9.75) is only one of a variety of possible controls (Balestrino, De Maria and Siciliano, 1985).

With $K_R = 0$ often we also have $K_I = B_I = 0$, so that the integrators are disregarded; the corresponding changes in the previous equations are quite obvious and are left to the reader.

9.8 Applications in robotics

A large variety of electromechanical systems can be modelled by a set of second order nonlinear differential equations as:

$$D(x^p(t))\dot{x}^v + c[x^p(t), x^v(t)] + g[x^p(t)] + d(t) = u(t) \quad (9.82a)$$

$$\dot{x}^p(t) = x^v(t) \quad (9.82b)$$

In a mathematical model of a general n-joint robot manipulator system $u \in R^n$ is a generalised torque vector; $x^p \in R^n$ and $x^v \in R^n$ are generalised joint co-ordinate (position and velocity) vectors; x^p is an angle or a distance, for a revolute or prismatic joint, respectively; $D(x^p)$ is the $n \times n$ symmetric definite positive matrix of inertial coefficients; $c(x^p, x^v)$ is the Coriolis and centrifugal force n-vector, and $g(x^p)$ is the gravitational force n-vector. $d(t)$ is an uncertainty n-vector whose components are smaller than the corresponding components of u; it takes into account unmodelled forces, such as friction, blacklash or environmental disturbances.

Hereafter we will apply the design procedure outlined in Section 9.6 to the plant described by (9.82).

Choosing $x_p = \begin{bmatrix} x_p \\ x_v \end{bmatrix}$ as the state vector, (9.82) can be rewritten as

$$\dot{x}_p = \begin{bmatrix} 0 & I_n \\ A_{1p} & A_{2p} \end{bmatrix} x_p + \begin{bmatrix} 0 \\ I_n \end{bmatrix} R(x_p)\, u_p + \begin{bmatrix} 0 \\ I_n \end{bmatrix} d_p \tag{9.83}$$

where $A_{1p} = A_{1p}(x^p)$, $A_{2p} = A_2(x^p, x^v)$, $R(x_p) = D^{-1}(x^p)$ is a positive definite matrix, $u_p = u$, and $d_p = -Rd$. The model equations are

$$\dot{x}_m = \begin{bmatrix} 0 & I_n \\ A_{1m} & A_{2m} \end{bmatrix} x_m + \begin{bmatrix} 0 \\ I_n \end{bmatrix} u_m \tag{9.84a}$$

$$y_m = (I_n \quad 0)\, x_m \tag{9.84b}$$

so that the perfect model-following conditions are automatically satisfied.

Moreover A_{1m} and A_{2m} are chosen as diagonal matrices in order to realise decoupled dynamics (Dubowski and Des Forges, 1979; Balestrino, De Maria and Sciavicco, 1983*b*). Introducing a bank of n integrators, with $B_I = I_n$,

$$\dot{x}_I = y_m - y_p = (C_1 C_2)(x_m - x_p) = (C_1 C_2)\, e \tag{9.85}$$

and assuming for the sake of simplicity

$$C_2 = I_n,\ K_R = I_n, \quad K_e = H(C_1 \quad I) = (HC_1 \quad H) \tag{9.86}$$

the error dynamics (9.40) become

$$\dot{e}_1 = e_2 \tag{9.87a}$$

$$\begin{aligned} \dot{e}_2 = {} & A_{1m}e_1 + A_{2m}e_2 - RMK_y e_1 + (A_{1m} - A_{1p} + RMK_p)\, x^v \\ & + (I - RMK_u)\, u_m - RMK_I x_I - RMN(v) - d_p \end{aligned} \tag{9.87b}$$

$$\dot{x}_I = C_I e_1 + e_2 \tag{9.87c}$$

$$v = HC_1 e_1 + He_2 + x_I \tag{9.87d}$$

Since H is a full rank matrix we can use the variables v instead of e_2 obtaining the equivalent system

$$\dot{e}_1 = H^{-1}v - H^{-1}x_I - C_1e_1 = e_2 \tag{9.88a}$$

$$\begin{aligned} v \quad & H(C_1 + H^{-1} + A_{1m})H^{-1}v \,\dot{-}\, HRMN(v) \\ & + (I - HRMK_I + HC_1 - HA_{2m}H^{-1})x_I + (HA_{1m} - HC_1^2 \\ & - HRMK_y - HA_{2m}C_1)e_1 + H(A_{1m} - A_{1p} + RMK_p)x^v \\ & + H(I - RMK_u)u_m - Hd \end{aligned} \tag{9.88b}$$

$$\dot{x}I = -H^{-1}x_1 + H^{-1}v \tag{9.88c}$$

If the system is in the sliding mode, i.e. $v =\dot{}\, v = 0$, then and (9.47) becom

$$\frac{d}{dt}(C_1e_{1s} + \dot{e}_{1s}) = -H^{-1}(C_1e_{1s} + \dot{e}_{1s}) \tag{9.89a}$$

$$\dot{x}_{Is} = -H^{-1}x_{Is} \tag{9.89b}$$

therefore choosing, for example, C_1 and H positive definite, the dynamics (9.89) are asymptotically stable and are arbitrarily assignable.

During the sliding mode from (9.88*b*) we can compute the equivalent control

$$\begin{aligned} w_{eq} = N(v)_s = & (HRM)^{-1}(I + HC_1 - HRMK_I - HA_{2m}H^{-1})x_{Is} \\ & + (HRM)^{-1}(HA_{1m} - HC_1^2 - HRMK_Y - HA_{2m}C_1)e_{1s} \\ & + (HRM)^{-1}[H](A_{1m} - A_{1p} + RMK_p)x^v \\ & + (HRM)^{-1}H(I - RMK_u)u_m - (HRM)^{-1}d \end{aligned} \tag{9.90}$$

In (9.90) we recognise several typical control actions:

(i) a proportional control (terms in e_{1s})
(ii) an integral action (terms in x_{Is})
(iii) a nonlinear compensation (feedback of x^v)
(iv) a feedfoward control (terms in u_m)
(v) a disturbance compensation (term in d)

Note that during the sliding mode we have also, see (9.88*c*),

$$x_{Is} = -C_1e_{1s} - \dot{e}_{1s} \tag{9.91}$$

therefore the integral action during the sliding motion is equivalent to being proportional derivative.

For $R(x^p)$ positive definite we suppose that for a given robot structure a matrix $M(x^p)$ exists such that (Kyriazov, 1987)

$$R(x^p)M(x^p) = N_0 + N_1(x^p) \tag{9.92a}$$

with

$$N_0 = \mathrm{diag}(n_{0i}), \quad n_{0i} > 0, \quad i = 1, \ldots\ldots, n \tag{9.92b}$$

$$\|N_0^{-1}N_1(x^p)\| < 1 \tag{9.92c}$$

It is in practice very convenient to implement the control in a decentralised setting; therefore we choose the matrices A_{1m}, A_{2m}, K_y, K_p, K_u, K_I and K_v all diagonal.

Putting

$$K_v = K_0 H^{-1} + K_I \tag{9.93a}$$

$$A_v = H^{-1} + H(C_1 + A_{2m}) H^{-1} - HN_0 K_0 H^{-1} \tag{9.93b}$$

$$B_0 = HN_0 \tag{9.93c}$$

we can rewrite (9.88*b*) in a more compact form:

$$\dot{v} = A_v v - B_0 N(v) + d_v + B_0 K_0 H^{-1} v \tag{9.94}$$

where d_v collects all remaining terms.

The structure of (9.94) is such that we can now apply the positive real lemma giving

$$A_v^T P_v + P_v A_v = -L_v L_v^T \tag{9.95}$$

$$B_0^T P_v = I \tag{9.95b}$$

Since A_v is diagonal, we can put $P_v = B_0^{-1}$, with P_v positive definite if the eigenvalues of A_v are real and negative; this can always be realised by a suitable choice of K_0. Hence the hyperstability of (9.94) is established if the passivity condition

$$v^T B_0 N(v) - v^T d_v - v^T B_0 K_0 H^{-1} v > 0 \tag{9.96}$$

is satisfied for some control law $N(v)$.

Using again, for the sake of simplicity, a generalised unit vector control law:

$$N(v) = K_0 H^{-1} v + D_1 v + D_v v/(v^T v)^{1/2}$$

$$= K_1 v + D_v v/(v^T v)^{1/2} \tag{9.97}$$

with K_1 and D_v positive definite matrices, not necessarily constant, (9.96) becomes

$$v^T B_0 K_1 v + v^T B_0 D_v v/(v^T v)^{1/2} > v^T d_v \tag{9.98}$$

Taking into account that B_0 is diagonal and positive definite, it is always possible to choose D_1 and D_v such that the inequality (9.98) is satisfied. If there are some constraints on the signal amplitudes then (9.98) can be satisfied only on a suitable neighbourhood of the state space including the origin for bounded disturbances.

In order to minimise the norm of d_v we can conveniently choose the matrices K_y, K_I, K_p and K_u. This choice is very easy if the robot manipulator works in a limited region of the space or if it moves with low speed; otherwise nonlinear adaptive compensation are needed. If there is limited computing power, special solutions such as piecewise constant matrices can be adopted.

Some examples of robotic applications are given in Balestrino, De Maria and Sciavicco (1984) and Lim and Eslami (1987), but we caution the reader to take into account some minor modifications to fit these papers into the present context.

9.9 A design example in aircraft control

In the previous Section the control law for a robotic manipulator has been designed using a model-following approach, but often the signals to be tracked by the plant are directly specified and there is no model to follow. In this case an inverse model-following technique is more appropriate (Landau, 1979); given the output to be tracked a model is used only to compute a suitable input u_m which is used for a feedfoward action (see Fig. 9.8).

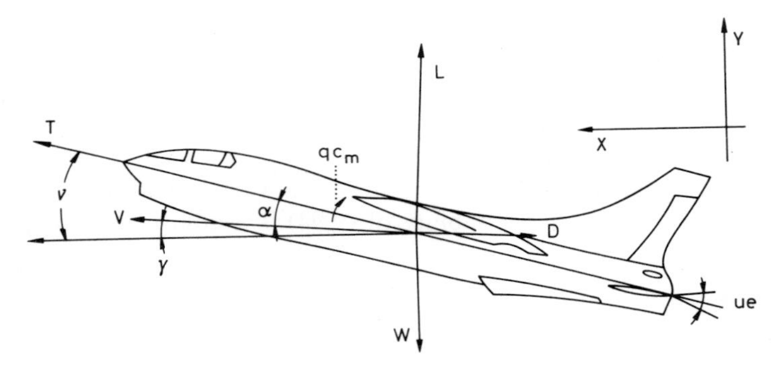

Fig. 9.9 *Aircraft dynamics*

The inverse model-following technique and the design approach of Section 9.7 are exemplified in this Section with reference to the control of high performance aircraft. Modelling such a plant as a linear time-invariant system and consequently designing the control via modern multivariable control techniques are standard procedures, but they prove to be of value only if the parametric variations are limited.

For high performance aircraft or missiles linearised dynamics lead to unsatisfactory designs, so that the analysis and the design must be carried out taking into account the nonlinearities and the parametric variations.

VSC theory can be applied successfully also in this case. We will apply the design procedure of Section 9.7 to obtain a suitable control law for the longitudinal dynamics of an aircraft depicted in Fig. 9.9. For the sake of simplicity we assume that the axis of zero lift and the longitudinal axis of the aircraft are coincident with the thrust axis.

The form of the dynamics equations (McRuer, Ashkenas and Graham, 1973) does not satisfy the perfect model-following conditions; therefore we rewrite the equations in the equivalent form

$$\frac{d}{dt}\begin{bmatrix}\dot{x}\\ \dot{h}\end{bmatrix} = -c_1\begin{bmatrix}C_L & C_D\\ -C_D & C_L\end{bmatrix}\begin{bmatrix}\dot{x}\\ \dot{h}\end{bmatrix} + c_2\begin{bmatrix}u_x\\ u_h\end{bmatrix} + \begin{bmatrix}0\\ -g\end{bmatrix} \quad (9.99)$$

$$\frac{d}{dt}\begin{bmatrix}\alpha\\ \dot{\alpha}\end{bmatrix} = c_3\begin{bmatrix}0 & 1\\ C_{m\alpha}(C_{m\dot{\alpha}} & + C_{mq})/2V\end{bmatrix}\begin{bmatrix}\alpha\\ \dot{\alpha}\end{bmatrix} + \begin{bmatrix}0\\ c_3 C_{me}\end{bmatrix}u_e + \begin{bmatrix}0\\ c_3 d(\alpha, \dot{\alpha})\end{bmatrix} \quad (9.100)$$

where

$$\theta = \alpha + \gamma, \qquad \dot{\theta} = q,$$

$$V = (\dot{x}^2 + \dot{h}^2)^{1/2}, \qquad \dot{x} = V\cos(\gamma), \qquad \dot{h} = V\sin(\gamma) \quad (9.101)$$

$$u_x = u_T\cos(\theta), \qquad u_h = u_T\sin(\theta), \qquad u_T = T/T_{max}$$

with T and T_{max} the effective and the maximum thrust respectively. The elevator angle is denoted by u_e and is assumed to be a control input disregarding the actuator dynamics. The parameters c_1, c_2 and c_3 are defined as

$$c_1 = \varrho V^2 S/2m, \quad c_2 = T_{max}/m, \quad c_3 = \varrho V^2 S\bar{c}/2I_y \quad (9.102)$$

where ρ is the air density and depends on the altitude h, and m and I_y are the mass and the moment of inertia of the aircraft, respectively. The aereodynamical coefficients have been written as follows:

$$\begin{aligned} C_L &= C_{L\alpha}(h, M)\,\alpha \\ C_D &= C_{D0}(h, M) + \mu(h, M)\,C_{L\alpha}(h, M)\,\alpha^2 \\ C_M &= C_{m0}(h, M) + C_{m\alpha}\alpha + (\bar{c}/2V)\,(C_{m\dot{\alpha}}(h, M)\,\dot{\alpha} \\ &\quad + C_{mq}(h, M)\,q) + C_{me}(h, M)\,u_e \end{aligned} \quad (9.103)$$

where the coefficients depend on the altitude h and on the Mach number M.

These coefficients are usually obtained experimentally and are functions of h and M. In (9.101) the term d includes all terms that are nonlinear and not present in the first two terms in the right hand of (9.103).

The dynamics (9.99) and (9.100) are nonlinear and are not decoupled, but usually they have quite different speeds of response. In what follows we will refer to the aircraft F4–C as a typical example. For the set of parameters

$$h \cong 0, \quad M = 0{\cdot}4, \quad \varrho = 2{\cdot}3769 \times 10^{-3}\,\mathrm{lb\,s^2/ft^4},$$

$$V = 446{\cdot}56\,\mathrm{ft/s}, \quad C_{L\alpha} = 3{\cdot}467, \quad C_{m\alpha} = -0{\cdot}3, C_{mq} = -2{\cdot}58,$$

$$C_{D0} = 1{\cdot}267 \times 10^{-2}, \quad \eta = 0{\cdot}533, \quad C_{m\alpha} = -1{\cdot}667,$$

$$\alpha_{trim} = 8{\cdot}8869 \times 10^{-2}\,\mathrm{rad}, \quad C_L = 0{\cdot}307, \quad C_{D\alpha} = 0{\cdot}0272$$

the eigenvalues of the dynamic matrix in (9.99) are; $-9{\cdot}507^{-3} \pm j2{\cdot}17 \times 10^{-2}$, while the eigenvalues of the dynamics matrix in (9.100) are: $-4{\cdot}96^{-1} \pm j6{\cdot}471$.

With

$$h = 3500\,\text{ft}, \quad M = 1{\cdot}5, \quad \varrho = 7{\cdot}382 \times 10^{-4}\,\text{lb sec}^2/\text{ft}^4,$$

$$V = 1459{\cdot}71\,\text{ft/sec}, \quad C_{L\alpha} = 2{\cdot}8, \quad C_{Do} = 3{\cdot}75 \times 10^{-2}, \quad \mu = 0{\cdot}91$$

$$C_{m\alpha} = -0{\cdot}77, \quad C_{mq} = -2{\cdot}6, \quad \alpha_{\text{trim}} = 3{\cdot}28 \times 10^{-2}\,\text{rad},$$

$$C_L = 9{\cdot}203 \times 10^{-2}, \quad C_D = 4{\cdot}025 \times 10^{-2},$$

the eigenvalues of the dynamic matrix in (9.1) are: $-9{\cdot}507^{-3} \pm j2{\cdot}17 \times 10^{-2}$, while the eigenvalues of the dynamics matrix in (9.2) are: $-4{\cdot}96^{-1} \pm j6{\cdot}471$.

We see that for extremely different flight regimes the linearised model eigenvalues are far apart from each other. Hence in the study of the dynamics of (9.100) the air speed and the air density can be assumed constant, while referring to the dynamics of (9.99) the variations of α and θ are assumed instantaneous with respect to the variations of $\dot{x}$ and $\dot{h}$. These assumptions can be substantiated by using the singular perturbation theory (Ioannou and Kokotovic, 1985).

Denoting $\dot{x}_m$ and $\dot{h}_m$ as the horizontal and vertical speeds to be tracked, and e_T the error vector,

$$e_T = (\dot{x}_m - \dot{x}, \quad \dot{h}_m - \dot{h})^T \tag{9.104}$$

from (9.99) we obtain

$$\begin{aligned} \dot{e}_T &= -c_1 A_T e_T + \begin{bmatrix} \ddot{x}_m \\ \ddot{h}_m \end{bmatrix} + c_1 A_T \begin{bmatrix} \dot{x}_m \\ \dot{h}_m \end{bmatrix} + \begin{bmatrix} 0 \\ g \end{bmatrix} - c_2 \begin{bmatrix} u_x \\ u_h \end{bmatrix} \\ &= -c_1 A_T e_T - I c_2 \begin{bmatrix} u_x \\ u_h \end{bmatrix} + d_T \end{aligned} \tag{9.105}$$

where

$$A_T = \begin{bmatrix} C_D & C_L \\ -C_L & C_D \end{bmatrix}$$

Using the positive real lemma

$$-c_1 A_T^T I + I(-c_1 A_T) = -2c_1 C_D I \tag{9.106a}$$

$$c_2 I = C \tag{9.106b}$$

Therefore the lemma is trivially satisfied because $c_1 C_D$ is positive if $V \neq 0$, and (9.106*b*) requires that the state variables (e_1, e_2) are available for measurement, as is usually the case. In order to assure asymptotic stability we require the hyperstability of the system (9.105), so the passivity condition

$$c_2(e_1 u_x + e_2 u_h) \geqslant e_T^T d_T = e_1 d_{T1} + e_2 d_{T2} \tag{9.107a}$$

must be satisfied.

From (9.107a) we can obtain the sufficient conditions

$$\begin{aligned} c_2 e_1 u_x &\geqslant (\ddot{x}_m + (\varrho VS/2m)(C_D \dot{x}_m + C_L \dot{h}_m))\, e_1 \\ c_2 e_2 u_h &\geqslant (\ddot{h}_m + (\varrho VS/2m)(C_D \dot{h}_m - C_L \dot{x}_m))\, e_2 \end{aligned} \quad (9.107b)$$

A possible choice of the control input is

$$\begin{aligned} u_x &= u_{xc} + u_{xu}\ \mathrm{sgn}\ (e_1) \\ u_h &= u_{hc} + u_{hu}\ \mathrm{sng}\ (e_2) \end{aligned} \quad (9.108)$$

where u_{xc} and u_{hc} are nonlinear compensation terms, while u_{xu} and u_{hu} are terms depending on the parameter uncertainty. Denoting for a given flight condition the estimated values of C_D and C_L as

$$\begin{aligned} C'_D &= C'_{D0} + \mu' C'_L\, \alpha^2 \\ C'_L &= C'_{L\alpha}\alpha \end{aligned} \quad (9.109)$$

and the maximum deviations from the nominal values as

$$\begin{aligned} C^*_{L\alpha} &= \max|C_{L\alpha} - C'_{L\alpha}| \\ C^*_{D0} &= \max|C_{D0} - C'_{D0}| \\ \eta^* &= \max|\eta - \eta'| \end{aligned} \quad (9.110)$$

we can choose u_x and u_h in (9.108) with

$$\begin{aligned} u_{xc} &= (\varrho VS/2T_{max})\,(C'_D \dot{x}_m + C'_L \dot{h}_m + 2m/(\varrho VS)\,\ddot{x}_m) \\ u_{hc} &= (\varrho VS/2T_{max})\,(C'_D \dot{h}_m - C'_L \dot{x}_m + 2m/(\varrho VS)\,(\ddot{h}_m + g)) \end{aligned} \quad (9.111)$$

$$\begin{aligned} u_{xu} &= (\varrho VS/2T_{max})\,(C^*_{L\alpha}|\alpha|\dot{h}_m + C^*_{D0}\dot{x}_m + C^*_{L\alpha}\eta^*\alpha^2\dot{x}_m) \\ u_{hu} &= (\varrho VS/2T_{max})\,(C^*_{L\alpha}|\alpha|\dot{x}_m + C^*_{D0}\dot{h}_m + C_L\eta^*\alpha^*\dot{h}_m) \end{aligned} \quad (9.112)$$

By comparing (9.108) with the last row terms in (9.101) we can compute

$$u_T = (u_x^2 + u_h^2)^{1/2} \quad (9.113)$$

and from

$$\tan(\theta) = \tan(\alpha + \gamma) = u_h/u_x, \qquad \tan(\gamma) = \dot{h}/\dot{x} \quad (9.114)$$

we get

$$\dot{x}\tan(\alpha) + \dot{h} = V^2 u_h/(\dot{x}u_h + \dot{h}u_x) \quad (9.115)$$

From this nonlinear equation, given $\dot{x}$, $\dot{h}$, h and M, we can compute the appropriate value of α.

An instantaneous variation of the angle of attack α cannot be obtained, because we can only modify (disregarding the actuators dynamics) the elevator angle u_e. Therefore a suitable control law must be designed so that the dynamic system (9.100) tracks a desired angle α_m with a settling time much shorter than the settling time of the system (9.105).

To this end, putting

$$e = \alpha_m - \alpha \tag{9.116}$$

from (9.100) we get

$$\ddot{e} = \ddot{\alpha}_m - \ddot{\alpha} = \ddot{\alpha}_m - c_3(C_{m\alpha}\alpha + (C_{m\dot{\alpha}} + C_{mq})\dot{\alpha}/2V) - c_3(C_{m\alpha}\gamma + (C_{m\dot{\alpha}} + C_{mq})\dot{\gamma}/2V) - c_3 d(\alpha, \dot{\alpha}) - c_3 C_{me} u_e \tag{9.117}$$

Putting

$$v = \dot{e} + c_3 k_0 e \tag{9.118}$$

$$u_e = [k_1 e + k_2 \dot{e} + k_3 \operatorname{sgn}(v) + k_4(\alpha_m, \gamma)] M_e \tag{9.119}$$

with

$$C_{me} M_e = N_0 + N_1 > 0, \quad N_0 > N_1 \geqslant 0 \tag{9.120}$$

where N_0 is constant and N_1 may be variable, and splitting any coefficient or variable z into its estimate z' and his variation z'', we can rewrite (9.117) as

$$\begin{aligned} \ddot{e} = {} & c_3[-N_0 k_1 e + C_{m\alpha} e - N_0 k_2 \dot{e} + (C_{m\dot{\alpha}} + C_{mq})\, \dot{e}/2V] \\ & - c_3 (N_0 + N_1)\, k_3 \operatorname{sgn}(v) + - c_3[N_0 k_4(\alpha, \gamma) + C'_{m\alpha} (\alpha_m, \gamma) \\ & + (C'_{m\dot{\alpha}} + C'_{mq})\, (\dot{\alpha} + \dot{\gamma})/2V + d'(\alpha, \dot{\alpha})] + - c_3 N_1 k_4(\alpha, \gamma) \\ & - c_3 C''_m (\alpha_m + \gamma) \\ & - c_3 (C''_{m\dot{\alpha}} + C''_{mq})\, (\dot{\alpha}_m + \dot{\gamma})/2V - c_3\, \mathrm{d}''(\alpha, \dot{\alpha}) \end{aligned} \tag{9.121}$$

Now we can choose $k_4(\alpha, \dot{\alpha})$ so that the contribution due to all nominal terms in (9.121) is annihilated, and from (9.118) and (9.121),

$$\dot{e} = - c_3 k_0 e + v \tag{9.122a}$$

$$v \quad c_3 [k_0 + (C_{m\dot{\alpha}} + C_{mq}/2V - N_0 k_2]\, v - c_3 w \tag{9.122b}$$

where

$$\begin{aligned} w = {} & (N_0 + N_1)\, k_3 \operatorname{sgn}(v) + N_1 k_4(\alpha, \dot{\alpha}) + C''_{m\alpha} (\alpha_m + \gamma) \\ & + (C''_{m\dot{\alpha}} + C''_{mq})\, (\dot{\alpha}_m + \dot{\gamma})/2V + d''(\alpha, \dot{\alpha}) + [k_0^2 + k_0 (C_{m\dot{\alpha}} \\ & + C_{mq})/2V + N_0 k_4 (\alpha_m, \gamma) - C_{m\alpha} - N_0 k_0 k_2] e \end{aligned} \tag{9.123}$$

By choosing k_2 in (9.122*b*) so that, with $w = 0$, the system is asymptotically stable with the desired settling time, the hyperstability of the global system is established by requiring that the passivity condition

$$vw \geqslant 0 \tag{9.124}$$

be satisfied.

This can be easily achieved by choosing k_3 as follows:

$$k_3 = k_{30} + k_{31}|\alpha_m| + k_{32}|\gamma| + k_{33}|\dot{\alpha}_m| + k_{34}|\dot{\gamma}| + k_{35}|e| \tag{9.125}$$

where all the coefficients k_{3i}, $i = 0, 1, \ldots, 5$ are positive and sufficiently large. However, if there are some constraints on the levels the comments given at the end of the Section 9.6 also valid in this case.

Whenever the system is in the sliding mode, i.e. $v = \dot{v} = 0$

$$\dot{e}_s = -c_3 k_0 e_s \tag{9.126}$$

and from (9.123) we can compute the equivalent control, putting $w = 0$.

Of course, the global control system is complex and complicated by issues

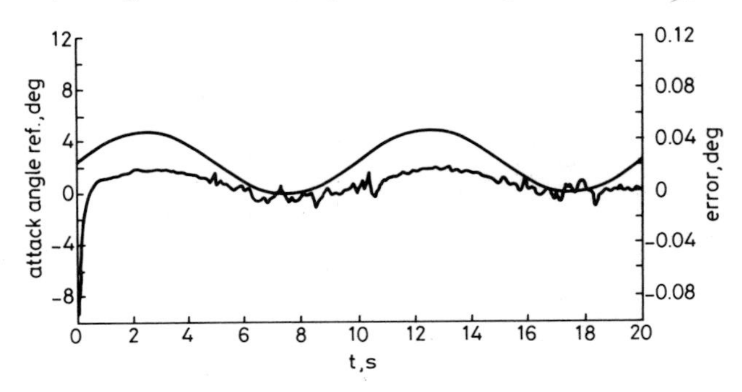

Fig. 9.10 *Attack angle reference and error*

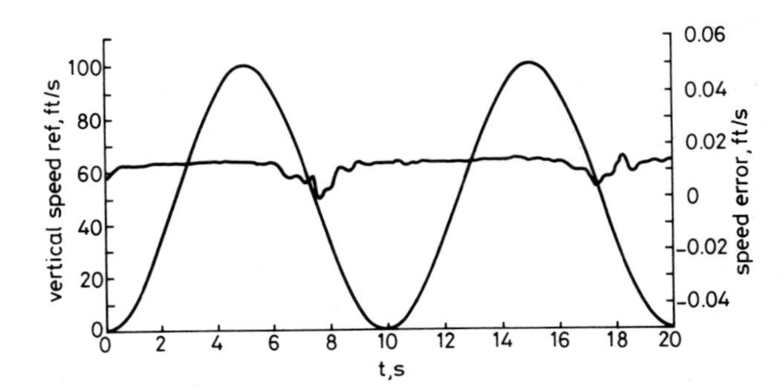

Fig. 9.11 *Vertical speed reference and error*

such as digital versus continuous implementation and by the physical constraints such as the dynamics and the saturation of the actuators and the flexural modes of the aircraft.

Some simulations have been successfully carried out for the aircraft F4–C in the descent and in the flare phases (Balestrino and Innocenti, 1988). In order to characterise the overall performance we also carried out some simulations with a sinusoidal vertical reference speed. Assuming the aircraft at a flight altitude of 30 000 ft with a speed of 950 ft/s at $t = 0$ the vertical speed to track varies from a minimum of zero to a maximum 100 ft/s, so that the aircraft is maneuvering inside the transonic region where the aereodynamical parameters undergo abrupt variations. The elevator angle and the thrust coefficient are bounded: $u_e \in (-20°, +20°)$, and $u_T \in (0, 1)$.

Referring to Figs. 9.10 and 9.11, taking into account that the vertical and horizontal speed errors are below 0·01 ft/s, we see that the frequency response from 0 to 0·1 Hz is very satisfactory. Increasing the frequency up to 0·2 Hz, the

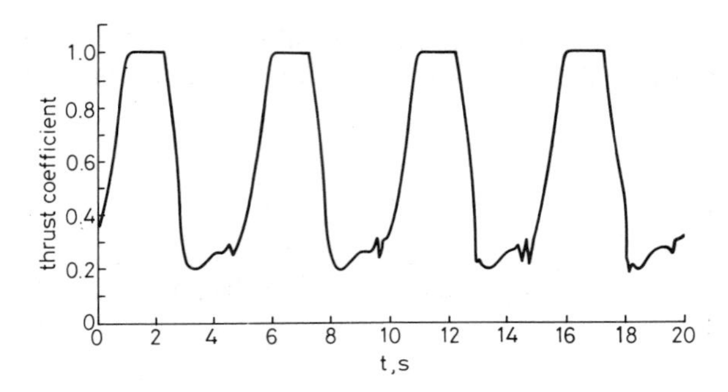

Fig. 9.12 *Thrust coefficient*

maximum vertical speed error rises to 0·2 ft/s, while the horizontal speed error after 2 s is 2·4 ft/s and is slightly diverging. In this case the thrust coefficient u_T periodically saturates; i.e. no sufficient thrust is available (see Fig. 9.12).

From these preliminary results it appears that the HVSC approach to robust control design is very promising also in flight control systems. Note that this approach allows us to consider the aircraft dynamics without making use of the linearised models, see Calise and Kramer (1984); therefore it is possible to design an efficient control for high performance aircraft or missiles for a wide range of parameter variations.

9.10 Conclusions

A systematic approach to the design of VSC systems has been given. Using the hyperstability theory we have established the stability of the global system and the existence of the sliding modes. Simple control laws can be used and high speed of response is obtainable by forcing the system with maximum allowable signals, e.g. by means of relay or unit-vector laws.

The sliding modes are typically chosen so that whenever the system attains such a regime it behaves like a linear time-invariant system with high degree of robustness and insensivity against parameter variations and environmental disturbances.

Some preliminary example have been reported in order to introduce the reader to the design technique, and the feasibility of the method has been proved both in robotics and in nonlinear flight control. We have deliberately assumed that all plant state variables were available for measurement, but have also indicated how it is possible to realise a nonlinear observer. The synthesis of VSC using nonlinear observers is outlined in Walcott and Żak (1988). Inaccurate modelling and output feedback pose a lot of problems, because in these cases the strong stability characteristics and the existence of the sliding modes are often destroyed.

A simple way to gain confidence with synthesis is an extensive and accurate simulation before implementing the control system. To this end it is very useful to realise effective CAD packages for variable structure control system design. A CAD package has been described in Dorling (1985) and Dorling and Zinober (1988).

9.11 References

ANDERSON, B. D. O., and VONGPANITLERD, S. (1973): 'Network analysis and synthesis' (Prentice Hall, Englewood Cliffs, NJ)

BALESTRINO, A., DE MARIA, G. and SCIAVICCO L. (1982): 'Adaptive hyperstable model following control of nonlinear plants', *Systems and Control Lett.*, **1**, pp. 232–236

BALESTRINO, A., DE MARIA, G. and SCIAVICCO, L. (1983*a*): 'Adaptive control design in servosystems'. 3rd IFAC Symp. on Contr. Power Elec. and Electron, Drives, Lausanne, pp. 125–131, (Pergamon Press)

BALESTRINO, A., DE MARIA, G. and SCIAVICCO, L. (1983*b*): 'An adaptive model following control for robotic manipulators', *J. Dyn. Syst. Measurement and Control*, **105**, pp. 143–151

BALESTRINO, A., DE MARIA, G. and SCIAVICCO, L. (1984): 'Robust control of robotic manipulators'. 6th IFAC World Congress, Budapest

BALESTRINO, A., DE MARIA, G. and SICILIANO, B. (1985): 'Hyperstable variable structure control for a class of uncertain systems'. 7th IFAC Symp. on Identification and Systems Parameter Estimation, York, Vol. II, pp. 1913–1919 (Pergamon Press)

BALESTRINO, A., GALLANTI, G. and KALAS, D. (1988): 'High performance DC drive using variable structure control theory', ICEM '88 Conference Proceedings, September 1988, Pisa

BALESTRINO, A. and INNOCENTI, M. (1988): 'Hyperstable variable structure control and aircraft flight control synthesis'. 12th IMACS World Congress on Scientific Computation, July 1988, Paris

BUHLER, H. (1979): 'Electronique de Réglage et de Cómmande' in 'Vol. XVI, Traité d'Electricité' (Editions Georgi, St. Saphorin, Lausanne)

BUHLER H. (1986): 'Réglage par mode de glissement' (Presses Polytechniques Ramandes, Lausanne)

CALISE, A. J. and KRAMER, F. S. (1984): 'A variable structure approach to robust control of VTOL aircraft', *J. Guidance*, **7**, pp. 620–625

CHEN, Y. H. (1987): 'On the Robustress of mismatched uncertain dynamical system' *J. Dyn. Syst. Measur. and Cont.*, **109**, pp. 29–35

CORLESS, M. J. and LEITMANN, G. (1981): 'Continous state feedback guaranteeing uniform ultimate boundedness for uncertain dynamical systems', *IEEE Trans*, **AC-26**, pp. 1139–1143

DORLING, C. M. (1985): 'The design of variable structure control systems'. Manual for the Vassyd CAD Package, Dept. of Applied and Computational Mathematics, University of Sheffield

DORLING, C. M. and ZINOBER, A. S. I. (1988): 'Two approaches to hyperplane design in multivariable variable structure control systems', *Int. J. Control*, **Vol. 48**, pp. 2043–54.

DUBOWSKY, S. and DESFORGES, D. T. (1979): 'The application of model reference adaptive control to robotic manipulators' *ASME J. Dynam. Syst., Meas. Contr.*, **101**, pp. 193–200.

EMELYANOV, S. V. (1985): 'Binary control systems,' Int. Res. Inst. for Management Sciences, Moscow

IOANNOU, P. A. and KOKOTOVIC, P. V. (1985): 'Adaptive systems with reduced models' (Springer Verlag, Berlin)

KALMAN, R. E. and BERTRAM, J. E. (1960): 'Control systems analysis and design via the 'second method' of Lyapunov: Continuous-time systems' *ASME J. Basic Eng.*, pp. 371–393

KYRIAZOV, P. and MARINOV, P. (1987): 'On the decoupled drive system design of industrial robots,' Proc. Control of Industrial Processes, BIAS '87, FAST, Milan, Italy, pp. 359–366

LANDAU, Y. D., (1979): 'Adaptive Control; The model reference approach' (Marcel Dekker, Basel)

LIM, K.Y and ESLAMI, M. (1987): 'Robust adaptive controller designs for robot manipulator systems' *IEEE J. Robotics and Automation*, **RA–3**, pp. 54–66

McRUER, D., ASHKENAS, I. and GRAHAM, D. (1973): 'Aircraft dynamics and automatic control' (Princeton University Pass, Princeton, N.J.)

POPOV, V. M. (1973): 'Hyperstability of control systems' (Springer-Verlag, Berlin)

PRALY, L. (1986): 'Approache entrée-sortie de la stabilité' *in* LANDAU, I. D. and DUGARD, L. (Eds.) 'Commande adaptative, aspects pratiques et Théoriques' (Masson, Paris) pp. 336–374

UTKIN, V. I. (1974): 'Sliding modes and their application in variable structure systems' (MIR, Moscow) (first English translation, 1978)

WALCOTT, B. L. and ŻAK, S. (1988): 'Combined observer-controller synthesis of uncertain dynamic systems with applications', *IEEE Trans.*, **SMC-18**, pp. 88–104

WILLEMS, J. C. (1972): 'Dissipative dynamical systems' Part I: 'General Theory'; Part II: 'Linear systems and quadratic supply rates' *Archiv for Rational Mechanics and Analysis*, **45**, pp. 321–393

YEDAVALLI, R. K. and LIANG, Z. (1987): 'Reduced conservatism in the ultimate boundedness control of mismatched uncertain linear systems' *J. Dyn. Syst. Meas. & Contr.*, **109**, pp. 1–6.

YOUNG, K. D., KOKOTOVIC, P. and UTKIN, V. (1977): 'A singular perturbation analysis of high-gain feedback systems,' *IEEE Trans.*, **AC-22**, pp. 931–938

Chapter 10

Nonlinear continuous feedback control for robust tracking

F. Garofalo and L. Glielmo

University of Naples

10.1 Introduction

One of the most interesting features of the VSC approach to the control of nominally linear systems is the insensitivity attained during the sliding mode to the uncertainties affecting the system both in the form of parameter variations and external disturbances (Utkin, 1978). In particular, provided that the uncertainties belong to a proper subspace (or, as it is often said, that the uncertainties are matched with the system), the motion on the sliding surface will depend on the surface itself, and virtually any desired (linear) behaviour can be obtained (Zinober, 1988). The control law is, by its very nature, discontinuous, and during sliding chatters at a (theoretically) infinite frequency. This may be suitable for some applications, but continuous nonlinear control, which yields motion remaining arbitrarily close to the sliding surface (see Ryan and Corless, 1984; and Burton and Zinober, 1986), is generally more desirable.

The description of the system behaviour in the sliding mode turns out to be very simple when using the 'equivalent control' technique (Utkin, 1978; Zinober, 1988). Another description of this behaviour can be found in Gutman (1979) where the theory of generalised dynamic systems has been used.

The remarkable simplification introduced by the equivalent control technique in the description of system dynamics in the sliding mode vanishes when dealing with control signals that are a continuous approximation of the discontinuous ones. A quantitative description of the behaviour of a system subject to any continuous approximation of a discontinuous control law is given in Leitmann (1981) and in Corless and Leitmann (1981; 1988). In particular, through the use of Lyapunov arguments, they evaluate the final precision that can be guaranteed under the worst realisation of the uncertainties affecting the system. This approach is often referred to as 'deterministic control of uncertain systems'. The objective achievable by the controller is the 'practical stability' (Corless and

Leitmann, 1990) of the closed loop rather than its asymptotic stability. It is interesting to note that a description of the effect of unmatched certainties is also possible (see Barmish and Leitmann, 1982).

Here, in the framework of the deterministic control of uncertain systems, we analyse the properties that an approximation of a discontinuous control must possess in order to preserve, to some extent, interesting features of a variable structure control. In particular we state the conditions under which a control law guarantees the attractivity of a certain region of the state space. Successively we analyse the conditions assuring that the motions of the system in this region lead to ultimate boundedness of the trajectories. We also prove that these conditions guarantee an arbitrary, exponential rate of convergence of the norm of the state vector (Ambrosino, Celentano and Garofalo, 1985*a*). After that we state a family of continuous laws (approximations of discontinuous ones) which satisfy these conditions. The analysis is carried out for a nominally linear system for which the uncertainties are structurally matched. Finally we exploit the possibility of using the degrees of freedom in the choice of the controller parameters in order to obtain simplified design procedures or simplified structures of the controllers (Ambrosino, Celentano and Garofalo, 1985*b*; Am}-brosino, Celentano and Garofalo, 1986).

10.2 Statement of the problem

We consider a multi-input multi-output system which can be described by the following differential equation

$$y^{(\nu)} = f(y, \ldots, y^{(\nu-1)}, q) + F(y, \ldots, y^{(\nu-1)}, q)u \qquad (10.1a)$$

$$y(t_0) = y_0, \ldots, y^{(\nu-1)}(t_0) = y_0^{(\nu-1)} \qquad (10.1b)$$

where $y \in R^m$ is the output, $u \in R^m$ is the input, and $q \in Q \subset R^\mu$ is the vector of the uncertainties affecting the model. A number of assumptions are made on the system (10.1).

Assumption 1: The set Q is compact and the function $q(\cdot)$: $R \to Q$ is Lebesgue measurable.

Assumption 2: Function $f(y, \ldots, y^{(\nu-1)}, q)$ is continuous in all its arguments and cone bounded with respect to $y, \ldots, y^{(\nu-1)}$; i,e, there exist constants k_i, $i = 1, \ldots, \nu$, such that

$$\| f(y, \ldots, y^{(\nu-1)}, q)\| \leqslant k_0 + k_1 \|y\| + \ldots + k_\nu \|y^{(\nu-1)}\|$$

Assumption 3: There exists a known matrix $W(y, \ldots, y^{(\nu-1)}) \in R^{mxm}$ and a scalar $\bar{\lambda}_D$ such that, for any $q \in Q$ and $(y, \ldots, y^{(\nu-1)}) \in R^{m\nu}$, the matrix

$$D(y, \ldots, y^{(\nu-1)}, q) \triangleq F(y, \ldots, y^{(\nu-1)}, q)\, W(y, \ldots, y^{(\nu-1)})$$

is positive/definite (PD in the sequel) and norm bounded, i.e.

$$\|D(y, \ldots, y^{(\nu-1)}, q)\| \leqslant \bar{\lambda}_D$$

As a consequence of assumptions 1 and 2, it is possible to find a function $g(\cdot):R^{mv} \rightarrow R^{+}$, cone bounded with respect to its arguments, satisfying

$$g(y, \ldots, y^{(v-1)}) \geqslant \max_{q \in Q} \| f(y, \ldots, y^{(v-1)}, q) \| \tag{10.2}$$

Assumption 3 enables us to look for a positive scalar λ_D satisfying

$$\lambda_D \leqslant \min \lambda_i \left(\frac{D + D^T}{2} \right) \tag{10.3}$$

where the minimum is taken for $q \in Q$, for $(y, \ldots, y^{(v-1)}) \in R^{mv}$, and $i = 1, \ldots, m$.

In (10.3) $\lambda_i\ (M)$ represents the ith eigenvalue of matrix M; in the following $\lambda_{max(min)}\ (M)$ will represent the operation of taking the maximum (minimum) eigenvalue of M. The arguments of a function will be omitted where no confusion is likely to occur.

Suppose that the reference trajectory $\hat{y}(\cdot):R \rightarrow R^m$ is given, satisfying the following

Assumption 4: The function $\hat{y}(\cdot)$ is continuous up to its vth derivative.

The objective of the control is to make the tracking error

$$\varepsilon \triangleq \hat{y} - y \tag{10.4}$$

ultimately bounded (Corless and Leitmann, 1990) in a ball around $\varepsilon = 0$, whose radius can be made arbitrarily small by a suitable choice of the parameters of the controller. Furthermore the controller should be able to ensure that the convergence of the error norm be arbitrarily close to an exponential convergence with desired time constant. The accomplishment of these objectives should be guaranteed for any realisation of the uncertain parameters $q(\cdot):R \rightarrow Q$ satisfying assumption 1.

Remark 1: Models of type (10.1) occur frequently when dealing with the control of complex mechanical structures like a robot manipulator (see, for instance, Ambrosino, Celentano and Garofalo, 1986).

Remark 2: The so-called 'matching assumption', regarding the way the uncertainties act on the system (Corless and Leitmann, 1988) are automatically satisfied in the case of equation (10.1), since matrix F is square and full rank in view of assumption 3.

10.3 Stability analysis

In this Section we state the dynamic equation of the tracking error and derive some stability results which will be used later for the synthesis of the controller.

Introducing the vector

$$w \triangleq f + Fu - \hat{y}^{(v)} - K_v \varepsilon - K_{v-1} \varepsilon^{(1)} - \ldots - K_1 \varepsilon^{(v-1)} \tag{10.5}$$

where K_i, $i = 1, \ldots, v$, are arbitrary $m \times m$ matrices, from (10.1) and (10.4) the error dynamic equation can be written as

$$\varepsilon^{(v)} + K_1 \varepsilon^{(v-1)} + \ldots + K_v \varepsilon = -w \tag{10.6a}$$

$$\varepsilon(t_0) = \varepsilon_0 \triangleq \hat{y}(t_0) - y(t_0)$$

$$\ldots$$

$$\varepsilon^{(v-1)}(t_0) = \varepsilon_0^{(v-1)} \triangleq \hat{y}^{(v-1)}(t_0) - y^{(v-1)}(t_0) \tag{10.6b}$$

Defining

$$\varepsilon^T \triangleq [\varepsilon^T \varepsilon^{(1)T} \ldots \varepsilon^{(v-1)T}], \tag{10.7}$$

a state space representation of eq. (10.6) can be given in the form

$$\dot{e} = Ee - B_1 w, \quad e(t_0) = e_0 \triangleq (\varepsilon_0 T \, \varepsilon_0^{(1)} T \ldots \varepsilon_0^{(v-1)} T)^T \tag{10.8}$$

where

$$E \triangleq \begin{bmatrix} 0 & I_m & \ldots\ldots & 0 & 0 \\ 0 & 0 & I_m \ldots & 0 & 0 \\ \vdots & \vdots & \vdots & \ddots & \vdots \\ 0 & \ldots & \ldots & 0 & I_m \\ -K_v & \ldots & \ldots & & -K_1 \end{bmatrix}, \quad B_1 \triangleq \begin{bmatrix} 0 \\ 0 \\ \vdots \\ I_m \end{bmatrix} \tag{10.9}$$

Suppose now that matrices K_i, $i = 1, \ldots, v$, are chosen so as to make matrix E in (10.8) asymptotically stable and let P be the solution of the Lyapunov equation

$$E^T P + PE = -Q, \qquad Q \text{ PD} \tag{10.10}$$

The following lemma can be stated:
Lemma 1: Choosing a real constant $\varrho_v > 0$, if signal w in 10.8 is such that

$$v^T [w - (B_1^T PB_1)^{-1} B_1^T PEe] \geqslant l(\|v\|) > 0 \tag{10.11}$$

$$\text{for any } e\text{: } \|v\| > \varrho_v$$

where

$$v \triangleq B_1^T Pe \tag{10.12}$$

and function $l(\cdot)$: $(\varrho_v, \infty) \to (0, \infty)$ is continuous, then the error $e(\cdot)$ is globally uniformly bounded and globally ultimately bounded in any ball of radius $\bar{\varrho}_e > \varrho_e$, $\underline{\varrho_e} = \varrho_e(\varrho_v)$.

Proof: First of all we give a different representation of system (10.8) in order to separately show the dynamics which take place in $\mathcal{N}(B_1^T P)$ and in a suitable complementary subspace. To this end we consider a matrix B_2 such that rank $(B_1| B_2) = mv$ and $B_1^T PB_2 = 0$, and we make the following change of basis:

$$\begin{bmatrix} z_1 \\ z_2 \end{bmatrix} = [B_1 \; B_2]^{-1} e \tag{10.13}$$

In the new co-ordinate system eqn. (10.8) takes the form

$$\dot{z}_1 = (B_1^T P B_1)^{-1} B_1^T PEB_1 z_1 + (B_1^T P B_1)^{-1} B_1^T PEB_2 z_2 - w \tag{10.14a}$$

$$\dot{z}_2 = (B_2^T P B_2)^{-1} B_2^T PEB_1 z_1 + (B_2^T P B_2)^{-1} B_2^T PEB_2 z_2 \tag{10.14b}$$

$$z_1(t_0) = z_{10}, \quad z_2(t_0) = z_{20} \tag{10.14c}$$

where z_{10} and z_{20} are obtained from e_0 through eqn. (10.13).

The analysis of the behaviour of the trajectories of system (10.14) is carried out in two steps. Firstly, through the use of equation (10.14*a*), we prove that, if signal w satisfies condition (10.11), then $z_1(\cdot)$ is ultimately bounded. Afterwards we prove, using equation (10.14*b*), that variable $z_2(\cdot)$ remains in turn ultimately bounded under the effect of the bounded input $z_1(\cdot)$.

For the first step of the proof we select the following Lyapunov candidate

$$V_1(z_1) \triangleq z_1^T (B_1^T P B_1) z_1 \tag{10.15}$$

Along the trajectories of system (10.14) we have

$$\dot{V}_1(z_1, t) = 2 z_1^T [\, B_1^T PE (B_1 z_1 + B_2 z_2) - (B_1^T P B_1) w \,] \tag{10.16}$$

Recalling (10.12) and (10.13), eqn. (10.16) can be rewritten as

$$\dot{V}(z_1, t) = 2 v^T [(B_1^T P B_1)^{-1} B_1^T PE_{a\,-\,w}] \tag{10.17}$$

which, by virtue of the hypothesis, is upper bounded by a negative function of $\|v\|$ for all z_1 such that $\|v\| > \varrho_v$.

Thus the trajectories $z_1(\cdot)$ are attracted toward the smallest ellipsoidal region of the family $Z(k) \triangleq \{\zeta_1 \in R^m : V_1(\zeta_1) \leqslant k\}$ containing $\{\zeta_1 \in R^m : \|B_1^T P B_1 \zeta_1\| \leqslant \varrho_v\}$.

The parameter k_ϱ which specifies that region can be found as

$$k_\varrho = \max_{\zeta 1} V(\zeta_1), \tag{10.18a}$$

s.t.

$$\|(B_1^T P B_1)\zeta_1\| = \varrho_v \tag{10.18b}$$

Straightforward application of the Lagrange multipler technique leads to

$$[(B_1^T P B_1) - \theta (B_1^T P B_1)^2]\zeta_1{}^* = 0 \tag{10.19}$$

where θ is the Lagrange multiplier. Since, by virtue of (10.18*b*), $\zeta_1{}^* \neq 0$, it is necessary that

$$\det [\theta I_m - (B_1^T P B_1)^{-1}] = 0 \tag{10.20}$$

so that

$$\theta^* = \lambda_i [(B_1^T P B_1)^{-1}], \quad i = 1, \ldots, m \tag{10.21}$$

are the possible solutions for θ.

Again, left-multiplying (10.19) by $\zeta_1{*}^T$, we obtain

$$V_1(\zeta_1{*}) = \theta^* \varrho_v^2 \tag{10.22}$$

and, since we are looking for the maximum of V_1,

$$k_\varrho = \lambda_{max}\,[(B_1^T P B_1)^{-1}]\varrho_v^2 \tag{10.23}$$

The smallest ball containing this ellipsoid has radius ϱ_1 given by

$$\varrho_1 = \left[\frac{k_\varrho}{\lambda_{min}\,(B_1^T P B_1)}\right]^{1/2} = \frac{\varrho_v}{\lambda_{min}\,(B_1^T P B_1)} \tag{10.24}$$

Now, with any $\bar{\varrho} > \varrho_1$, using the results contained in Corless and Leitmann (1988) (see also Leitmann, 1981; and Garofalo and Leitmann, 1989) we can conclude that, for any real $R_1 > 0$ and z_{10}: $\|z_{10}\| \leqslant R_1$

(i) The trajectories $z_1(\cdot)$ are globally uniformly bounded

(ii) They are globally ultimately bounded; i.e. there exists a finite time $t_1 \triangleq t_0 + T(R_1)$ at which they enter the ball of radius $\bar{\varrho}_1$ remaining in it thereafter.

As a second step of the proof we analyse the behaviour of variable z_2. We consider equation (10.14) as describing a linear time invariant system subject to a disturbance $z_1(\cdot)$ which has been proved to be uniformly bounded; its trajectories will be bounded as well if the dunamic matrix

$$F \triangleq (B_2^T P B_2)^{-1} B_2^T P E B_2 \tag{10.25}$$

is asymptotically stable (see, for instance, Desoer and Vidvasagar. 1975). Indeed, choosing $S = B_2^T P B_2$, from the Lyapunov equation (10.10) we have

$$SF + F^T S = -L \tag{10.26}$$

with $L \triangleq B_2^T Q B_2$ a PD matrix.

This proves uniform boundedness of the trajectories $z_2(\cdot)$. In order to prove their ultimate boundedness we note that, since matrix F in (10.25) is asymptotically stable, positive real constants c and d exist (Kalman and Bertram, 1960) such that

$$\|e^{Ft}\| \leqslant ce^{-dt} \tag{10.27}$$

Defining

$$G \triangleq (B_2^T P B_2)^{-1} B_2^T P E B_1 \tag{10.28}$$

from (10.14*b*), (10.25) and (10.27), we have

$$\begin{aligned} \|z_2(t)\| &= \left\| e^{F(t-t_1)}\left[e^{F(t_1-t_0)} z_2(t_0) + \int_{t_0}^{t_1} e^{F(t_1-\tau)} Gz_1(\tau)\, d\tau \right] \right. \\ &\quad \left. + \int_{t_1}^{t} e^{F(t-\tau)} Gz_1(\tau)\, d\tau \right\| \\ &\leqslant ce^{-d(t-t_1)} \|z_2(t_1)\| + \|G\|\bar{\varrho}_1 (c/d\rho)(1 - e^{-d(t-t_1)}) \end{aligned} \tag{10.29}$$

From expression (10.29) global ultimate boundedness of $z_2(\cdot)$ follows, for any $\bar{\varrho}_2 > \underline{\varrho}_2$ with

$$\underline{\varrho}_2 \triangleq (c/d)\|G\|\underline{\varrho}_1 \tag{10.30}$$

The proof of the lemma is concluded by noting, from (10.13), that

$$\|e\| \leqslant \|B_1\| \; \|z_1\| + \|B_2\| \; \|z_2\| \tag{10.31}$$

and hence vector $e(\cdot)$ is ultimately bounded in any ball of radius $\bar{\varrho}_e > \underline{\varrho}_e$ with

$$\begin{aligned} \underline{\varrho}_e &\triangleq \|B_1\|\underline{\varrho}_1 + \|B_2\|\underline{\varrho}_2 \\ &= \frac{\|B_1\| + (c/d)\|G\| \; \|B_2\|}{\lambda_{min}\,(B_1^T P B_1)^{-1}} \varrho_v \end{aligned} \tag{10.32}$$

Lemma 2: If in addition to the hypotheses of Lamma 1, we have:

$$v^T w \geqslant 0 \qquad \text{for any } e\text{: } \|v\| > \varrho_v \tag{10.33}$$

then the norm of the trajectories of system (10.8), while converging toward any ball of ultimate boundedness, are upper bounded by a time function which is arbitrarily close to an exponential one with a time constant τ given by

$$\tau \triangleq \frac{2}{\lambda_{min}\,(QP^{-1})} \tag{10.34}$$

Proof: Choose

$$V(e) \triangleq e^T P e \tag{10.35}$$

Along the trajectories of (10.8) we have

$$\begin{aligned} \dot{V}(e,t) &= -e^T Q e - 2e^T P B_1 w \\ &= -e^T Q e - 2v^T w \end{aligned} \tag{10.36}$$

so that, from (10.33),

$$-\frac{\dot{V}}{V} \geqslant \frac{e^T Q e}{e^T P e} \geqslant \lambda_{min}\,(QP^{-1}) = \frac{2}{\tau} \tag{10.37}$$

and then

$$V(e) \leqslant V(e_0)\, e^{-2(t-t_0)/\tau} \tag{10.38}$$

from which

$$\|e(t)\| \leqslant \left[\frac{\lambda_{max}(P)}{\lambda_{min}(P)}\right]^{1/2} \|e_0\| e^{-(t-t_0)/\tau}$$

$$\text{for any } t\text{: } t \geqslant t_0 \text{ and } \|v(t)\| > \varrho_v \tag{10.39}$$

Since $\{\zeta_1 \in R^m : \|B_1^T P B \zeta_1\| > \varrho_v\} \supset S_1 \triangleq \{\zeta_1 \in R^m : \|\zeta_1\| > \underline{\varrho}_1\}$, then (10.39) holds in the region S_1 and so any ball of ultimate boundedness of radius $\bar{\varrho}_1 > \underline{\varrho}_1$ is approached with that rate in the z_1 space.

On the other hand, the evolution of variable z_2, described by (10.14), can always be considered as sum of free and forced response, say z_{2l} and z_{2f} respectively. Since the former is governed by

$$z_{2l} = Fz_{2l}, \tag{10.40}$$

choosing

$$V_2(z_{2l}) = z_{2l}^T (B_2^T PB_2)z_{2l} \tag{10.41}$$

we easily obtain that, along the free response trajectories,

$$-\frac{\dot{V}_2}{V_2} = \frac{z_{2l}^T (B_2^T QB_2)z_{2l}}{z_{2l}^T (B_2^T PB_2)z_2} \geqslant \lambda_{min} (QP^{-1}) = \frac{2}{\tau} \tag{10.42}$$

and hence

$$\|z_{2l}(t)\| \leqslant \left[\frac{\lambda_{max}(P)}{\lambda_{min}(P)}\right]^{1/2} \|z_2(t_1)\|e^{-(t-t_1)/\tau} \quad t \geqslant t_1$$

From (10.29) it follows that, for $t \geqslant t_1,$ $\|z_{2f}(t)\| \leqslant \bar{\varrho}_1\alpha(t)$, where

$$\alpha(t) \triangleq \|G\| \text{ (c/d) } [1 - e^{-d(t-t_1)}] \tag{10.43}$$

so that

$$\|z_2(t)\| \leqslant \left[\frac{\lambda_{max}(P)}{\lambda_{min}(P)}\right]^{1/2} \|z_2(t_1)\|e^{-(t-t_1)/\tau} + \bar{\tau}_1\alpha(t) \tag{10.44}$$

In conclusion, for $t \geqslant t_1$ we have

$$\begin{aligned} \|e(t)\| &= \|B_1 z_1 + B_2 z_2\| \\ &\leqslant \|B_1\|\bar{\varrho}_1 + \|B_2\|\left\{\left[\frac{\lambda_{max}(P)}{\lambda_{min}(P)}\right]^{1/2} \|z_2(t_1)\|e^{-(t-t_1)/\tau} + \bar{\varrho}_1\alpha(t)\right\} \\ &= \|B_2\|\left[\frac{\lambda_{max}(P)}{\lambda_{min}(P)}\right]^{1/2} \|z_2(t_1)\|e^{-(t-t_1)/\tau} + \bar{\varrho}_1 [\|B_1\| + \|B_2\|\alpha(t)] \end{aligned} \tag{10.45}$$

so that, by a suitable choice of $\bar{\varrho}_1$ (that is, ϱ_v), it is possible to arbitrarily approach the exponential decreasing function in the right-hand side of (10.45).

Remark 3: It is interesting to note that in the z_1 space, owing to the structure of the Lyapunov function (10.15) and the definition of v in (10.12), the ball of radius ϱ_1, which is the smallest ball containing the attractive Lyapunov ellipsoid, is also the smallest ball containing the ellipsoid $\{z_1: \|B_1^T PB_1 z_1\| = \varrho_v\}$, so that no loss of 'final precision' of the tracking error occurs in passing from the region in which hypotheses (10.1) applies to that one where the attractive behaviour can be observed.

Remark 4: A different evaluation of the region of ultimate boundedness in the z_2 space can be obtained using the following result given in Desoer and Vidyasagar (1975) and Vidyasagar (1980), based on the notion of measure of a matrix.

Given an induced matrix norm $\|\cdot\|$, the corresponding 'measure' of a square matrix $A \in R^{n\times n}$ is defined as

$$\mu(A) \triangleq \lim_{\zeta\to 0} \frac{\|I_n + \zeta A\| - 1}{\zeta} \tag{10.46}$$

The value of the measure strongly depends on the vector norm considered; moreover in general the computation of (10.46) is not an easy task, but for some canonical norms the following correspondences turn out:

$$\|x\|_1 = \sum_{i=1}^{n} |x_i| \Rightarrow \mu_1(A) = \max_j \left[a_{jj} + \sum_{i\neq j} |a_{ij}| \right] \tag{10.47a}$$

$$\|x\|_2 = \left[\sum_{i=1}^{n} |x_i|^2 \right]^{1/2} \Rightarrow \mu_2(A) = \lambda_{max} \left[\frac{A + A^T}{2} \right] \tag{10.47b}$$

$$\|x\|_\infty = \max_i |x_i| \Rightarrow \mu_\infty(A) = \max_i \left[a_{ii} + \sum_{j\neq i} |a_{ij}| \right] \tag{10.47c}$$

The measure has several interesting properties; here in particular we recall that (see Desoer and Vidyasagar, 1980)

$$-\mu(-A) \leqslant Re\,[\lambda_i(A)] \leqslant \mu(A) \quad i = 1, \ldots, n \tag{10.48}$$

Considering equation (10.14) rewritten as (see also (10.25) and (10.28))

$$\dot{z}_2 = Fz_2 + Gz_1 \tag{10.49}$$

the following inequality applies for any $t \geqslant t_1$:

$$\|z_2(t)\| \leqslant e^{\mu(F)(t-t_1)} \|z_2(t_1)\| + \int_{t_1}^{t} e^{\mu(F)(t-\tau)} \|Gz_1(\tau)\| d\tau$$

and then, considering the upper bound on $\|z_1\|$,

$$\|z_2(t)\| \leqslant \begin{cases} e^{\mu(F)(t-t_1)} \|z_2(t_1)\| - \dfrac{\|G\|\bar{\varrho}_1}{\mu(F)} [1 - e^{\mu(F)(t-t_1)}] \\ \qquad\qquad\qquad\qquad (\text{if } \mu(F) \neq 0) \\ \|z_2(t)\| + \|G\|\bar{\varrho}_1(t - t_1). \quad (\text{if } \mu(F) = 0) \end{cases} \tag{10.50}$$

When

$$\mu(F) < 0 \tag{10.51}$$

we can use (10.50*a*) to obtain a different expression for $\underline{\varrho}_2$; that is

$$\underline{\varrho}_2 = \frac{\|G\|}{|\mu(F)|} \underline{\varrho}_1 \tag{10.52}$$

We should emphasise that:

(1) Inequality (10.51) is a sufficient condition for asymptotic stability of matrix F, but not a necessary one (see (10.48));
(ii) Since inequality (10.50) applies for any matrix measure (that is, induced by any vector norm), one can use the minimum of those given in (10.47) for the best result.

Remark 5: The second part of the proof of lemma 1 closely resembles a result obtained by Utkin (1971) while treating the problem of discontinuities in differential equations; this seems worthwhile to briefly recall here.

Consider the system

$$\dot{x} = \phi(x,t) + \Psi(x,t)u \qquad x(t_0) = x_0 \tag{10.53}$$

with $x \in R^n$, $u \in R^P$, and the surface

$$\Gamma \triangleq \{x : s(x) = 0\} \tag{10.54}$$

where $s(x)$ is a sufficiently smooth p-dimensional vector values function. Let $J(x)$ be the Jacobian of s with respect to x; that is

$$J(x) \triangleq \frac{\partial s}{\partial x} \tag{10.55}$$

The control function $u^*(x,t)$ which ensures that a trajectory starting on Γ remains on it, it defined to be the 'equivalent control' and its expression can easily be found by imposing $\dot{s} = 0$, thus obtaining

$$u^*(x, t) = -(J\Psi)^{-1}J\phi \tag{10.56}$$

as long as $J\Psi$ is nonsingular. Let $x^*(\cdot)$ be the trajectory corresponding to the equivalent control. Choosing a real positive number Δ, consider the region defined as

$$S_\Delta \triangleq \{x : \|s(x)\| < \Delta\} \tag{10.57}$$

and suppose that

(i) A control $\tilde{u}$ and time instants $\tilde{t}$ and $\tilde{T}$ exist, such that

$$x(t) \in S_\Delta, \qquad \text{for any } t \in [\tilde{t}, \tilde{T}] \tag{10.58}$$

(ii) A trajectory $x^*(\cdot)$ and a positive real number L exist such that

$$\|x(t) - x^*(t)\| \leqslant L\Delta \tag{10.59}$$

then, under some further technical structural conditions (see Utkin, 1971), a real $H > 0$ exists such that

$$\|x(t) - x^*(t)\| \leqslant H\Delta \qquad \text{for any } t \in [\tilde{t}, \tilde{T}] \tag{10.60}$$

In other words, if a trajectory is 'trapped' for some finite time inside the region

S_Δ, it remains close to a suitable equivalent trajectory and, if Δ tends to zero, $x(\cdot)$ approaches arbitrarily close to $x^*(\cdot)$.

Apparently in our framework the linear equation (10.8) corresponds to equation (10.53) and the surface Γ is the hyperplane corresponding to the null space of $B_1^T P$. The trajectory corresponding to the equivalent control, obtained by (10.13) and (10.14) with $z_1 \equiv 0$, is asymptotically stable and, thanks to the linearities, Utkin's result can be extended on an infinite time interval.

Remark 6: When lemmas 1 and 2 apply, an upper bound on the norm of e can be computed in the following way. First of all we observe that, from (10.39) we have

$$\|e(t)\| \leqslant \left[\frac{\lambda_{max}(P)}{\lambda_{min}(P)}\right]^{1/2} \|e_0\|, \quad \text{for any } t \in (t_0, t_1) \tag{10.61}$$

while inequality (10.45) leads to

$$\|e(t)\| \leqslant \|B_2\| \left[\frac{\lambda_{max}(P)}{\lambda_{min}(P)}\right]^{1/2} \|z_2(t_1)\| + \bar{\varrho}_1 [\|B_1\| + \|B_2\| \|G\| (c/d)] \quad \text{for any } t \geqslant t_1 \tag{10.62}$$

In turn, from (10.61),

$$\|B_1 z_1(t_1) + B_2 z_2(t_1)\| \leqslant \left[\frac{\lambda_{max}(P)}{\lambda_{min}(P)}\right]^{1/2} \|e_0\| \tag{10.63}$$

and from the definition of t_1

$$\|z_1(t_1)\| \leqslant \bar{\varrho}_1, \tag{10.64}$$

using the property

$$|\|a\| - \|b\|| \leqslant \|a + b\|, \quad \text{for any } a, b \in R^\Omega \tag{10.65}$$

one obtains

$$\|B_2 z_2(t_1)\| \leqslant \|B_2\| \|z_2(t_1)\| \leqslant \left[\frac{\lambda_{max}(P)}{\lambda_{min}(P)}\right]^{1/2} \|e_0\| + \|B_1 z_1(t_1)\|$$

$$\leqslant \left[\frac{\lambda_{max}(P)}{\lambda_{min}(P)}\right]^{1/2} \|e_0\| + \|B_1\| \bar{\varrho}_1 \tag{10.66}$$

Inequality (10.62) thus becomes

$$\|e(t)\| \leqslant \frac{\lambda_{max}(P)}{\lambda_{min}(P)} R_e + \bar{\varrho}_1 \left[\|B_1\| \left\{1 + \left[\frac{\lambda_{max}(P)}{\lambda_{min}(P)}\right]^{1/2}\right\} + \|B_2\| \|G\| (c/d)\right] \tag{10.67}$$

assuming that a bound R_e on $\|e_0\|$ is known.

10.4 Design of the controller

In this Section we show that feedback control functions u exist such that the hypotheses of lemmas 1 and 2 are satisfied. The main result is stated in the following:

Theorem 1. Consider the tracking system described by eqs (10.8), (10.9), and (10.5), and assume that matrices K_i, $i = 1, \ldots, \nu$, are chosen so that matrix E is asymptotically stable. Define the following function:

$$\gamma(y, \ldots, y^{(\nu-1)}, e, \hat{y}^{(\nu)}) \triangleq g(y, \ldots, y^{(\nu-1)}) + \max[\|K\|; \|K + (B_1^T P B_1)^{-1} B_1^T P E\|]\|e\| + \|\hat{y}^{(\nu)}\| \tag{10.68}$$

where

$$K = [K_\nu \ldots K_1] \tag{10.69}$$

and the real constants

$$\delta > 0 \tag{10.70}$$

$$h > \frac{1 + \delta/\varrho_\nu}{\lambda_D} \tag{10.71}$$

$$0 > \gamma \leqslant \gamma(y, \ldots, y^{(m-1)}, e, \hat{y}^{(m)}), \text{ for any } y, \ldots, y^{(\nu-1)}, \hat{y}^{(\nu)} \text{ for any } e: \|v\| > \varrho_\nu \tag{10.72}$$

Then the continuous feedback function

$$u(t) = p(e(t), t) \tag{10.73}$$

$$p(e, t) = h\gamma W \frac{v}{\|v\| + \delta'} \tag{10.73b}$$

determines a signal w in (10.5) which satisfies conditions (10.11) and (10.33) of lemmas 1 and 2.

Proof: Using (10.5), (10.7), (10.69), and (10.73) and assumption 3, we have

$$\begin{aligned} v^T w &= v^T \left[Fh\gamma W \frac{v}{\|v\| + \delta} + f - Ke - \hat{y}^{(\nu)} \right] \\ &= h\gamma \frac{v^T D v}{\|v\| + \delta} + v^T [f - Ke - \hat{y}^{(\nu)}] \end{aligned} \tag{10.74}$$

Since the following inequalities apply:

$$v^T D v \geqslant \lambda_D \|v\|^2 \qquad \text{(from (10.3))} \tag{10.75a}$$

$$-v^T f \leqslant \|v\| g \qquad \text{(from (10.2))} \tag{10.75b}$$

$$v^T K e \leqslant \max[\|K\|; \|K + (B_1^T P B_1)^{-1} B_1^T P E\|]\|e\| \; \|v\|; \tag{10.75c}$$

$$v^T \hat{y}^{(\nu)} \leqslant \|v\| \; \|\hat{y}^{(\nu)}\| \tag{10.75d}$$

we obtain

$$v^T w \geqslant h\gamma\lambda_D \frac{\|v\|^2}{\|v\| + \delta} - g\|v\| - \max\,[\|K\|;\ \|K + (B_1^T PB_1)^{-1} B_1^T PE\|]$$

$$\|e\|\ \|v\| - \|\hat{y}^{(v)}\|\ \|v\| \geqslant$$

$$\geqslant \gamma\,(1 + \delta/\varrho_v) \frac{\|v\|^2}{\|v\| + \delta} - \gamma\|v\|$$

$$= \frac{\gamma\|v\|\delta}{\varrho_v\,(\|v\| + \delta)}\,[\|v\| - \varrho_v]$$

$$= l\,(\|v\|) > 0, \text{ for any } e\text{: } \|v\| > \varrho_v \qquad (10.76)$$

which satisfies (10.33). Similarly (10.11) can be proved.

The preceding proof, and in particular eqn. (10.74), suggest the following: *Corollary 1:* Another class of feedback control functions satisfying the conditions for lemmas 1 and 2 is given by

$$p'(e,t) = h\gamma_s W v \qquad (10.77)$$

where γ_s can be any real function such that:

$$\gamma_s \geqslant \frac{\gamma}{\|v\| + \delta}. \qquad (10.78)$$

Remark 7: The following functions are included in the class defined by corollary 1:

$$p_1 = \frac{h\gamma W v}{\max |v_i| + \delta}; \qquad (10.79)$$

$$P_2 = \begin{cases} \dfrac{h\gamma W v}{\delta} & \text{for } \|v\| < \delta \\ & \qquad \text{(saturation control)} \\ \dfrac{h\gamma W v}{\|v\|} & \text{for } \|v\| > \delta \end{cases} \qquad (10.80)$$

$$p_3' = \frac{h\gamma W v}{\delta} \qquad \text{(proportional control)} \qquad (10.81)$$

Remark 8: In remark 6 we have seen that, if an upper bound on $\|e_0\|$ is known, and the system is stabilised via lemmas 1 and 2, it is possible to determine a compact set, say S_a, to which the error $e(t)$ belongs for any t. If also $\hat{y}(t), \ldots \hat{y}^{(v)}(t)$ range in compact sets $S_y, \ldots S_{y(v)}$, a control complying with conditions (10.11) and (10.33) can be given as

$$p''(e,t) = h\bar{\gamma} W \frac{v}{\|v\| + \delta} \qquad (10.82)$$

where the constant gain $\bar{\gamma}$ substitutes for the function $\gamma(y, \ldots y^{(\nu-1)}, e, \hat{y}^{(\nu)})$ in (10.68) and must satisfy the inequality

$$\bar{\gamma} \geqslant \max \left[\| f - Ke - \hat{y}^{(\nu)} \|; \; \| f - (K + (B_1^T P B_1)^{-1} B_1^T P E) e - \hat{y}^{(\nu)} \| \right]$$

$$\text{s.t.} \quad q \in Q$$

$$e \in S_e$$

$$\hat{y} \in S_y, \ldots \hat{y}^{(\nu)} \in S_{y(\nu)} \tag{10.83}$$

Remark 9: Note that the proposed feedback controller in (10.73) is continuous and differentiable. Moreover, defining the vector $Y \triangleq [y^T, \ldots y^{(\nu-1)T}]^T$, system (101.$a$) subject to the feedback controller can be rewritten as

$$Y = H(Y, q, t) \tag{10.84}$$

Assumptions 2 and 3 ensure that vector function H is cone bounded with respect to Y, and this enables us to apply a 'global existence theorem' (see Driver, 1977) to prove the continuability of the solution of (10.84) over an infinite time interval.

10.5 Design consideration

In this Section we show how the degrees of freedom so far left in the choice of some parameters can be exploited in order to easily get a desired convergence rate in addition to a decoupled structure of the controller. The first useful result is stated in the following:

Theorem 2: Assume that the eigenvalues $\lambda_1, \ldots \lambda_{m\nu}$, of the asymptotically stable matrix E are real and distinct, and let Z be a nonsingular transformation matrix such that

$$Z^{-1} E Z = \Lambda \triangleq \operatorname{diag} [\lambda_1, \ldots \lambda_{m\nu}] \tag{10.85}$$

Then, choosing

$$Q = -2 (Z^T)^{-1} \Lambda Z^{-1} \tag{10.86}$$

the solution P of the Lyapunov equation (10.10) and the time constant τ in (10.34) are given by

$$P = (Z Z^T)^{-1} \tag{10.87}$$

$$\tau = \frac{2}{\lambda_{\min}(Q P^{-1})} = \frac{1}{\lambda_{\max}(E)} \tag{10.88}$$

Proof: This involves only simple algebraic manipulation and is therefore omitted.

Theorem 2 enables the designer to directly select the rate of convergence by suitably placing the eigenvalues of matrix E. In this regard, the particular structure of that matrix leads to another interesting result:

Theorem 3: If matrices K_i, $i = 1, \ldots v$, in (10.5) are chosen to have the form

$$K_i = \text{diag}\,[k_{i1}, \ldots, k_{im}] \tag{10.89}$$

then the eigenvalues of E are the roots λ_{ij}, $i = 1, \ldots v$, $j = 1, \ldots m$ of the polynomials

$$P_j(\lambda) \triangleq \lambda^v + k_{1j}\lambda^{v-1} + \ldots + k_{vj}, \quad j = 1, \ldots m \tag{10.90}$$

Moreover, if the roots of each polynomial are distinct, then the matrix P in (10.87) is given by

$$P = \text{block}\,\{P_{ij}\}_{\substack{i=1,\ldots v\\ j=1,\ldots v}} \tag{10.91a}$$

$$P_{ij} = \text{diag}\,[p_{1ij}, \ldots p_{mij}] \tag{10.91b}$$

where the elements P_{lij} are entries of $v \times v$ matrices obtained as follows:

$$\{p_{lij}\}_{\substack{i=1,\ldots v\\ j=1,\ldots v}} \begin{bmatrix} v & \sum_k \lambda_{kl} & \cdots & \sum_k \lambda_{kl}^{v-1} \\ \sum_k \lambda_{kl} & \sum_k \lambda_{kl}^2 & \cdots & \sum_k \lambda_{kl}^{v} \\ \cdots & \cdots & \cdots & \cdots \\ \sum_k \lambda_{kl}^{v-1} & \cdots & \cdots & \sum_k \lambda_{kl}^{2v-2} \end{bmatrix}^{-1} \tag{10.92}$$

$l = 1, \ldots m$

Proof: Consider the transformation matrix $T \in R^{(mv)\times(mv)}$ given by

$$T = \begin{bmatrix} l_1^T & 0 & 0 & \cdots & 0 \\ 0 & l_1^T & 0 & \cdots & 0 \\ \cdot\cdot & \cdot\cdot & \cdot & \cdot\cdot & \cdot \\ 0 & 0 & \cdot\cdot & \cdot\cdot & l_1^T \\ l_2^T & 0 & \cdot\cdot & \cdot\cdot & 0 \\ \cdot\cdot & \cdot & \cdot\cdot & \cdot\cdot & \cdot \\ 0 & \cdot\cdot & \cdot\cdot & \cdot\cdot & l_v^T \end{bmatrix} \tag{10.93}$$

where l_i denotes the ith column of the identity matrix I_v. It is easy to show that

$$\hat{E} \triangleq T^{-1}ET = \text{diag}\,[C_1, \ldots C_m] \tag{10.94}$$

where

$$C_j \triangleq \begin{bmatrix} 0 & 1 & 0 & \cdots & 0 \\ 0 & 0 & 1 & \cdots & 0 \\ \cdot & \cdot & \cdot & \cdots & \cdot \\ 0 & 0 & \cdots & \cdots & 1 \\ -k_{vj} & -k_{v-1,j} & \cdots & \cdots & -k_{1j} \end{bmatrix} \quad j = 1,\ldots m \tag{10.95}$$

Therefore the characteristic polynomial of E is given by

$$p(\lambda) = \det(\lambda I_{mv} - \hat{E}) = \prod_{i=1}^{m} P_j(\lambda) \tag{10.96}$$

Furthermore, since the eigenvectors of the companion matrix C_j are of the form

$$u_i^{(C_j)} = [1 \ \lambda_{ij} \ldots \lambda_{ij}^{v-1}]^{\mathrm{T}}, \qquad i = 1, \ldots v, \tag{10.97}$$

the eigenvectors of E (which are the columns of the Z matrix in (10.85) will be

$$u_{ij} = T\,[0 \ldots 0 \ u_i^{(C_j)T} \ 0 \ldots 0]^T \tag{10.98}$$

Hence, standard but tedious manipulations lead to

$$P = \left[\text{block} \left\{ \text{diag} \left[\textstyle\sum_k \lambda_{k1}^{i+j-2}, \ldots, \sum_k \lambda_{km}^{i+j-2} \right] \right\}_{\substack{i=1,\ldots v \\ j=1,\ldots v}} \right]^{-1} \tag{10.99}$$

and then to (10.91) and (10.92).

Using the structure of the K_i matrices suggested in theorem 3, from (10.7), (10.9) and (10.91) it turns out that

$$v = B_1^T P e = \begin{bmatrix} \sum_{j=1}^{v} p_{1vj} \quad e_1^{(j-1)} \\ \cdot \\ \cdot \\ \cdot \\ \sum_{j=1}^{v} p_{mvj} \quad e_m^{(j-1)} \end{bmatrix} \tag{10.100}$$

that is, the ith component of v is made up of a linear combination of the derivatives of the ith component of the error, and thus the signal v can be obtained by means of a bank of generalised decoupled PD controllers. The design of the controller via (10.100) requires the matrix inversion of eq. (10.92). If, however, one is not interested in fixing the rate of convergence of the tracking error through the placement of the eigenvalues of matrix E, but wants only the asymptotic stability of E to be guaranteed, it is possible to avoid the matrix inversion by using the following:

Theorem 4: The parameters p_{ivj}, $i = 1, \ldots m$ in (10.100) can be chosen equal to the coefficients of m arbitrary, strictly Hurwitz polynomials

$$d_i(s) \triangleq p_{iv1} + p_{iv2}s + \ldots + p_{ivv}s^{v-1} \tag{10.101}$$

with distinct roots.

Proof: It must be proved that, given m polynomials of the form (10.101), it is always possible to find an asymptotically stable matrix E, and a PD matrix Q, such that the solution P of the Lyapunov eq. (10.10) determines, through (10.100), the desired PD coefficients. To this aim, first of all we note that (10.100) can be rewritten as

$$v = \begin{bmatrix} d_1^T \boldsymbol{e}_1 \\ \cdot \\ \cdot \\ \cdot \\ d_m^T \boldsymbol{e}_m \end{bmatrix} \tag{10.102}$$

with

$$d_i \triangleq [p_{iv1}\, p_{iv2} \ldots p_{ivv}]^{\mathrm{T}}, \qquad i = 1, \ldots, m \tag{10.103}$$

$$\boldsymbol{e}_i = [e_i\, e_i^{(1)} \ldots e_i^{(v-1)}]^{\mathrm{T}}, \qquad i = 1, \ldots m \tag{10.104}$$

Secondly, we recall that the Lyapunov equation for the matrix $\hat{E}$ in (10.94) can be written as

$$\hat{E}^T \hat{P} + \hat{P}\hat{E} = -\hat{Q}, \tag{10.105}$$

with

$$\hat{P} = T^T P T \tag{10.106}$$

and

$$\hat{Q} = T^T Q T \tag{10.106}$$

Taking into account the diagonal form of $\hat{E}$ and choosing

$$\hat{P} = \text{blockdiag}\,[P_1, \ldots, P_m] \tag{10.107}$$

$$\hat{Q} = \text{blockdiag}\,[Q_1, \ldots, Q_m] \tag{10.107}$$

equation (10.102) splits into the following m equations:

$$C_i^T P_i + P_i C_i = -Q_i, \qquad i = 1, \ldots m \tag{10.108}$$

Furthermore

$$\hat{B}_1 \triangleq T^{-1} B_1 = \begin{bmatrix} l_v & 0 & \ldots & 0 \\ 0 & l_v & \ldots & 0 \\ \cdot & \cdot & \cdot\;\cdot\;\cdot & \cdot \\ 0 & 0 & \ldots & l_v \end{bmatrix} \in R^{(mv) \times m} \tag{10.109}$$

where again l_v is the vth column of I_v, and

$$T^{-1} e = [\boldsymbol{e}_1^T \ldots \boldsymbol{e}_m^T]^T \tag{10.110}$$

Now, using (10.106), (10.107), (10.109) and (10.110), we have

$$\begin{aligned} v &= B_1^T P e \\ &= \hat{B}_1 \hat{P}(T^{-1} e) \\ &= \begin{bmatrix} l_v^T & P_1 & e_1 \\ & \cdot & \\ & \cdot & \\ & \cdot & \\ l_v^T & P_m & e_m \end{bmatrix} \end{aligned} \qquad (10.111)$$

It remains to be shown that, for any given vector d_i, an asymptotically stable matrix C_i, as in (10.95), and PD matrices Q_i and P_i exist such that

(i) eqn (10.108) is satisfied
(ii) $l_v^T P_i = d_i^T$

This is easily proved by application of the Kalman–Yacubovitch lemma to the transfer function

$$W_i(s) = d_i^T (s\mathrm{I}_v - C_i)^{-1} l_v = d_i(s)/c_i(s) \qquad (10.112)$$

noting that, for any given strictly Hurwitz polynomial $d_i(s)$, there exists a strictly Hurwitz polynomial $c_i(s)$ which makes $W_i(s)$ strictly positive real.

10.6 Conclusions

A robust tracking control system has been presented. The model, affected by bounded parameter uncertainties, has a structure often encountered in the analysis of complex mechanical systems. The continuous nonlinear control law which has been proposed (and which is an approximation of discontinuous laws) ensures the attractiveness of a certain region of the state space, while leaving the possibility of choosing the rate of convergence toward any ball of ultimate boundedness or just a decoupled structure of the controller. The method of analysis and design merges ideas of the VSC and Lyapuncv approaches. Some of those ideas have been developed here and enriched for the tracking problem.

10.7 References

AMBROSINO, G., CELENTANO, G., and GAROFALO, F., (1985a): 'Robust model tracking control for a class of nonlinear plants', *IEEE Trans.*, **AC-30,** 275–279

AMBROSINO, G., CELENTANO, G., and GAROFALO, F. (1985b): 'Decentralized PD controllers for tracking control of uncertain multivariable systems', Proc. 7th IFAC/IFORS Symp. on Ident. and Sys. Param. Estim., York, England, pp. 1907–1911

AMBROSINO, G., CELENTANO, G., and GAROFALO, F. (1986): 'Tracking control of high performance robots via stabilizing controllers for uncertain systems', *J. Opt. Th. Appl.*, **50,** pp. 239–255

BARMISH, B. R., and LEITMANN, G. (1982): 'On ultimate boundedness control of uncertain systems in the absence of matching assumption', *IEEE Trans.*, **AC-27,** pp. 153–158

BURTON, J.A., and ZINOBER, A. S. I. (1986): 'Continuous approximation of variable structure control', *Int. Jnl. Systems Science,* **17**, pp. 876–885

CORLESS, M., and LEITMANN, G. (1981) 'Continuous state feedback guaranteeing uniform ultimate boundedness for uncertain dynamic systems', *IEEE Trans.*, **AC-26,** pp. 1139–1144.

CORLESS, M., and LEITMANN, G., (1990), Chap. 11 in this book

DESOER, C. A., and VIDYASAGAR, M, (1975): 'Feedback systems: input-output properties' (Academic Press, NY)

DRIVER, R. D. (1977): 'Ordinary and delay differential equation' (Berlin: Springer-Verlag, Berlin)

GAROFALO, F., and LEITMANN, G. (1987): 'Guaranteeing ultimate boundedness and exponential rate of convergence for a class of nominally linear uncertain systems', *ASME J. Dyn. Sys. Meas. Control.* (to appear)

GUTMAN, S. (1979): 'Uncertain dynamic systems: A Lyapunov min–max approach', *IEE Trans.*, **AC-24,** pp. 437–443

KALMAN, R. E., and BERTRAM, J. E. (1960): 'Control system analysis and design via the second method of Lyapunov: Continuous-time systems', *ASME J. Basic Eng.*, pp. 371–393

LEITMANN, G. (1981): 'On the efficacy of nonlinear control in uncertain linear systems', *ASME J. Dyn. Sys. Meas. Control,* **103,** pp. 95–102

RYAN, E. P., and CORLESS, M. (1984): 'Ultimate boundedness and asymptotic stability of a class of uncertain dynamical systems via continuous and discontinuous feedback control', *IMA J. Math. Control Inf.*, **1,** pp. 223–242

UTKIN, V. I. (1971): 'Equation of sliding mode in discontinuous systems I', *Automat. Remote Control,* pp. 1897–1907

UTKIN, V. I. (1978): 'Sliding modes and their applicators to variable structure systems', MIR (Moscow)

VIDYASAGAR, M. (1980) 'Nonlinear systems' (Academic Press, NY)

ZINOBER, A. S. I. (1990), Chap. 1 in this book

Chapter 11

Deterministic control of uncertain systems: A Lyapunov theory approach†

M. Corless

Purdue University

G. Leitmann

University of California

11.1 Introduction

In order to control the behaviour of a system in the 'real' world, the system analyst seeks to capture the system's salient features in a mathematical model. This abstraction of the 'real' system usually contains uncertain elements; for example, uncertainties due to parameters, constant or varying, which are unknown or imperfectly known, or uncertainties due to unknown or imperfectly known inputs into the system. Despite such imperfect knowledge about the chosen mathematical model, one often seeks to devise controllers which will 'steer' the system in some desired fashion, for example, so that the system response will approach or track a desired reference response; by suitable definition of the system (state) variables such a problem can usually be cast into that of stabilising a prescribed state.

Two main avenues are open to the analyst seeking to control an uncertain dynamical system. He may choose a stochastic approach in which information about the uncertain elements as well as about the system response is statistical

† Supported by the US National Science Foundation and the US Air Force Office of Scientific Research under grants MSM-8706927 and ECS-8602524

in nature; for example, see Åström (1970) and Kushner (1966). Loosely speaking, when modelling via random variables, one is content with desirable behaviour on the average. The other approach to the control of uncertain systems, and the one for which we shall opt in the present discussion, is deterministic. Available, or assumed, information about uncertain elements is deterministic in nature. Here one seeks controllers which assure the desired response of the dynamical system.

We consider the problem of obtaining memoryless stabilising feedback controllers for uncertain dynamical systems described by ordinary differential equations. Various classes of controllers are presented. The design of all these controllers is based on Lyapunov theory.

We utilise the results to obtain tracking controllers for a general class of uncertain mechanical systems. These controllers are illustrated by application to a model of the Manutec r3 robot which has an uncertain payload.

Before proceeding with the problem, we introduce some basic notions and results for ordinary differential equations.

11.2 Basic notions

Let $T = (\underline{t}, \infty)$ where $\underline{t} \in [-\infty, \infty)$; let X be a non-empty open subset of $\mathscr{R}^n$; and let $f: T \times X \to \mathscr{R}^n$. Consider the first order ordinary differential equation (ODE)

$$\dot{x}(t) = f(t, x(t)) \tag{11.1}$$

where $\dot{x}(t)$ denotes the derivative of the function $x(\cdot)$ at t. By a solution of (11.1) we shall mean an absolutely continuous function $x(\cdot)$: $[t_0, t_1) \to X$, where $t_0 \in T$ and $t_1 \in (t_0, \infty]$, which satisfies (11.1) almost everywhere† on $[t_0, t_1)$.

When considering a system described by an equation of the form (11.1), we shall refer to X as the 'state space', a member of X as a 'state', eqn. (11.1) as the 'state equation', and a solution of (11.1) as a 'state evolution', 'state motion', or 'state history'.

11.2.1 Existence and extension of solutions

Since we consider here systems described by ODEs, the two properties introduced in this section are of fundamental importance.

Definition 2.1: Eqn. (11.1) has (global) existence of solutions iff, given any pair $(t_0, x^0) \in T \times X$, there exists a solution $x(\cdot)$: $[t_0, t_1) \to X$ of (11.1) with $x(t_0) = x^0$.

The following theorem (see Coddington and Levinson, 1955; or Hale, 1980 for a proof) yields sufficient conditions for existence of solutions.

† That is, everywhere except possibly on a set of Lebesgue measure zero.

Theorem 2.1: If f is a Caratheodory[‡] function, eqn. (11.1) has global existence of solutions.

Definition 2.2: Eqn. (11.1) has indefinite extension of solutions iff, given any solution $x(\cdot)$: $[t_0, t_1) \to X$ of (11.1), there exists a solution $x^e(\cdot)$: $[t_0, \infty) \to X$ of (11.1) with $x^e(t) = x(t)$ for all $t \in [t_0, t_1)$.

The following theorem, which can be deduced from the results presented in Hale (1980), chapter 1, provides useful sufficient conditions for indefinite extension of solutions.

Theorem 2.2: Suppose f is Caratheodory and for each solution $x(\cdot)$: $[t_0, t_1) \to X$ of (11.1) with $t_1 < \infty$, there exists a compact subset C of X such that $x(t) \in C$ for all $t \in [t_0, t_1)$. Then, eqn. (11.1) has indefinite extension of solutions.

11.2.2 Boundedness and stability

In this section, we formalise the notion of a system described by (11.1) exhibiting 'desirable' behaviour with respect to a state $x^* \in \bar{X}$, where $\bar{X}$ is the closure of X.

Definition 2.3: The solutions of (11.1) are globally uniformly bounded (GUB) iff, given any compact subset C of X, there exists $d(C) \in \mathscr{R}_+$ such that, if $x(\cdot)$: $[t_0, t_1) \to X$ is any solution of (11.1) with $x(t_0) \in C$, then $\|x(t)\| \leqslant d(C)$ for all $t \in [t_0, t_1)$.

Definition 2.4: x^* is uniformly stable (US) for (11.1) or (11.1) is uniformly stable about x^* iff, given any neighbourhood[†] B of x^*, there exists a neighbourhood B_0 of x^* such that, if $x(\cdot)$: $[t_0, t_1) \to X$ is any solution of (11.1) with $x(t_0) \in B_0$, then $x(t) \in B$ for all $t \in [t_0, t_1)$.

Definition 2.5: x^* is a global uniform attractor (GUA) for (11.1) iff, given any neighbourhood B of x^* and any compact subset C of X, there exists $T(C, B) \in \mathscr{R}_+$ such that, if $x(\cdot)$: $[t_0, \infty) \to X$ is any solution of (11.1) with $x(t_0) \in C$, then $x(t) \in B$ for all $t \geqslant t_0 + T(C, B)$.

Definition 2.6: x^* is globally uniformly asymptotically stable (GUAS) for (11.1), or (11.1) is globally uniformly asymptotically stable about x^*, iff:

(i) The solutions of (11.1) are GUB.
(ii) x^* is US for (11.1).
(iii) x^* is a GUA for (11.1).

‡ See Appendix 11.9.1, or just note that, if f is continuous, it is Caratheodory.

† By a neighbourhood of x^*, we mean a set containing an open set which contains x^*.

Remark 2.1: Frequently, in the definition of uniform stability of x^* in the literature, x^* is assumed to be an equilibrium state for (11.1), i.e. $x^* \in X$ and $f(t, x^*) = 0$‡ for all $t \in T$, or, equivalently, the function $x(\cdot): T \to X, x(t) \equiv x^*$, is a solution of (11.1). However, one may readily show that, if a state $x^* \in X$ is uniformly stable and if $x(\cdot)$: $[t_0, t_1) \to X$ is any solution of (11.1) with $x(t_0) = x^*$, it necessarily follows that $x(t) = x^*$ for all $t \in [t_0, t_1)$.

11.2.3 Lyapunov functions and a sufficient condition for GUAS

In this section, we restrict the discussion to differential equations of the form (11.1) with $X = \mathscr{R}^n$, i.e.

$$\dot{x}(t) = f(t, x(t)) \tag{11.2}$$

where $f: T \times X \to \mathscr{R}^n$, $X = \mathscr{R}^n$, and $T = (\underline{t}, \infty)$ with $\underline{t} \in [-\infty, \infty)$. In particular, we present a theorem (Theorem 2.3) which yields a sufficient condition assuring that (11.2) is GUAS about the zero state. The condition utilises the notion of a Lyapunov function which we shall define presently.

Definition 2.7: A function $V: T \times \mathscr{R}^n \to \mathscr{R}_+$ is a candidate Lyapunov function iff it is continuously differentiable and there exist functions $\gamma_1, \gamma_2: \mathscr{R}_+ \to \mathscr{R}_+$ of class KR† such that, for all $(t, x) \in T \times \mathscr{R}^n$,

$$\gamma_1(\|x\|) \leqslant V(t, x) \leqslant \gamma_2(\|x\|) \tag{11.3}$$

Remark 2.2: Suppose $V: T \times \mathscr{R}^n \to \mathscr{R}$ is given by

$$V(t, x) = W(x)$$

for all $(t, x) \in T \times \mathscr{R}^n$, where $W: \mathscr{R}^n \to \mathscr{R}$ is a continuously differentiable function satisfying

$$W(0) = 0,$$

$$x \neq 0 \Rightarrow W(x) > 0$$

$$\lim_{\|x\| \to \infty} W(x) = \infty$$

for all $x \in \mathscr{R}^n$. Then, V is a candidate Lyapunov function. To see this, define γ_1, $\gamma_2: \mathscr{R}_+ \to \mathscr{R}_+$ by

$$\gamma_1(r) = \inf_{\|x\| \geqslant r} W(x), \qquad \gamma_2(r) = \sup_{\|x\| \leqslant r} W(x),$$

for all $r \in \mathscr{R}_+$.

Definition 2.8: A function $V: T \times \mathscr{R}^n \to \mathscr{R}_+$ is a Lyapunov function for (11.2) iff it is a candidate Lyapunov function and there exists a function

‡ We use '0' to denote a zero vector.

† See Appendix 11.9.2

$\gamma_3:\mathscr{R}_+ \to \mathscr{R}_+$ of class $K^\dagger$ such that for all $(t, x) \in T \times \mathscr{R}^n$

$$\frac{\partial V}{\partial t}(t, x) + \frac{\partial V}{\partial x}(t, x) f(t, x) \leqslant -\gamma_3(\|x\|) \tag{11.4}$$

We can now introduce a sufficient condition for (11.2) to be GUAS about zero.

Theorem 2.3: If there exists a Lyapunov function for (11.2), then (11.2) is GUAS about zero.

Proofs of various versions of Theorem 2.3 can be found in the literature; see, for example, Kalman and Bertram (1960), LaSalle and Lefschetz (1961), Hahn (1967), Cesari (1971), and LaSalle (1976). Also, Theorem 2.3 is a corollary of Theorem 6.1 of which there is a proof in Corless (1984). The following corollary is readily deduced from Theorems 2.1, 2.2 and 2.3.

Corollary 2.1: If $f:T \times \mathscr{R}^n \to \mathscr{R}^n$ is Caratheodory and there exists a Lyapunov function for (11.2), then (11.2) has existence and indefinite extension of solutions and is GUAS about zero.

11.2.4 Systems with control

Consider now a system with control; i.e. a system whose state evolution depends not only on an initial state but also on an externally applied control input. For some non-empty set $U \subset \mathscr{R}^m$, the set of control values, and some function $F:T \times X \times U \to \mathscr{R}^n$ (T and X are as before), such a system is described by

$$\dot{x}(t) = F(t, x(t), u(t)) \tag{11.5}$$

where $u(t) \in U$ is the control value at t.

Here we shall consider control to be given by a memoryless feedback controller; i.e. we shall consider

$$u(t) = p(t, x(t)) \tag{11.6}$$

for some feedback control function $p:T \times X \to U$. Substituting (11.6) into (11.5), a feedback controlled system can be described by

$$\dot{x}(t) = F(t, x(t), p(t, x(t))) \tag{11.7}$$

i.e., it can be described by (11.1) with

$$f(t, x) = F(t, x, p(t, x)) \tag{11.8}$$

Since we shall be considering GUAS as a criterion for desirable system behaviour, we introduce the following definitions.

Definition 2.9: A feedback control function $p:T \times X \to U$ stabilises (11.5) about x^* iff (11.7) has existence and indefinite extension of solutions and is GUAS about x^*.

Definition 2.10: (11.5) is stabilisable about x^* iff there exists a $p: T \times X \rightarrow U$ which stabilises (11.5) about x^*.

11.3 Uncertain systems

In the previous section, we introduced the notion of a system with control, i.e. a system described by (11.5) for some function $F\colon T \times X \times U \rightarrow \mathscr{R}^n$. When modelling a 'real' system, one usually does not have, or cannot obtain, an 'exact' model. The model usually contains uncertain elements; for example, uncertainties due to parameters, constant or varying, which are unknown or imperfectly known, or uncertainties due to unknown or imperfectly known inputs into the system. Our model of an uncertain system is of the form

$$\dot{x}(t) = \bar{F}(t, x(t), u(t), \omega) \tag{11.9}$$

$$\omega \in \Omega \tag{11.10}$$

where Ω is some known, non-empty, set and $\bar{F}: T \times X \times U \times \Omega \rightarrow \mathscr{R}^n$ is known. All the uncertainty in the system is represented by the lumped uncertain element ω. This uncertain element could be an element of $\mathscr{R}^q$ representing unknown constant parameters and inputs; it could be a function from $\mathscr{R}$ into $\mathscr{R}^q$ representing unknown time varying parameters and inputs; it could also be a function from $\mathscr{R} \times \mathscr{R}^n$ into $\mathscr{R}^q$ representing nonlinear elements which are difficult to characterize.

As a simple example of an uncertain system, consider a scalar system subject to an uncertain Lebesgue measurable input disturbance $v: \mathscr{R} \rightarrow \mathscr{R}$

$$\dot{x}(t) = -x(t) + v(t) + u(t) \tag{11.11}$$

Table 11.1 *Three different possible assumptions on the knowledge of v*

1	$v(t) = \omega \;\; \forall\, t \in \mathscr{R}$; $\omega \in \mathscr{R}$ unknown
2	$v\colon \mathscr{R} \rightarrow [\underline{\varrho}, \bar{\varrho}]$; $\underline{\varrho}, \bar{\varrho}$ known
3	$v\colon \mathscr{R} \rightarrow \mathscr{R}$, bounded; bounds unknown

Table 11.1 lists three different possible assumptions on the knowledge of v.

In case 1, the disturbance is simply an unknown constant and the system model is given by (11.9) – (11.11) with $\Omega = \mathscr{R}$ and

$$\bar{F}(t, x, u, \omega) = -x + \omega + u. \tag{11.12}$$

In case 2, the disturbance is an unknown Lebesgue measurable function with known upper and lower bounds, $\bar{\varrho}$ and $\underline{\varrho}$, respectively. In this case,

$$\Omega = \{\omega: T \rightarrow [\underline{\varrho}, \bar{\varrho}] \,|\, \omega \text{ is Lebesgue measurable}\}$$

and

$$\bar{F}\ (t, x, u, \omega) \quad = \quad -x + \omega(t) + u. \tag{11.13}$$

In case 3, which requires less information on the uncertainty than cases 1 and 2, the disturbance is modelled by a bounded measurable function with no assumption on the knowledge of its bounds. In this case,

$$\Omega \quad = \quad \{\omega{:}T \rightarrow \mathscr{R} \mid \omega \text{ is Lebesgue measurable and bounded}\}$$

and $\bar{F}$ is given by (11.13).

As a second example, consider a scalar system

$$\dot{x}(t) \quad = \quad v(t)x(t) + u(t) \tag{11.14}$$

where the Lebesgue measurable function $v{:}\mathscr{R} \rightarrow \mathscr{R}$ models an uncertain parameter. Again, Table 11.1 lists some possible assumptions on the knowledge of v.

An uncertain system described by case 2 of the previous example is an example of the type of uncertain system we shall be considering in this chapter. Basically, the uncertain systems considered here are specified by specifying for each $(t, x, u) \in T \times X \times U$, the set of possible values which $\bar{F}(t, x, u, \omega)$ may assume. For cases 1 and 3, see Corless (1984) and Corless and Leitmann (1983b).

11.4 Initial problem statement: Stabilisation

11.4.1 Problem statement

Basically, the type of problem we shall consider initially in this chapter is as follows:

Given an uncertain system described by

$$\dot{x}(t) \quad = \quad \bar{F}(t, x(t), u(t), \omega) \tag{11.15}$$

$$\omega \in \Omega \tag{11.16}$$

where Ω is a known non-empty set and $\bar{F}{:}T \times X \times U \times \Omega \rightarrow \mathscr{R}^n$ is known, and a 'desirable' state $x^* \in \bar{X}$, obtain a feedback control function $p{:}T \times X \rightarrow U$ which stabilises (11.15) about x^*.

Since the only information available on ω is a set Ω to which ω belongs, we attempt to solve the above problem by looking for a feedback control function which stabilises (11.15) about x^* for all $\omega \in \Omega$.

We now introduce:

Definition 4.1: A feedback control function $p{:}T \times X \rightarrow U$ stabilises (11.15) – (11.16) about x^* iff p stabilises (11.15) about x^* for each $\omega \in \Omega$.

The problem we shall consider is that of obtaining a feedback control function which stabilises (11.15) – (11.16) for a given $\bar{F}$ and Ω.

Definition 4.2: (11.15) – (11.16) is stabilisable about x^* iff there exists $p: T \times X \to U$ which stabilises (11.15) – (11.16) about x^*.

Remark 4.1: Note that stabilisability of (11.15) for each $\omega \in \Omega$ does not imply stabilisability of (11.15) – (11.16). It might be the case that, for each $\omega \in \Omega$, there exists p (dependent on ω) which stabilises (11.15), but there does not exist p which stabilises (11.15) for all $\omega \in \Omega$. For example, consider the uncertain scalar system

$$\begin{aligned} \dot{x}(t) &= \omega u(t) \\ \omega &\in \{-1, 1\} \end{aligned} \tag{11.17}$$

Although (11.17) is stabilisable for each ω (e.g. let $p(t, x) = x$ and $p(t, x) = -x$, for $\omega = -1$ and $\omega = 1$, respectively) it is unlikely that there exists a single feedback control function which stabilizes (11.17) for each ω. However, there does exist a non-memoryless or dynamic controller which stabilises (11.17) for each ω; see Nussbaum (1983).

11.4.2 A useful theorem for the synthesis of stabilising controllers

In this section, we present a theorem (Theorem 4.1) which is useful in the synthesis of zero-state stabilising feedback control functions for uncertain systems whose state space is $\mathscr{R}^n$. For a given uncertain system, the theorem yields criteria which, if satisfied by a feedback control function, ensure that the feedback control function is a stabilising controller.

Theorem 4.1: Consider an uncertain system described by (11.15) – (11.16) with $X = \mathscr{R}^n$ and suppose that $p: T \times X \to U$ is such that

$$\dot{x}(t) = \bar{F}(t, x(t), p(t, x(t)), \omega) \tag{11.18}$$

has existence and indefinite extension of solutions for all $\omega \in \Omega$. If, for each $\omega \in \Omega$, there exists a Lyapunov function for (11.18), then p stabilises (11.15) – (11.16) about zero.

Proof: The proof follows readily from Theorem 2.3.

In the next section, we consider a particular class of uncertain systems. For each member of the class, we present a class of candidate stabilising feedback control functions whose design is based on meeting the Lyapunov criterion in Theorem 4.1.

11.5 L–G controllers

In this section, we consider first a class of uncertain systems which have been treated previously in the literature; see Gutman and Leitmann (1976), Gutman (1979), Corless and Leitmann (1981), and Barmish *et al.* (1983). For each member of this class, we present a class of candidate stabilising controllers. We then enlarge the class of systems for which the presented controllers are candidate stabilising controllers. Finally, we present a theorem which yields some properties of systems subject to the controllers presented.

11.5.1 Original class of uncertain systems

A member of the class of uncertain systems under consideration here is described by (11.9) – (11.10)

$$\dot{x}(t) = \bar{F}(t, x(t), u(t), \omega) \tag{11.19}$$

$$\omega \in \Omega \tag{11.20}$$

where $X = \mathscr{R}^n$, $U = \mathscr{R}^m$, and $\bar{F}$ satisfies assumptions A1 and A2.

Assumption A1: ($\bar{F}(t, x, \cdot, \omega)$ is affine) There exist functions $\bar{f}: T \times X \times \Omega \to \mathscr{R}^n$ and $\bar{B}: T \times X \times \Omega \to \mathscr{R}^{n \times m}$ such that for all $(t, x, u, \omega) \in T \times X \times U \times \Omega$

$$\bar{F}(t, x, u, \omega) = \bar{f}(t, x, \omega) + \bar{B}(t, x, \omega) u \tag{11.21}$$

If A1 is satisfied, then $\bar{F}$ has a unique representation in the form of (11.21); i.e. there exists a unique pair $(\bar{f}, \bar{B})$ for which (11.21) is satisfied for all $(t, x, u, \omega) \in T \times X \times U \times \Omega$; this pair is given by

$$\bar{f}(t, x, \omega) = \bar{F}(t, x, 0, \omega), \tag{11.22}$$

$$\bar{B}(t, x, \omega) u = \bar{F}(t, x, u, \omega) - \bar{F}(t, x, 0, \omega) \quad \forall\, u \in U \tag{11.23}$$

for all $(t, x, \omega) \in T \times X \times \Omega$.

Assumption A2: There exist Caratheodory functions $f: T \times X \to \mathscr{R}^n$ and $B: T \times X \to \mathscr{R}^{n \times m}$, a candidate Lyapunov function $V: T \times X \to \mathscr{R}_+$, a strongly Caratheodory† function $\varrho^0: T \times X \to \mathscr{R}_+$, and a constant $c \in \mathscr{R}_+$ such that

(i) V is a Lyapunov function for

$$\dot{x}(t) = f(t, x(t)). \tag{11.24}$$

(ii) There exist functions $e: T \times X \times \Omega \to \mathscr{R}^m$ and $E: T \times X \times \Omega \to \mathscr{R}^{m \times m}$ such that $e(\cdot, \omega)$ and $E(\cdot, \omega)$ are strongly Caratheodory for all $\omega \in \Omega$,

$$\bar{f} = f + Be \tag{11.25}$$

$$\bar{B} = B + BE \tag{11.26}$$

and

$$\|e(t, x, \omega)\| \leqslant \varrho^0(t, x) \tag{11.27}$$

$$\|E(t, x, \omega)\| \leqslant c < 1 \tag{11.28}$$

† See Appendix 11.9.1 or just note that, if a function is continuous, it is strongly Caratheodory.

Thus, utilising (11.19) – (11.21), (11.25) and (11.26), any system under consideration here is described by

$$\begin{aligned} \dot{x}(t) &= f(t, x(t)) + B(t, x(t))[e(t, x(t), \omega) \\ &\quad + E(t, x(t), \omega)\, u(t) + u(t)], \end{aligned} \tag{11.29}$$

where f has a Lyapunov function V, and e and E satisfy (11.27) and (11.28) for some strongly Caratheodory function ϱ^0 and constant $c \geqslant 0$.

Remarks 5.1

(i) For a certain or completely known system, i.e. for a system for which ω is known, it should be clear that, under condition (11.28), (11.29) is stabilisable. If one lets

$$\begin{aligned} u(t) &= p(t, x(t)) \\ p(t, x) &= -[I + E(t, x, \omega)]^{-1}\, e(t, x, \omega) \end{aligned} \tag{11.30}$$

then (11.29) reduces to (11.24) which, as a consequence of f being Caratheodory, part (i) of assumption A2, and corollary 2.1, has existence and indefinite extension of solutions and is GUAS about zero.

(ii) In the literature, conditions (11.25) and (11.26) are sometimes referred to as matching conditions; see Gutman (1979), Corless and Leitmann (1981), Ryan and Corless (1984), Barmish and Leitmann (1982), and Chen and Leitmann (1987); the latter two references concern the relaxation of the matching conditions.

When Assumption A1 is satisfied, the existence of f, B satisfying (11.25) and (11.26) is equivalent to either of the following two conditions:

Condition C1: There exist Caratheodory functions $f: T \times X \to \mathscr{R}^n$ and $B: T \times X \to \mathscr{R}^{n \times m}$ such that for all $(t, x, u, \omega) \in T \times X \times U \times \Omega$,

$$\bar{F}(t, x, u, \omega) - f(t, x) \in R(B(t, x)) \tag{11.31}$$

where $R(B(t, x))$ denotes the range space of $B(t, x)$.

Condition C2: There exist Caratheodory functions $f: T \times X \to \mathscr{R}^n$ and $B: T \times X \to \mathscr{R}^{n \times m}$ such that for all $(t, x, u, \omega) \in T \times X \times U \times \Omega$,

$$[I - B(t, x)\, B^{\dagger}(t, x)]\, [\bar{F}(t, x, u, \omega) - f(t, x)] = 0 \tag{11.32}$$

where $B^{\dagger}(t, x)$ denotes the pseudoinverse‡ of $B(t, x)$; see Luenberger (1966).

(iii) Suppose one has an uncertain system described by (11.19) – (11.20) which satisfies A1 and A2(ii) but for which A2(i) is relaxed to:

Assumption A3: There exists a strongly Caratheodory function

‡ If $B \in \mathscr{R}^{n \times m}$ and rank $(B) = m$, then $B^{\dagger} = [B^T B]^{-1} B^T$.

$p^0: T \times X \to U$ such that V is a Lyapunov function for

$$\dot{x}(t) = \tilde{f}(t, x(t)) \tag{11.33}$$

where

$$\tilde{f} = f + Bp^0 \tag{11.34}$$

Assumption A3 ensures that

$$\dot{x}(t) = f(t, x(t)) + B(t, x(t))u(t) \tag{11.35}$$

is stabilisable about zero.

Letting

$$\tilde{u}(t) = u(t) - p^0(t, x(t)) \tag{11.36}$$

one has

$$u(t) = p^0(t, x(t)) + \tilde{u}(t) \tag{11.37}$$

and, utilizing assumptions A1 and A2(ii), one can obtain

$$\dot{x}(t) = \tilde{F}(t, x(t), \tilde{u}(t), \omega) \tag{11.38}$$

where

$$\tilde{F}(t, x, u, \omega) = \tilde{f}(t, x) + B(t, x)[\tilde{e}(t, x, \omega) + E(t, x, \omega)\tilde{u} + \tilde{u}] \tag{11.39}$$

$$\tilde{e}(t, x, \omega) = e(t, x, \omega) + E(t, x, \omega)p^0(t, x) \tag{11.40}$$

$$\|\tilde{e}(t, x, \omega)\| \leqslant \tilde{\rho}^0(t, x) \tag{11.41}$$

$$\tilde{\varrho}^0(t, x) = \rho^0(t, x) + c\|p^0(t, x)\| \tag{11.42}$$

for all $(t, x, u, \omega) \in T \times X \times U \times \Omega$. Hence, one may obtain a new system description which satisfies A1 and A2; $\tilde{f}$, B, V, $\tilde{\rho}^0$, and c assure A1 and A2.

11.5.2 L–G controllers

Consider any uncertain system described by (11.19) – (11.20) (with $X = \mathscr{R}^n$) which satisfies Assumptions A1 and A2. An *L–G* (Leitmann–Gutman) controller for such a system is any function $p: T \times X \to U$ which satisfies

$$p(t, x) = -\varrho^c(t, x)\|\alpha(t, x)\|^{-1}\alpha(t, x) \quad \text{if} \quad \alpha(t, x) \neq 0, \tag{11.43}$$

$$\varrho^c(t, x) \geqslant \varrho(t, x) \tag{11.44}$$

where

$$\alpha(t, x) = B^T(t, x)\frac{\partial V^T}{\partial x}(t, x) \tag{11.45}$$

$$\varrho(t, x) = \varrho^0(t, x)/(1 - c) \tag{11.46}$$

for all $(t, x) \in T \times X$, and (f, B, V, ϱ^0, c) assure satisfaction of A2.

For previous literature on the above controllers, see, for example, Gutman (1979) and Gutman and Leitmann (1976).

11.5.3 Extension of original system class

In this section, we present a class of uncertain systems which is a generalisation of the class presented in Section 11.5.1. An uncertain system in this class is described by (11.19) – (11.20) where $X = \mathscr{R}^n$, $U = \mathscr{R}^m$, and $\bar{F}$ satisfies the following assumption.

Assumption A4: There exist a Caratheodory function $B: T \times X \to \mathscr{R}^{n \times m}$, a candidate Lyapunov function $V: T \times X \to \mathscr{R}_+$ and a strongly Caratheodory function $\varrho: T \times X \to \mathscr{R}_+$ such that

$$\bar{F}(t, x, u, \omega) = f^s(t, x, \omega) + B(t, x)\, g(t, x, u, \omega) \tag{11.47}$$

for all $(t, x, u, \omega) \in T \times X \times U \times \Omega$, for some functions $f^s: T \times X \times \Omega \to \mathscr{R}^n$ and $g: T \times X \times U \times \Omega \to \mathscr{R}^m$ which satisfy:

(i) For each $\omega \in \Omega$, $f^s(\cdot, \omega)$ is Caratheodory and V is a Lyapunov function for

$$\dot{x}(t) = f^s(t, x(t), \omega) \tag{11.48}$$

(ii) For each $\omega \in \Omega$, $g(\cdot, \omega)$ is strongly Caratheodory and

$$\|u\| \geqslant \varrho(t, x) \Rightarrow u^T g(t, x, u, \omega) \geqslant 0 \tag{11.49}$$

for all $(t, x, u) \in T \times X \times U$.

An L–G controller for a system in this class is any function $p: T \times X \to U$ which satisfies (11.43) – (11.45), where (B, V, ϱ) is any triple which assures A4.

Remark 5.2: Letting

$$\varrho(t, x) = \varrho^0(t, x)/(1 - c)$$

one can readily show that a member of the class of systems considered in Section 11.5.1 is a member of the class treated here; see Corless (1984).

11.5.4 Properties of systems with L–G controllers

We have the following theorem:

Theorem 5.1: Consider any uncertain system described by (11.19) – (11.20) (with $X = \mathscr{R}^n$) which satisfies assumption A4. If p is any corresponding L–G controller (as given by (11.43) – (11.45)) for which

$$\dot{x}(t) = \bar{F}(t, x(t), p(t, x(t)), \omega) \tag{11.50}$$

has existence and indefinite extension of solutions for all $\omega \in \Omega$, then p stabilises (11.19) – (11.20) about zero.

Proof: A proof is given in Corless (1984).

Looking at Theorem 5.1, it can be seen that we have not completely solved the original problem for an uncertain system presented in Section 11.5.3. To do so we need to exhibit an *L–G* controller p which assures existence and indefinite extension of solutions to (11.50) (as defined in definitions 2.1 and 2.2) for all $\omega \in \Omega$. This, however, is not possible in general. Except in special cases‡ it is not possible to obtain a function p, satisfying (11.43) and (11.44) for all $(t, x) \in T \times X$, which is continuous in x. Thus, one cannot assure that $\bar{f}: T \times X \times \Omega \to \mathscr{R}^n$, given by

$$\bar{f}(t, x, \omega) = \bar{F}(t, x, p(t, x), \omega) \quad \forall\ (t, x) \in T \times X$$

is continuous in x. Hence (11.50) does not satisfy the usual requirements for existence of solutions; see Theorem 2.1.

In view of the above, we need to relax the requirements of the original problem statement. Here are two possible relaxations:

(i) Relax the requirements which must be met by a function in order to be considered a solution of (11.50). This is the approach taken in (for example) Gutman (1979) and Gutman and Leitmann (1976) where the notion of a generalised solution is introduced and the *L–G* controllers solve the relaxed problem for the systems considered there.

(ii) Relax the requirement of GUAS of (11.50) about zero. This is what is done in (for example) Barmish *et al.* (1983), Corless and Leitmann (1981), and Leitmann (1978*b*) and what we do in the next section by introducing the notion of practical stabilisation. In this approach, one may solve the relaxed problem with controllers which are continuous in the state; hence they are more desirable from the viewpoint of practical implementation as well.

11.6 Relaxed problem statement: Practical stabilisation

Before introducing a relaxed problem statement, we need some new notions. Consider a system described by (11.1), i.e.

$$\dot{x}(t) = f(t, x(t)) \tag{11.51}$$

where $f: T \times X \to \mathscr{R}^n$ and suppose that $x^* \in \bar{X}$. For any subset B of $\mathscr{R}^n$, we have the following definition.

Definition 6.1: The solutions of (11.51) are globally uniformly ultimately bounded (GUUB) within B iff, given any compact subset C of X, there exists $T(C) \in \mathscr{R}_+$ such that, if $x(\cdot): [t_0, \infty) \to X$ is any solution of (11.51) with $x(t_0) \in C$, $x(t) \in B$ for all $t \geqslant t_0 + T(C)$.

If $B \subset \mathscr{R}^n$ is a neighbourhood of x^*, we have the following definition.

‡ For example, $\varrho(t, x) = \gamma(t, x)\|\alpha(t, x)\|$ for all $(t, x, u) \in T \times X \times U$, where $\gamma: T \times X \to \mathscr{R}_+$ is Caratheodory.

Definition 6.2: System (11.51) B-tracks x^* or tracks x^* to within B iff:
(i) The solutions of (11.51) are GUB.
(ii) There exists a neighbourhood B_0 of x^* such that, if $x(\cdot): [t_0, t_1) \to X$ is any solution of (11.51) with $x(t_0) \in B_0$, then $x(t) \in B$ for all $t \in [t_0, t_1)$.
(iii) The solutions of (11.51) are GUUB within B.

Remark 6.1: If (11.51) B-tracks x^* for any neighbourhood B of x^*, it is GUAS about x^*.

The following theorem yields sufficient conditions for B-tracking of the zero state when $X = \mathscr{R}^n$.

Theorem 6.1: Consider any system described by (11.51) with $X = \mathscr{R}^n$ and suppose there exist a candidate Lyapunov function $V: T \times X \to \mathscr{R}_+$, a class K function $\gamma_3: \mathscr{R}_+ \to \mathscr{R}_+$, and a constant $\varepsilon \in \mathscr{R}_+$, which satisfy

$$\lim_{r \to \infty} \gamma_3(r) > \varepsilon \tag{11.52}$$

such that for all $(t, x) \in T \times X$,

$$\frac{\partial V}{\partial t}(t, x) + \frac{\partial V}{\partial x}(t, x) f(t, x) \leqslant -\gamma_3(\|x\|) + \varepsilon. \tag{11.53}$$

Then, (11.51) tracks the zero state to within any neighbourhood[‡] B of $S(\delta)$, where

$$S(\delta) = \{x \in \mathscr{R}^n : \|x\| \leqslant \delta\} \tag{11.54}$$

$$\delta = \bar{\gamma}_1^{-1}(\gamma_2(r)), \tag{11.55}$$

$$r = \bar{\gamma}_3^{-1}(\varepsilon), \tag{11.56}$$

and $\gamma_1, \gamma_2: \mathscr{R}_+ \to \mathscr{R}_+$ are any class KR functions for which

$$\gamma_1(\|x\|) \leqslant V(t, x) \leqslant \gamma_2(\|x\|) \tag{11.57}$$

for all $(t, x) \in T \times X$.

Proof: Corless (1984) contains a proof.

We now introduce:

Definition 6.3: A collection P of feedback control functions $p: T \times X \to U$ practically stabilises (11.9)–(11.10) about x^* iff, given any neighbourhood B of x^*, there exists $p \in P$ such that for all $\omega \in \Omega$,

$$\dot{x}(t) = F(t, x(t), p(t, x(t)), \omega)$$

has existence and indefinite extension of solutions and B-tracks x^*.

[‡] If $S \subset \mathscr{R}^n$, then a neighbourhood of S is any subset of $\mathscr{R}^n$ which contains an open set containing S. For the definition of $\bar{\gamma}_1^{-1}$ and $\bar{\gamma}_3^{-1}$ and some of their properties see Appendix 11.9.2.

The relaxed problem we shall consider is that of obtaining a collection P of feedback control functions which practically stabilises (11.9) – (11.10) about x^* for a given $\bar{F}$.

Definition 6.4: (11.9) – (11.10) is practically stabilisable about x^* iff there exists a collection P of feedback control functions which practically stabilises (11.9) – (11.10) about x^*.

We now present a theorem which is useful in the synthesis of zero-state practically stabilising sets of feedback controllers for uncertain systems whose state space is $\mathscr{R}^n$.

For a given uncertain system, the theorem yields criteria which, if satisfied by a collection of feedback control functions, assure that the collection is a practically stabilising collection.

Theorem 6.2: Consider an uncertain system described by (11.9) – (11.10) with $X = \mathscr{R}^n$ and suppose that P is a collection of feedback control functions $p: T \times X \to U$. If there exists a candidate Lyapunov function $V: T \times X \to \mathscr{R}_+$ and a class K function $\gamma_3: \mathscr{R}_+ \to \mathscr{R}_+$ such that given any $\varepsilon > 0$ there exists $p \in P$ which assures that for all $\omega \in \Omega$,

$$\dot{x}(t) = \bar{F}(t, x(t), p(t, x(t)), \omega)$$

has existence and indefinite extension of solutions and

$$\frac{\partial V}{\partial t}(t, x) + \frac{\partial V}{\partial x}(t, x)\, \bar{F}(t, x, p(t, x), \omega) \leqslant -\gamma_3(\|x\|) + \varepsilon \quad (11.58)$$

for all $(t, x) \in T \times X$, then P practically stabilises (11.9) – (11.10) about zero.

Proof: This theorem follows from Theorem 6.1 and the fact that, given any neighbourhood B of the origin in $\mathscr{R}^n$, there exists $\varepsilon > 0$ which satisfies (11.52) and assures that B is a neighbourhood of the corresponding $S(\delta)$ as given by (11.54) – (11.56).

In the next section, we present some practically stabilising controller sets whose design is based on meeting the criteria in the above theorem.

11.7 Modified L–G controllers

11.7.1 Systems under consideration

The class of uncertain systems under consideration here is more general than the original class presented in Section 11.5.1, but not as general as the class presented in Section 11.5.3. An uncertain system considered here is described by (11.19) – (11.20), where $X = \mathscr{R}^n$, $U = \mathscr{R}^m$ and $\bar{F}$ satisfies the following assumption.

Assumption A5: There exist a Caratheodory function $B: T \times X \to \mathscr{R}^{n \times m}$, a candidate Lyapunov function $V: T \times X \rightarrow \mathscr{R}_+$, a class K function $\gamma_3: \mathscr{R}_+ \rightarrow \mathscr{R}_+$, and strongly Caratheodory functions $\kappa, \varrho: T \times X \rightarrow \mathscr{R}_+$ such that

$$\bar{F}(t, x, u, \omega) = f^s(t, x, \omega) + B(t, x)\, g(t, x, u, \omega) \tag{11.59}$$

for all $(t, x, u, \omega) \in T \times X \times U \times \Omega$, for some functions $f^s: T \times X \times \Omega \rightarrow \mathscr{R}^n$ and $g: T \times X \times U \times \Omega \rightarrow \mathscr{R}^m$ which satisfy:

(i) For each $\omega \in \Omega$, $f^s(\cdot, \omega)$ is Caratheodory and

$$\frac{\partial V}{\partial t}(t, x) + \frac{\partial V}{\partial x}(t, x) f^s(t, x, \omega) \leqslant -\gamma_3(\|x\|) \tag{11.60}$$

for all $(t, x) \in T \times X$.

(ii) For each $\omega \in \Omega$, $g(\cdot, \omega)$ is strongly Caratheodory and there exist functions $\beta_1, \beta_2: T \times X \times \Omega \rightarrow \mathscr{R}_+$ such that for all $(t, x, u) \in T \times X \times U$,

$$u^T g(t, x, u, \omega) \geqslant -\beta_1(t, x, \omega)\|u\| + \beta_2(t, x, \omega)\|u\|^2 \tag{11.61}$$

and

$$\beta_1(t, x, \omega) \leqslant \kappa(t, x) \tag{11.62}$$

$$\beta_1(t, x, \omega) \leqslant \beta_2(t, x, \omega)\, \varrho\,(t, x) \tag{11.63}$$

11.7.2 Modified L–G controllers

In this section we present some zero-state practically stabilising controller sets for the systems considered in the previous section. Each controller presented is a continuous-in-state approximation to some *L–G* controller presented in Section 11.5.2.

Consider any uncertain system described in Section 11.7.1 and let $(B, V, \gamma_3, \kappa, \varrho)$ be any quintuple which assures the satisfaction of Assumption A5. Choose any strongly Caratheodory functions $\kappa^c, \varrho^c: T \times X \rightarrow \mathscr{R}_+$ which satisfy

$$\kappa^c(t, x) \geqslant \kappa(t, x), \quad \varrho^c(t, x) \geqslant \varrho(t, x) \tag{11.64}$$

for all $(t, x) \in T \times X$ and define

$$\alpha(t, x) = B(t, x)^T \frac{\partial V}{\partial x}(t, x)^T \tag{11.65}$$

$$\mu(t, x) = \kappa^c(t, x)\, \alpha\,(t, x) \tag{11.66}$$

A proposed set of modified *L–G* controllers for practical stabilisability is the set

$$P = \{p^\varepsilon : \varepsilon > 0\}$$

where $p^\varepsilon: T \times X \rightarrow U$ is any strongly Caratheodory function which satisfies

$$p^\varepsilon(t, x) = -\varrho^c(t, x) s(t, x) \tag{11.67}$$

$$\|\alpha(t, x)\| s(t, x) = \|s(t, x)\| \alpha(t, x) \quad (11.68)$$

$$\|\mu(t, x)\| > 0 \Rightarrow \|s(t, x)\| \geqslant 1 - \|\mu(t, x)\|^{-1}\varepsilon \quad (11.69)$$

for all $(t, x) \in T \times X$.

As an example of a function satisfying the above requirements on p^ε, consider p^ε given by

$$p^\varepsilon(t, x) = \begin{cases} -\varrho^c(t, x)\|\mu(t, x)\|^{-1}\mu(t, x) & \text{if } \|\mu(t, x)\| > \varepsilon \\ -\varrho^c(t, x)\varepsilon^{-1}\mu(t, x) & \text{if } \|\mu(t, x)\| \leqslant \varepsilon \end{cases} \quad (11.70)$$

for all $(t, x) \in T \times X$; see Corless and Leitmann (1981).

As a further example, consider

$$p^\varepsilon(t, x) = -\varrho^c(t, x)[\|\mu(t, x)\| + \varepsilon]^{-1} \mu(t, x) \quad (11.71)$$

for all $(t, x) \in T \times X$; this function is suggested by the controllers presented in Ambrosino *et al.* (1985).

We have now the following theorem.

Theorem 7.1: Consider any uncertain system described by (11.19) – (11.20) (with $X = \mathscr{R}^n$ and $U = \mathscr{R}^m$) which satisfies Assumption A5; let P be any corrresponding set of modified L–G controllers as defined above; and suppose that $\gamma_1, \gamma_2 : \mathscr{R}_+ \to \mathscr{R}_+$ are KR functions which assure that V is a candidate Lyapunov function. Then, for each $p^\varepsilon \in P$ for which

$$\lim_{r\to\infty} \gamma_3(r) > \varepsilon \quad (11.72)$$

and for each $\omega \in \Omega$,

$$\dot{x}(t) = \bar{F}(t, x(t), p^\varepsilon(t, x(t)), \omega) \quad (11.73)$$

has existence and indefinite extension of solutions and tracks the zero state to within any neighbourhood B of $S(\delta_\varepsilon)$ where

$$S(\delta_\varepsilon) = \{x \in \mathscr{R}^n : \|x\| \leqslant \delta_\varepsilon\}, \quad (11.74)$$

$$\delta_\varepsilon = \bar{\gamma}_1^{-1} (\gamma_2(r_\varepsilon)), \quad (11.75)$$

$$r_\varepsilon = \bar{\gamma}_3^{-1} (\varepsilon). \quad (11.76)$$

Proof: The proof is similar to that of Theorem 7.1 in Corless (1984).

From Theorem 7.1, we can deduce the following corollary:

Corollary 7.1: Consider any uncertain system described by (11.19) – (11.20) (with $X = \mathscr{R}^n$ and $U = \mathscr{R}^m$) which satisfies Assumption A5 and let P be any corresponding set of modified L–G controllers as defined above. Then P practically stabilises (11.19) – (11.20) about zero.

Proof: This corollary follows from Theorem 7.1 and the fact that, given any neighbourhood B of the origin in $\mathscr{R}^n$, there exists $\varepsilon > 0$ which satisfies (11.72) and assures that B is a neighbourhood of $S(\delta_\varepsilon)$ as given by (11.74) – (11.76).

11.8 Tracking controllers for uncertain mechanical systems

In this section we consider uncertain mechanical systems, i.e. systems which are modelled using the laws of mechanics. Examples include robots and spacecraft. The systems treated here have a finite number of degrees of freedom (DOFs) and their main characterisation is that the number of independent scalar control inputs is the same as the number of DOFs of the system.

Utilising the results of the previous section, we obtain a set of controllers which assure that the trajectory of the controlled system can track a prespecified desired trajectory to within any desired degree of accuracy.

The controllers are illustrated by application to a 3-DOF model of the Manutec r3 robot which has an uncertain payload.

11.8.1 Systems under consideration and problem statement

Consider an uncertain mechanical system with N DOFs described by

$$J(t, q(t), \omega)\ddot{q}(t) = U(t, q(t), \dot{q}(t), \omega) + W(t, q(t), \dot{q}(t))\hat{u}(t) \tag{11.77}$$

$$\omega \in \Omega \tag{11.78}$$

where $t \in \mathscr{R}$ represents time; $q(t) \in \mathscr{R}^N$ is a vector of generalised co-ordinates which describe the configuration of the system; $\hat{u}(t) \in \mathscr{R}^N$ is a vector of control inputs; and ω is the lumped uncertain element.

The set Ω and the functions $J: \mathscr{R} \times \mathscr{R}^N \times \Omega \to \mathscr{R}^{N\times N}$, $U: \mathscr{R} \times \mathscr{R}^N \times \mathscr{R}^N \times \Omega \to \mathscr{R}^N$, and $W: \mathscr{R} \times \mathscr{R}^N \times \mathscr{R}^N \to \mathscr{R}^{N\times N}$ are known.

The following assumption is satisfied.

Assumption A6

(i) For each $\omega \in \Omega$, the functions $J(\cdot, \omega): \mathscr{R} \times \mathscr{R}^N \to \mathscr{R}^{N\times N}$ and $U(\cdot, \omega): \mathscr{R} \times \mathscr{R}^N \times \mathscr{R}^N \to \mathscr{R}^N$ are strongly Caratheodory.

(ii) For each $(t, q, \dot{q}, \omega) \in \mathscr{R} \times \mathscr{R}^N \times \mathscr{R}^N \times \Omega$, $J(t, q, \omega)$ is symmetric and $W(t, q, \dot{q})$ is non-singular.

(iii) There exist strongly Caratheodory functions $\hat{\beta}_0: \mathscr{R} \times \mathscr{R}^N \times \mathscr{R}^N \to \mathscr{R}_+$ and $\hat{\beta}_1, \hat{\beta}_2: \mathscr{R} \times \mathscr{R}^N \to \mathscr{R}_+$ such that for all $(t, q, \dot{q}) \in \mathscr{R} \times \mathscr{R}^N \times \mathscr{R}^N$

$$\| U(t, q, \dot{q}, \omega) \| \leqslant \hat{\beta}_0(t, q, \dot{q}) \tag{11.79}$$

$$\lambda_{max}[J(t, q, \omega)] \leqslant \hat{\beta}_1(t, q) \qquad (11.80)^\ddagger$$

‡ If $M \in \mathscr{R}^{N\times N}$ has all real eigenvalues, $\lambda_{max}[M]$ ($\lambda_{min}[M]$) is the maximum (minimum) eigenvalue of M.

$$\lambda_{min}[J(t, q, \omega)] \geqslant \hat{\beta}_2(t, q) > 0 \tag{11.81}$$

for all $\omega \in \Omega$.

For a prescribed, desired, piecewise C^2 motion $\bar{q}(\cdot):\mathcal{R} \to \mathcal{R}^N$, we consider the problem of obtaining a set $\hat{P}$ of feedback control functions with the following property. For each $\delta > 0$, there exists $\hat{p} \in \hat{P}$ such that by letting

$$\hat{u}(t) = \hat{p}(t, q(t), \dot{q}(t))$$

in (11.77) the resulting controlled system has the property that for any initial condition $(t_0, q^0, \dot{q}^0) \in \mathcal{R} \times \mathcal{R}^N \times \mathcal{R}^N$ and any $\omega \in \Omega$, $\|q(\cdot) - \bar{q}(\cdot)\|, \|\dot{q}(\cdot) - \bar{q}(\cdot)\|$ are bounded and

$$\limsup_{t\to\infty} \|q(t) - \bar{q}(t)\| \leqslant \delta, \tag{11.82}$$

where $q(t_0) = q^0$ and $\dot{q}(t_0) = \dot{q}^0$. Roughly speaking, we wish to assure asymptotic tracking of $\bar{q}(\cdot)$ by $q(\cdot)$ to within any desired degree of accuracy.

11.8.2 Proposed controllers

First we reformulate the problem into one already considered in Sections 11.6 and 11.7. Towards that end introduce the state tracking error

$$x(t) = \begin{bmatrix} q(t) - \bar{q}(t) \\ \dot{q}(t) - \dot{\bar{q}}(t) \end{bmatrix} \tag{11.83}$$

also choose any nonsingular matrix $T \in \mathcal{R}^{N\times N}$ and define a new control input

$$\tilde{u}(t) = T^T W(t, q(t), \dot{q}(t))\hat{u}(t) \tag{11.84}$$

Thus

$$\hat{u}(t) = [T^T W(t, q(t), \dot{q}(t))]^{-1}\tilde{u}(t) \tag{11.85}$$

and the evolution of $x(\cdot)$ is described by

$$\dot{x}(t) = Ax(t) + B[h(t, x(t), \omega) + G(t, x(t), \omega)\tilde{u}(t)] \tag{11.86}$$

$$\omega \in \Omega$$

with

$$A = \begin{bmatrix} 0 & \mathrm{I} \\ 0 & 0 \end{bmatrix}, \quad B = \begin{bmatrix} 0 \\ T \end{bmatrix} \tag{11.87}$$

where 0 and I are the zero and identity matrices in $\mathcal{R}^{N\times N}$, respectively, and‡

$$h = T^{-1}J^{-1}U - T^{-1}\ddot{\bar{q}} \tag{11.88}$$

$$G = \tilde{J}^{-1}, \quad \tilde{J} = T^T J T \tag{11.89}$$

‡ For ease of exposition, arguments have been omitted in the following expressions.

The original problem can be solved now by obtaining a zero-state practically stabilising set $\tilde{P}$ of controllers (which generate $\tilde{u}$) for (11.86).

To obtain a system description which satisfies Assumption A5, let

$$\tilde{u}(t) = p^1(t, x(t)) + p^2(t, x(t)) + u(t) \tag{11.90}$$

which yields

$$\dot{x}(t) = f^s(t, x(t), \omega) + Bg(t, x(t), u(t), \omega) \tag{11.91}$$

$$\omega \in \Omega$$

where

$$f^s(t, x, \omega) = Ax + BG(t, x, \omega)p^1(t, x) \tag{11.92}$$

$$g(t, x, u, \omega) = h(t, x, \omega) + G(t, x, \omega)p^2(t, x) + G(t, x, \omega)\, u \tag{11.93}$$

The function $p^1: \mathscr{R} \times \mathscr{R}^n \to \mathscr{R}^N$ $(n = 2N)$ is chosen so that the system

$$\dot{x}(t) = f^s(t, x(t), \omega) \tag{11.94}$$

is GUAS about 0. Specifically, choose any positive definite, symmetric matrix $Q \in \mathscr{R}^{n \times n}$, any positive scalar $\sigma > 0$, and any strongly Caratheodory function $\gamma: \mathscr{R} \times \mathscr{R}^n \to \mathscr{R}_+$ which satisfies

$$\gamma(t, x) \geqslant \tfrac{1}{2}\sigma\, \lambda_{max}[\tilde{J}(t, x, \omega)] \tag{11.95}$$

for all $(t, x, \omega) \in \mathscr{R} \times \mathscr{R}^n \times \Omega$; the existence of γ is guaranteed by Assumption A6(iii). Then

$$p^1(t, x) = -\gamma(t, x)B^T Px \tag{11.96}$$

where $P \in \mathscr{R}^{n \times n}$ is the unique positive definite, symmetric solution of the algebraic Riccati equation

$$PA + A^T P - \sigma PBB^T P + 2Q = 0 \tag{11.97}$$

With p^1 chosen as indicated above, consider the function $V: \mathscr{R} \times \mathscr{R}^n \to \mathscr{R}_+$, given by

$$V(t, x) = \tfrac{1}{2}\, x^T Px \tag{11.98}$$

as a candidate Lyapunov function for (11.94). Then,

$$\frac{\partial V}{\partial t}(t, x) + \frac{\partial V}{\partial x}(t, x) f^s(t, x, \omega) \leqslant -x^T Qx;$$

hence inequality (11.60) of Assumption A5 is satisfied with

$$\gamma_3(\|x\|) = \lambda_{min}[Q]\|x\|^2 \tag{11.99}$$

The function $p^2: \mathscr{R} \times \mathscr{R}^n \to \mathscr{R}^N$ is strongly Caratheodory and is chosen to reduce the magnitude of the uncertain term $h + Gp^2$.

Utilising (11.93) we note that

$$u^T g(t, x, u, \omega) \geqslant -\beta_1(t, x, \omega)\|u\| + \beta_2(t, x, \omega)\|u\|^2$$

where

$$\beta_1(t, x, \omega) = \|h(t, x, \omega) + G(t, x, \omega)p^2(t, x)\| \tag{11.100}$$

$$\beta_2(t, x, \omega) = \lambda_{min}[G(t, x, \omega)] = [\lambda_{max}[\tilde{J}(t, x, \omega)]]^{-1} > 0 \tag{11.101}$$

It follows from Assumption A6(iii) that there exist strongly Caratheodory functions $\kappa, \varrho : \mathscr{R} \times \mathscr{R}^n \to \mathscr{R}_+$ such that for all $(t, x) \in \mathscr{R} \times \mathscr{R}^n$

$$\beta_1(t, x, \omega) \leqslant \kappa(t, x) \tag{11.102}$$

$$\beta_1(t, x, \omega)/\beta_2(t, x, \omega) \leqslant \varrho(t, x) \tag{11.103}$$

for all $\omega \in \Omega$; thus (11.61) – (11.63) are satisfied and hence A5 is satisfied.

Utilising the results of Section 11.7.2, a practically stabilising set of controllers for (11.91) is the set

$$P = \{p^\varepsilon : \varepsilon > 0\} \tag{11.104}$$

where $p^\varepsilon : \mathscr{R} \times \mathscr{R}^n \to \mathscr{R}^N$ is any strongly Caratheodory function which satisfies (11.67) – (11.69) where

$$\alpha(t, x) = B^T P x \tag{11.105}$$

and κ^c, ϱ^c, μ are chosen to satisfy (11.64), (11.66).

Thus, a practically stabilising set of controllers for (11.86) is

$$\tilde{P} = \{\tilde{p} : \tilde{p} = p^1 + p^2 + p^\varepsilon, p^\varepsilon \in P\}$$

where p^1, p^2 are as specified previously. The controllers generating $\hat{u}(t)$ are given by (11.85) with

$$\tilde{u}(t) = \tilde{p}(t, x(t))$$

and $\tilde{p} \in \tilde{P}$.

11.8.3 Application to a Manutec r3 robot

In this section, we apply the results of the previous section to a 3-DOF model of a Manutec r3 robot as illustrated in Fig. 11.1. A complete description of the robot is given in Otter and Turk (1987). Torques are applied to the arms by three electric motors, one for each arm.

Letting

$$q = [q_1\ q_2\ q_3]^T$$

$$\hat{u} = [\hat{u}_1\ \hat{u}_2\ \hat{u}_3]^T$$

where $\hat{u}_i$ is the control voltage applied to motor i, the robot can be modelled by

(11.77). In this model, ω is the mass of an uncertain payload modelled as a point mass located at P. The only information available on ω is an upper bound $\bar{m}$. Thus

$$\omega \in \Omega = [0, \bar{m}]$$

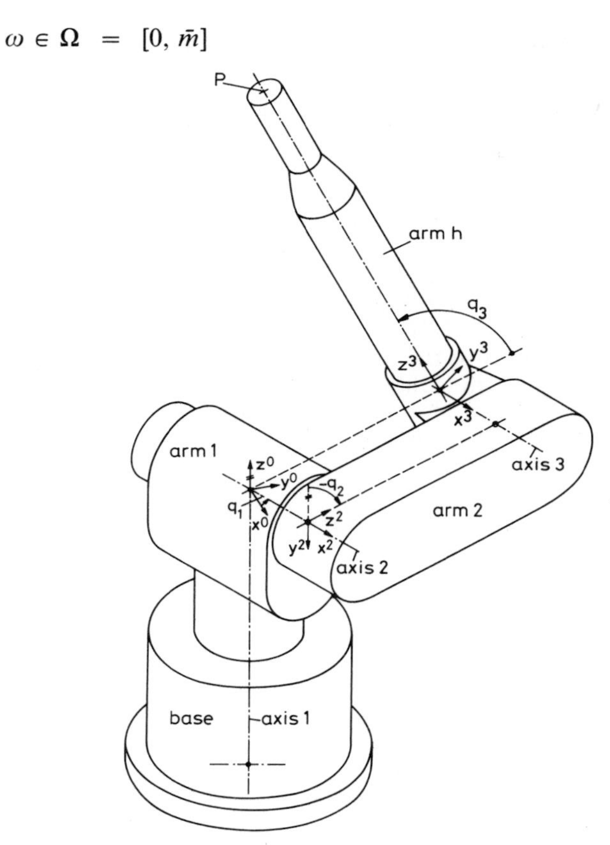

Fig. 11.1 *Manutec r3 robot (from Otter and Turk, 1987)*

The mass matrix $J(t, q, \omega)$ is symmetric and W is constant, diagonal, and non-singular. The generalised force term $U(t, q, \dot{q}, \omega)$ is of the form

$$U(t, q, \dot{q}, \omega) = \chi^g(q, \omega) + \chi^d(q, \dot{q}, \omega)$$

Utilising the expressions given in Otter and Turk (1987) for J, χ^g, χ^d, one can show that there exist scalars $c_1, c_2, c_3, c_4 \in \mathscr{R}_+$ such that

$$\|\chi^g(q, \omega)\| \leqslant c_1$$

$$\|\chi^d(q, \dot{q}, \omega)\| \leqslant c_2\|\dot{q}\|^2$$

$$\lambda_{max}[J(t, q, \omega)] \leqslant c_3$$

$$\lambda_{min}[J(t, q, \omega)] \geqslant c_4 > 0$$

for all $(t, q, \dot{q}) \in \mathscr{R} \times \mathscr{R}^3 \times \mathscr{R}^3$ and $\omega \in \Omega$. Thus Assumption A6 is satisfied; let

$$\hat{\beta}_0(t, q, \dot{q}) = c_1 + c_2\|\dot{q}\|^2$$

$$\hat{\beta}_1(t, q) = c_3$$

$$\hat{\beta}_2(t, q) = c_4$$

Suppose $\bar{q}(\cdot):\mathscr{R} \to \mathscr{R}^3$ is a given desired trajectory and define

$$x = \begin{bmatrix} q - \bar{q}(t) \\ \dot{q} - \dot{\bar{q}}(t) \end{bmatrix}$$

Now choose any nonsingular matrix $T \in \mathscr{R}^{3\times 3}$, any positive definite symmetric matrix $Q \in \mathscr{R}^{6\times 6}$, any positive scalar $\sigma > 0$, and let $P \in \mathscr{R}^{6\times 6}$ be the unique positive definite solution of (11.97).

Then, utilising the results of the previous section, a set of controllers which yield tracking of $\bar{q}(\cdot)$ by $q(\cdot)$ to within any desired degree of accuracy is given by

$$\hat{u}(t) = (T^T W)^{-1}\,\tilde{p}^{\varepsilon}(t, q(t), \dot{q}(t))$$

where, for any $\varepsilon > 0$,

$$\tilde{p}^{\varepsilon}(t, q, \dot{q}) = p^1(x) + p^2(t) + p^{\varepsilon}(t, x)$$

with

$$p^1(x) = -\gamma\alpha(x)$$

$$p^2(t) = \gamma_2 T^{-1}\ddot{\bar{q}}(t)$$

$$p^{\varepsilon}(t, x) = \varrho^c(t, \dot{q})[\varepsilon + \|\mu(t, x)\|]^{-1}\mu(t, x)$$

Table 11.2

$\tilde{J} = T^T J T$
$\lambda_{min}[\tilde{J}] \geqslant \boldsymbol{\sigma} > 0$
$\lambda_{max}[\tilde{J}] \leqslant \bar{\sigma}$
$\gamma = \frac{1}{2}\sigma\bar{\sigma}$
$\gamma_2 = 2(\boldsymbol{\sigma}^{-1} + \bar{\sigma}^{-1})^{-1}$
$\|T^{-1}J^{-1}\chi^g\| \leqslant \kappa_1$
$\|T^{-1}J^{-1}\chi^d\| \leqslant \kappa_2\|\dot{q}\|^2$
$\|\gamma_2\tilde{J}^{-1} - I\| \leqslant \kappa_3$
$\|\tilde{J}\|\,\|T^{-1}J^{-1}\chi^g\| \leqslant \rho_1$
$\|\tilde{J}\|\,\|T^{-1}J^{-1}\chi^d\| \leqslant \rho_2\|\dot{q}\|^2$
$\|\tilde{J}\|\,\|\gamma_2\tilde{J}^{-1} - I\| \leqslant \rho_3$

where

$$\alpha(x) = B^T P x$$

$$\mu(t, x) = \kappa^c(t, \dot{q})\alpha(x);$$

the bounding functions κ^c and ϱ^c are chosen to be of the form

$$\kappa^c(t, \dot{q}) = \kappa_1 + \kappa_2 \|\dot{q}\|^2 + \kappa_3 \|T^{-1}\ddot{\bar{q}}(t)\|$$

$$\varrho^c(t, \dot{q}) = \varrho_1 + \varrho_2 \|\dot{q}\|^2 + \varrho_3 \|T^{-1}\ddot{\bar{q}}(t)\|$$

The scalars γ, γ_2, κ_i, ϱ_i, $i = 1, 2, 3$, are chosen to satisfy the inequalities given in Table 11.2 for all $(t, q, \dot{q}, \omega) \in \mathscr{R} \times \mathscr{R}^3 \times \mathscr{R}^3 \times \Omega$.

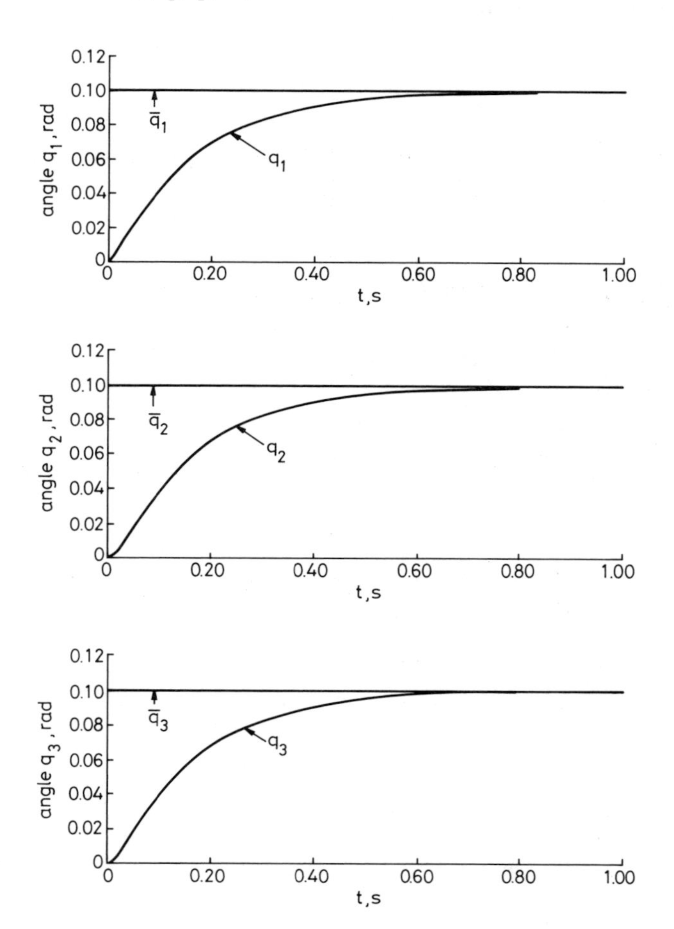

Fig. 11.2 *Numerical simulation results for $\omega = 0$*

Numerical simulation results: In numerical simulations we used the system parameter values given in Otter and Turk (1987); in particular, we chose

$$\bar{m} = 15\,\text{kg}$$

We let

$$T = \begin{bmatrix} 1{\cdot}4142 & 0 & 0 \\ 0 & 1 & 0 \\ 0 & 0 & 1{\cdot}8708 \end{bmatrix}, \quad \sigma = 0{\cdot}5$$

Q was chosen so that

$$B^T P = -2\sigma^{-1} T^{-1} \underline{A}$$

where

$$\underline{A} = \begin{bmatrix} -a_1 & 0 & 0 & -a_2 & 0 & 0 \\ 0 & -a_1 & 0 & 0 & -a_2 & 0 \\ 0 & 0 & -a_1 & 0 & 0 & -a_2 \end{bmatrix}, \quad a_1 = 72, \quad a_2 = 12.$$

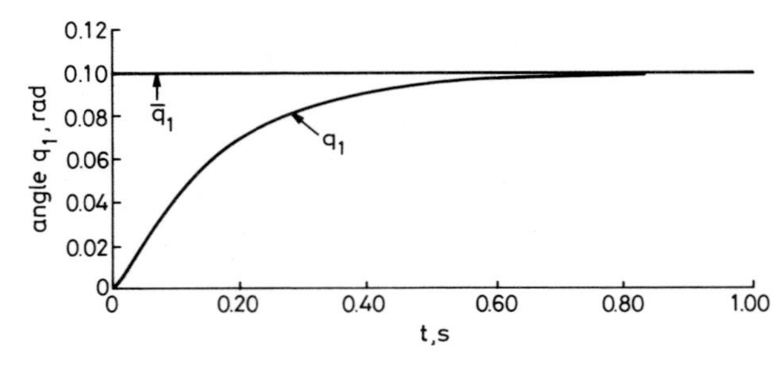

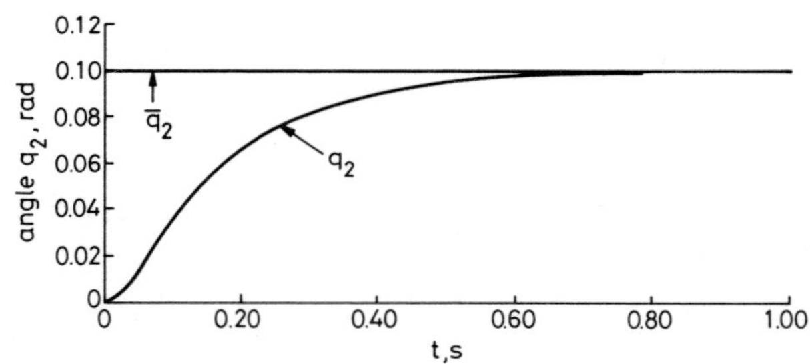

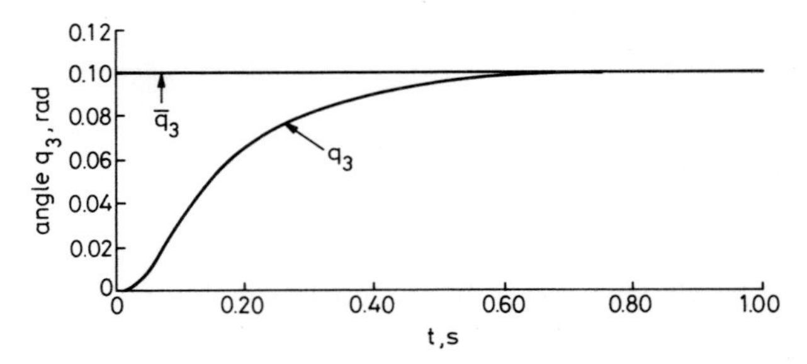

Fig. 11.3 *Results for* $\omega = 15\,kg$

The rest of the controller parameters were as follows:

$$\varepsilon = 5, \quad \underline{\sigma} = 33{\cdot}1, \quad \bar{\sigma} = 182,$$

$$\kappa_1 = 7{\cdot}29, \quad \kappa_2 = 0{\cdot}744, \quad \kappa_3 = 1{\cdot}25,$$

$$\varrho_1 = 941, \quad \varrho_2 = 136, \quad \varrho_3 = 226$$

Figs. 11.2 – 11.5 illustrate the results obtained for desired motion

$$\bar{q}_i(t) \equiv 0{\cdot}1 \text{ rad}, \quad i = 1, 2, 3$$

and initial conditions

$$q_i(0) = \dot{q}_i(0) = 0, \quad i = 1, 2, 3$$

For Figs. 11.2 and 11.4, $\omega = 0$; for Figs. 11.3 and 11.5, $\omega = 15\,\text{kg}$.

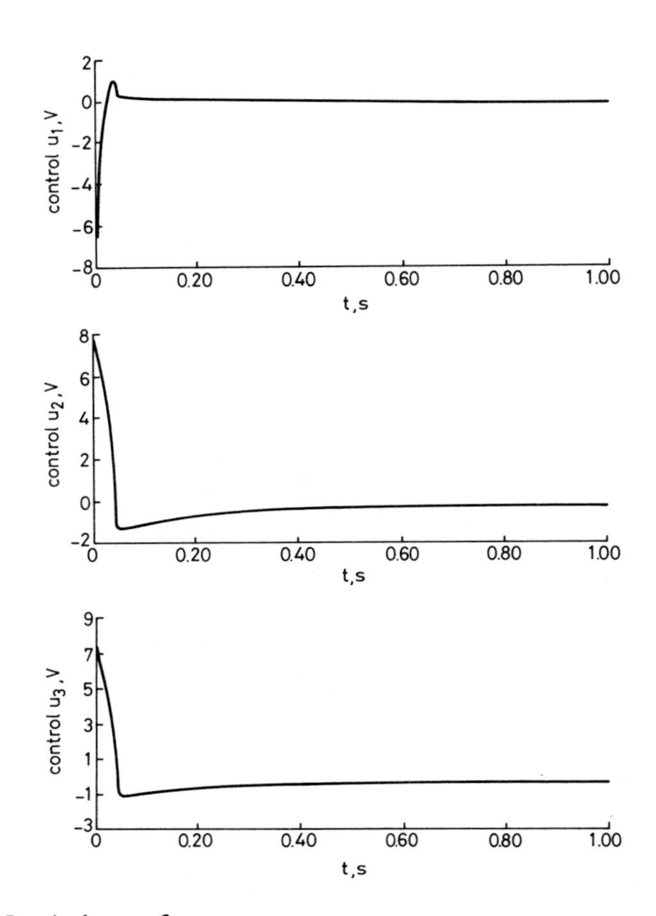

Fig. 11.4 *Results for $\omega = 0$*

11.9 Appendix

11.9.1 Caratheodory functions

In Section 11.9.1, T is any non-empty Lebesgue measurable subset of $\mathscr{R}$ and X is any non-empty subset of $\mathscr{R}^n$.

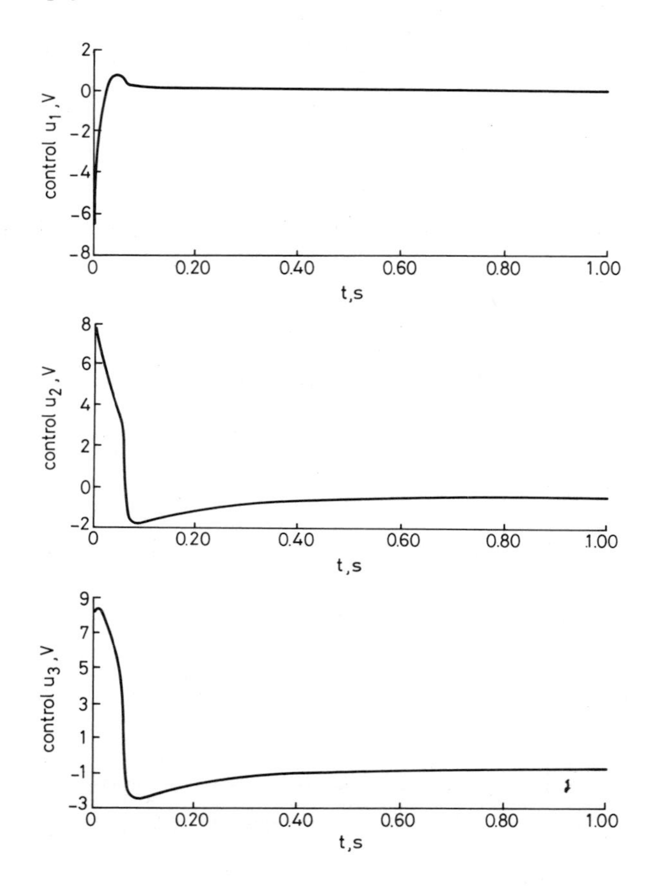

Fig. 11.5 *Results for ω_l = 15 kg*

Definition 9.1

(i) A function $f: T \times X \rightarrow \mathscr{R}^p$ is Caratheodory iff: for each $t \in T$, $f(t, \cdot)$ is continuous; for each $x \in X, f(x, \cdot)$ is Lebesgue measurable; and for each compact subset C of $T \times X$ there exists a Lebesgue integrable function $M_C(\cdot)$ such that, for all $(t, x) \in C$,

$$\|f(t, x)\| \leqslant M_C(t),$$

(ii) A function $f: T \times X \rightarrow \mathscr{R}^p$ is strongly Caratheodory iff it satisfies (i) with $M_C(\cdot)$ replaced by a constant M_C.

11.9.2 K and KR functions

Definition 9.2

(i) A function $\gamma: \mathscr{R}_+ \rightarrow \mathscr{R}_+$ belongs to class K iff it is continuous and satisfies

$$r_1 \leqslant r_2 \Rightarrow \gamma(r_1) \leqslant \gamma(r_2) \; \forall \; r_1, r_2 \in \mathscr{R}_+,$$

$$\gamma(0) = 0, \quad r > 0 \Rightarrow \gamma(r) > 0.$$

(ii) A function $\gamma: \mathscr{R}_+ \rightarrow \mathscr{R}_+$ belongs to class KR iff it belongs to K and

$$\lim_{r \rightarrow \infty} \gamma(r) = \infty$$

Lemma 9.1: If γ belongs to K, then there exist functions $\underline{\gamma}^{-1}, \bar{\gamma}^{-1}: [0, l) \rightarrow \mathscr{R}_+$, where $l = \lim_{r \rightarrow \infty} \gamma(r)$, such that

$$\underline{\gamma}^{-1}(s) = \inf \{r \in \mathscr{R}_+ : \gamma(r) = s\} \; \forall \; s \in [0, l)$$

$$\bar{\gamma}^{-1}(s) = \sup \{r \in \mathscr{R}_+ : \gamma(r) = s\} \; \forall \; s \in [0, l)$$

and these functions are strictly increasing and satisfy

$$\gamma(\underline{\gamma}^{-1}(s)) = s = \gamma(\bar{\gamma}^{-1}(s)) \; \forall \; s \in [0, l)$$

$$\underline{\gamma}^{-1}(\gamma(r)) \leqslant r \leqslant \bar{\gamma}^{-1}(\gamma(r)) \; \forall \; r \in \mathscr{R}_+$$

Proof: Corless (1984) contains a proof.

11.10 References

AMBROSINO, G., CELENTANO, G., and GAROFALO, F. (1985): 'Robust model tracking control for a class of nonlinear plants', *IEEE Trans.*, **AC-30**, pp. 275–279

ASTRÖM, J. K. (1970): 'Introduction to stochastic control theory' (Academic Press, NY)

BARMISH, B. R., CORLESS, M., and LEITMANN, G. (1983): 'A new class of stabilizing controllers for uncertain dynamical systems', *SIAM J. Contr. Optimiz.*, **21**, pp. 246–255

BARMISH, B. R., and LEITMANN, G. (1982): 'On ultimate boundedness control of uncertain systems in the absence of matching conditions', *IEEE Trans.*, **AC-27**, pp. 153–158

BARMISH, B. R., PETERSEN, I. R., and FEUER, A. (1983): 'Linear ultimate boundedness control of uncertain dynamical systems', *Automatica*, **19**, pp. 523–532

BASS, R. W. (1957): Discussion of LETOV, A. M.: 'Die Stabilität von Regelsystemen mit nachgebender Rückführung'. Proc. Heidelberg Conf. Automatic Control, pp. 209–210.

BREINL, W., and LEITMANN, G. (1983): 'Zustandsrückführung für dynamische Systeme mit Parameterunsicherheiten', *Regelungstechnik*, **31**, pp. 95–103

CESARI, L. (1971): 'Asymptotic behavior and stability problems in ordinary differential equations' (Springer-Verlag, NY)

CHANG, S. S. L., and PENG, T. K. C. (1972): 'Adaptive guaranteed cost control of systems with uncertain parameters', *IEEE Trans.*, **AC-17**, pp. 474–483

CHEN, Y. H. (1986): 'On the deterministic performance of uncertain dynamical systems', *Int. J. Control*, **43**, pp. 1557–1579

CHEN, Y. H. (1987*a*): 'Robust output feedback controller: Direct design', *Int. J. Control*, **46**, pp. 1083–1091

CHEN, Y. H. (1987*b*): 'Robust output feedback controller: Indirect design', *Int. J. Control*, **46**, pp. 1093–1103

CHEN, Y. H. and LEE, C. S. (1987): 'On the control of an uncertain water quality system', *Optimal Control Appl. Methods*, **8**

CHEN, Y. H. and LEITMANN, G. (1987): 'Robustness of uncertain systems in the absence of matching assumptions', *Int. J. Control*, **45**, pp. 1527–1542

CODDINGTON, E. A., and LEVINSON, N. (1955): 'Theory of ordinary differential equations', (McGraw-Hill, NY)

CORLESS, M. (1984): 'Control of uncertain systems', Ph.D. Dissertation, University of California, Berkeley

CORLESS, M. (1987): 'Robustness of a class of feedback-controlled uncertain nonlinear systems in the presence of singular perturbations', Proc. American Control Conference, Minneapolis

CORLESS, M., GOODALL, D. P., LEITMANN, G., and RYAN, E. P. (1985): 'Model-following controls for a class of uncertain dynamical systems', Proc. IFAC Conf. on Identification and System Parameter Estimation, York University, England

CORLESS, M., and LEITMANN, G. (1981): 'Continuous state feedback guaranteeing uniform ultimate boundedness for uncertain dynamic systems', *IEEE Trans.*, **AC-26**, pp. 1139–1144

CORLESS, M., and LEITMANN, G. (1983*a*): 'Erratum to 'Continuous state feedback guaranteeing uniform ultimate boundedness for uncertain dynamic systems', *IEEE Trans.*, **AC-28**, p. 249

CORLESS, M., and LEITMANN, G. (1983*b*): 'Adaptive control for systems containing uncertain functions and unknown functions with uncertain bounds, *J. Optimiz. Theory Applic.*, **41**, pp. 155–168

CORLESS, M., and LEITMANN, G. (1984): 'Adaptive control for uncertain dynamical systems', *in* BLAQUIERE, A., and LEITMANN, G. (Eds.): 'Dynamical systems and microphysics: control theory and mechanics', (Academic Press, NY)

CORLESS, M., and LEITMANN, G. (1985): 'Adaptive long-term management of some ecological systems subject to uncertain disturbances', *in* FEICHTINGER, G. (ed.): 'Optimal control theory and economic analysis 2' (Elsevier Science Publishers, Amsterdam)

CORLESS, M., LEITMANN, G., and RYAN, E. P. (1984): 'Tracking in the presence of bounded uncertainties', Proc. 4th IMA Int. Conf. Control Theory, Cambridge University, England

CORLESS, M., and MANELA, J. (1986): 'Control of uncertain discrete-time systems,' Proc. American Control Conf., Seattle, Washington

DOYLE, J. C. and STEIN, G. (1981): 'Multivariable feedback design: Concepts for a classical/modern synthesis', *IEEE Trans.*, **AC-26**, pp. 4–16

EL-GHEZAWI, O. M. E., ZINOBER, A. S. I., and BILLINGS, S. A. (1983): 'Analysis and design of variable structure systems using a geometric approach', *Int. J. Control*, **38**, pp. 657–671

ESLAMI, M., and RUSSELL, D. L. (1980): 'On stability with large parameter variations: Stemming from the direct method of Lyapunov', *IEEE Trans.*, **AC-25**, pp. 1231–1233

GRAYSON, L. P. (1964): 'Two theorems on the second method', *IEEE Trans.*, **AC-9**, p. 587

GRAYSON, L. P. (1965): 'The status of synthesis using Lyapunov's method', *Automatica*, **3**, pp. 91–121

GUTMAN, S. (1975): 'Differential games and asymptotic behavior of linear dynamical systems in the presence of bounded uncertainty', Ph.D. Dissertation, University of California, Berkeley

GUTMAN, S. (1976): 'Uncertain dynamical systems – A differential game approach', NASA TMX-73

GUTMAN, S. (1979): 'Uncertain dynamical systems – Lyapunov min–max approach', *IEEE Trans.*, **AC-24**, pp. 437–443

GUTMAN, S., and LEITMANN, G. (1975*a*): 'On a class of linear differential games', *J. Optimiz. Theory Applic.*, **17**, pp. 511–522

GUTMAN, S., and LEITMANN, G. (1975*b*): 'Stabilizing control for linear systems with bounded parameter and input uncertainty', Proc. 2nd IFIP Conf. Optimiz. Techniques, (Springer-Verlag, Berlin)

GUTMAN, S., and LEITMANN, G. (1976): 'Stabilizing feedback control for dynamical systems with bounded uncertainty', Proc. IEEE Conf. Decision Control

GUTMAN, S., and PALMOR Z. (1982): 'Properties of min–max controllers in uncertain dynamical systems', *SIAM J. Control Optimization*, **20**, 850–861

HA, I. J., and GILBERT, E. G. (1987): 'Robust tracking in nonlinear systems,' *IEEE Trans.*, **AC-32**, pp. 763–771

HAHN, W. (1967): 'Stability of motion' (Springer-Verlag, Berlin)

HALE, J. K. (1980): 'Ordinary differential equations' (Krieger)

HIRSCH, M. W., and SMALE, S. (1974): 'Differential equations, dynamical systems, and linear algebra, (Academic Press, NY)

HOLLOT, C. V., and BARMISH, B. R. (1980): 'Optimal quadratic stabilizability of uncertain linear systems', Proc. 18th Allerton Conf. Communications Contr. Computing

JOHNSON, G. W. (1964): 'Synthesis of control systems with stability constraints via the direct method of Lyapunov', *IEEE Trans.*, **AC-9**, pp. 270–273

KALMAN, R. E. and BERTRAM, J. E. (1960): 'Control system analysis and design via the 'second method' of Lyapunov. I: Continuous-time systems', *J. Basic Engineering*, **82**, pp. 371–393

KUSHNER, H. J. (1986): 'On the status of optimal control and stability for stochastic systems', IEE Int. Conv. Rec., **14**, pp. 143–151

LASALLE, J. P. (1962): 'Stability and control', *SIAM J. Control*, **1**, pp. 3–15

LASALLE, J. P. (1976): 'The stability of dynamical systems', SIAM

LASALLE, J. P., and LEFSCHETZ, S. (1961): ' Stability by Lyapunov's direct method with applications' (Academic Press, NY)

LEE, C. S., and LEITMANN, G. (1983): 'On optimal long-term management of some ecological systems subject to uncertain disturbances', *Int. J. Syst. Sci.*, **14**, pp. 979–994

LEITMANN, G. (1968): 'A simple differential game', *J. Optimiz. Theory Applic.*, **2**, pp. 220–225

LEITMANN, G. (1974): 'On stabilizing a linear system with bounded state uncertainty', *in* 'Topics in contemporary mechanics' (Springer-Verlag, Vienna)

LEITMANN, G. (1976): 'Stabilization of dynamical systems under bounded input disturbance and parameter uncertainty', *in* ROXIN, E. O., LIU, P.-T. and STERNBERG, R. L. (Eds): 'Differential games and control theory. II' (Marcel Dekker, NY) pp. 47–64

LEITMANN, G. (1978*a*): 'Guaranteed ultimate boundedness for a class of uncertain linear dynamical systems', *in* LIU, P. T., and ROXIN, E. O. (Eds.): 'Differential games and control theory. III' (Marcel Dekker, NY) pp. 29–49

LEITMANN, G. (1978*b*): 'Guaranteed ultimate boundedness for a class of uncertain linear dynamical systems', *IEEE Trans.*, **AC-23**, pp. 1109–1110

LEITMANN, G. (1979*a*): 'Guaranteed asymptotic stability for a class of uncertain linear dynamical systems', *J. Optimiz. Theory Applic.*, **27**, 99–106

LEITMANN, G. (1979*b*): 'Guaranteed asymptotic stability for some linear systems with bounded uncertainties', *J. Dynam. Syst. Meas. Contr.*, **101**, pp. 212–216

LEITMANN, G. (1980): 'Deterministic control of uncertain systems', *Acta Astronautica*, **7**, pp. 1457–1461

LEITMANN, G. (1981): 'On the efficacy of nonlinear control in uncertain linear systems', *J. Dynam. Syst. Meas. Control*, **102**, pp. 95–102

LEITMANN, G., LEE, C. S., and CHEN, Y. H. (1986): 'Decentralized control for a large scale uncertain river system', Proc. IFAC Workshop on Modelling, Decisions and Games For Social Phenomena, Beijing, China

LEITMANN, G., RYAN, E. P., and STEINBERG, A. (1986): 'Feedback control of uncertain systems: Robustness with respect to neglected actuator and sensor dynamics,' *Int. J. Control*, **43**, pp. 1243–1256

LEITMANN, G., and WAN, H. Y. Jr., (1978): 'A stabilization policy for an economy with some unknown characteristics,' *J. Franklin Inst.*, **306**, pp. 23–33

LEITMANN, G., and WAN, H.Y., Jr. (1979*a*): 'Macro-economic stabilization policy for an uncertain dynamic economy', *in* 'New trends in dynamic system theory and economics' (Academic Press, NY)

LEITMANN, G., and WAN, H. Y., Jr. (1979*b*): 'Performance improvement of uncertain macro-economic systems' in LIU, P.-T. (Ed.); 'Dynamic optimization and mathematical economics' (Plenum Press, NY)

LETOV, A. M. (1955): 'Stability of nonlinear regulating systems' (Izdatel'stvo Technichesko-Teoreticheskoi Literatury, Moscow) (in Russian)

LUENBERGER, D. G. (1966) 'Optimization by vector space methods' (Wiley, NY)

LUR'E, A. I. (1951): 'Some nonlinear problems in the theory of automatic control (Gostekhizdat, Moscow) (in Russian); Akademie Verlag, 1957, German translation, HM's Stationery Office, 1957; English translation

MADANI-ESFAHANI, S. M., DECARLO, R. A., CORLESS, M. J., and ZAK, S. H. (1986): 'On deterministic control of uncertain nonlinear systems', Proc. American Control Conf., Seattle, Washington

MOLANDER, P. (1979): 'Stabilisation of uncertain systems', LUTFD2/(TFRT-1020)/1–111, Lund Institute of Technology

MONOPOLI, R. V. (1965): 'Synthesis techniques employing the direct method', *IEEE Trans.*, **AC-10**, pp. 369–370

MONOPOLI, R. V. (1966*a*): 'Discussion on Two theorems on the second method', *IEEE Trans.*, **AC-11**, pp. 140–141

MONOPOLI, R. V. (1966*b*): 'Corrections to: 'Synthesis techniques employing the direct method', *IEEE Trans.*, **AC-11**, p. 631

MONOPOLI, R. V. (1966*c*): 'Engineering aspects of control system design via the 'direct method' of Lyapunov', CR-654, NASA

NUSSBAUM, R. D. (1983): 'Some remarks on a conjecture in parameter adaptive control', *Systems and Control Letters*, **3**, pp. 243–246

OTTER, M., and TURK, S. (1987): 'Mathematical model of the Manutec r3 robot', Institut für Dynamik der Flugsysteme, DFVLR, W. Germany

PATEL, R. V., TODA, M., and SRIDHAR, B. (1977): 'Robustness of linear quadratic state feedback designs in the presence of system uncertainty', *IEEE Trans.*, **AC-22**, pp. 945–949

PETERSEN, I. R., and BARMISH, B. R. (1984): 'Control effort considerations in the stabilisation of uncertain dynamical systems', Proc. Amer. Contr. Conf.

PETERSEN, I. R. and HOLLOT, C. (1986): 'A Riccati equation approach to the stabilization of uncertain linear systems', *Automatica*, **22**, pp. 397–411

RYAN, E. P. (1983): 'A variable structure approach to feedback regulation of uncertain dynamical systems', *Int. J. Control*, **38**, pp. 1121–1134

RYAN, E. P., and CORLESS, M. (1984): 'Ultimate boundedness and asymptotic stability of a class of uncertain dynamical systems via continuous and discontinuous feedback control', *IMA J. Math. Contr. Inf.*, **1**, pp. 223–242

RYAN, E. P., LEITMANN, G., and CORLESS, M. (1985): 'Practical stabilizability of uncertain dynamical systems: Application to robotic tracking', *J. Optimiz. Theory. Applic.*, **47**, pp. 235–252

SAFONOV, M. G. (1980): 'Stability and robustness of multivariable feedback systems' (MIT Press, Cambridge)

SHOURESHI, R., CORLESS, M. J., and ROESLER, M. D. (1987): 'Control of industrial manipulators with bounded uncertainties', *J. Dynam. Syst. Meas. Contr.*, **109**, pp. 53–59

SLOTINE, J. J., and SASTRY, S. S. (1983): 'Tracking control of non-linear systems using sliding surfaces, with application to robot manipulators', *Int. J. Control*, **38**, pp. 465–492

SPONG, M. W., THORP, J. S., and KLEINWAKS, J. M. (1984): 'The control of robot manipulators with bounded input. Part II: Robustness and disturbance rejection', Proc. IEEE Conf. on Decision Control

THORP, J. S., and BARMISH, B. R. (1981): 'On guaranteed stability of uncertain systems via linear control', *J. Optimiz. Theory Applic.*, **35**, pp. 559–579

UTKIN, V. I. (1977): 'Variable structure systems with sliding mode', *IEEE Trans.*, **AC-22**, pp. 212–222

VINKLER, A., and WOOD, L. J. (1979): 'Multistep guaranteed cost control of linear systems with uncertain parameters', *J. Guidance Control*, **2**, pp. 449–456

WIBERG, D. M. (1971): 'State space and linear systems' (McGraw-Hill, NY)

YOSHIZAWA, T., (1960): 'Lyapunov's function and boundedness of solutions', *Bol. Soc. Mat. Mex. Ser. 2,* **5**, pp. 146–151

YOUNG, K.-K. D. (1978): 'Design of variable structure model-following control systems', *IEEE Trans.*, **AC-23**, pp. 1079–1085

Chapter 12

Control of uncertain systems with neglected dynamics†

M. Corless

Purdue University

G. Leitmann

University of California

E. P. Ryan

University of Bath

12.1 Introduction

The prototype for the class of systems considered in this chapter is depicted in Fig. 12.1 and consists of a dynamical process P (imperfectly known) controlled by a (judiciously designed) feedback law (operator F) acting on state data generated by sensor S and implemented via actuator A.

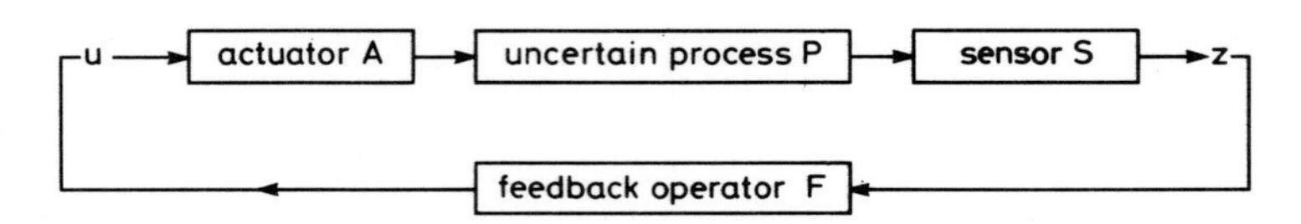

Fig. 12.1 *Prototype system*

† This chapter is based on research partially supported by the US National Science Foundation under grant MSM-8706927, by the US National Science Foundation and the US Air Force Office of Scientific Research under grant ECS-8602524, and by the UK Science and Engineering Research Council under grant GR/D/60638.

We assume (realistically) that the sensor and actuator are dynamic elements of the feedback loop; furthermore, we adopt the viewpoint that these dynamics are 'fast' relative to those of the process P to be controlled. If this is not the case, then, at the modelling stage, the sensor and actuator should be explicitly incorporated as an integral part of the process to be controlled.

We recognise, of course, that in the context of nonlinear systems, the concept of 'fastness' is difficult to quantify. Here we use the term loosely to indicate that the overall system exhibits a 'two time scale' structure as described in the next section.

12.2 The full-order system

The above prototype typifies a general class of singularly perturbed uncertain systems which can be decomposed, by means of a scalar parameter μ, into two coupled subsystems which henceforth will be referred to as the 'slow' subsystem (with state $x(t)$) and the 'fast' subsystem (with state $y(t)$). The parameter μ, henceforth referred to as the singular perturbation parameter, can be interpreted as some measure of the ratio of characteristic times of the fast and slow subsystems.

We model this general class of systems by the following coupled pair of differential equations:

$$\dot{x}(t) = X(t, x(t), y(t), u(t)), \quad x(t) \in \mathscr{R}^n, \; u(t) \in \mathscr{R}^m \tag{12.1a}$$

$$\mu\dot{y}(t) = Y(t, x(t), y(t), u(t), \mu), \quad y(t) \in \mathscr{R}^p, \; \mu \in (0, \infty) \tag{12.1b}$$

with measured output

$$z(t) = Sx(t) + Ty(t), \quad z(t) \in \mathscr{R}^n \tag{12.1c}$$

where X and Y are uncertain functions with the following structure:

$$X(t, x, y, u) = A_{11}x + A_{12}y + B_1u + g_1(t, x, y, u) \tag{12.2a}$$

$$Y(t, x, y, u, \mu) = C(t)[A_{21}x + y + B_2u] + g_2(t, x, y, u, \mu) \tag{12.2b}$$

A_{ij}, B_i, S and T are known constant real matrices; C is an uncertain measurable matrix-valued function; g_1 and g_2 are uncertain Carathéodory functions (i.e. measurable in their first argument, continuous in their other arguments and integrably bounded on compact sets).

Note that we require that the dimension of the output space coincides with the dimension of the slow subsystem state space. We refer to system (12.1)–(12.2) as the full-order system (a dynamical system on $\mathscr{R}^{n+p}$).

Now suppose that the dynamics of the fast subsystem are neglected, i.e. suppose that μ is set to zero, in which case (12.1b) reduces to an algebraic constraint on (12.1a). This procedure yields the reduced-order system (a dynamical system on $\mathscr{R}^n$). Suppose further that a feedback strategy is designed which guarantees some stability property $\mathscr{P}$ for the uncertain reduced-order system. One such design is proposed in Section 12.5 and analysed in Section 12.6, using the deterministic framework developed in, for example, Gutman (1979), Leitmann (1980, 1981), Corless & Leitmann (1981), Barmish & Leitmann (1982), Barmish, Corless & Leitmann (1983) and Ryan & Corless (1984). Then the essential question to be addressed is that of structural stability of property $\mathscr{P}$ with respect to a singular perturbation; i.e. does property $\mathscr{P}$ persist when the fast dynamics are re-introduced? More usefully, does there exist a calculable threshold value $\mu^* > 0$ such that property $\mathscr{P}$ persists for all values of the singular perturbation parameter in the interval $(0, \mu^*)$?

Our objective is to answer such questions affirmatively, under additional hypotheses on the full-order system. The first of these is an assumption which ensures that a well-defined reduced-order system results from setting $\mu = 0$ in (12.1*b*).

Assumption A1:

(i) $C(\cdot) = C_0 + \Delta C(\cdot)$, where $C_0 \in \mathscr{R}^{p \times p}$ is known with spectrum $\sigma(C_0) \subset \mathscr{C}^-$ (the open left half complex plane) and $\Delta C: \mathscr{R} \to \mathscr{R}^{p \times p}$ is an unknown measurable function with known bound κ_c (sufficiently small), viz. for all t, $\|\Delta C(t)\| \leqslant \kappa_c < 1/2\|P\|^{-1}$, where $P > 0$ (symmetric) solves the Lyapunov equation $PC_0 + C_0^T P + I = 0$;
(ii) $g_2(\cdot, \cdot, \cdot, \cdot, 0) = 0$.

12.3 The reduced-order system

Solving the algebraic equation $Y(t, x, y, u, 0) = 0$ for y (uniquely in view of Assumption A1) determines the function

$$(x, u) \mapsto H(x, u) := -[A_{21}x + B_2 u] \tag{12.3}$$

The reduced-order system associated with (12.1) is now defined as

$$\dot{x}(t) = X_r(t, x(t), u(t)), \quad x(t) \in \mathscr{R}^n \tag{12.4a}$$

with output

$$z(t) = Sx(t) + TH(x(t), u(t)), \quad z(t) \in \mathscr{R}^n \tag{12.4b}$$

where

$$X_r(t, x, u) := X(t, x, H(x, u), u) = \bar{A}x + \bar{B}u + \bar{g}(t, x, u) \tag{12.5a}$$

and

$$\bar{A} := A_{11} - A_{12}A_{21}, \; \bar{B} := B_1 - A_{12}B_2, \; \bar{g}(t, x, u) := g_1(t, x, H(x, u), u) \tag{12.5b}$$

At this stage, we loosely define our preliminary goal as that of rendering, by feedback, some acceptably small compact neighbourhood of the zero state of (12.4) globally attractive. Thus, it is not unreasonable to require the following:

Assumption A2:

(i) $(\bar{A}, \bar{B})$ is a stabilisable pair
(ii) $S\text{-}TA_{21}$ is invertible.

Now, let $(Q, \gamma_0) \in \mathscr{R}^{n \times n} \times \mathscr{R}^+ (\mathscr{R}^+ := [0, \infty))$ be a pair of design parameters with the properties (i) Q is symmetric and positive definite, (ii) $\gamma_0 > 0$ if $\sigma(\bar{A}) \not\subset \mathscr{C}^-$. These properties, in conjunction with A2, ensure that the Riccati equation

$$K\bar{A} + \bar{A}^T K + Q - 2\gamma_0 K\bar{B}\bar{B}^T K = 0 \tag{12.6}$$

admits a unique real positive-definite symmetric solution $K > 0$. Hence, for example, in the absence of uncertainty ($\bar{g} \equiv 0$) and if $S = I$ and $T = 0$, the output feedback law $u = -\gamma_0 \bar{B}^T Kz$ renders the zero state of (12.4) asymptotically stable.

We now impose some additional structure and bounds on the system uncertainty.

Assumption A3:

There exist a known subspace $\mathscr{S} \subset \mathscr{R}^m$, known non-negative real numbers α_1, α_2, α_3, α_4, β_1, β_2, a known continuous function ξ: $\mathscr{R} \times \mathscr{R}^n \to \mathscr{R}^+$, a known matrix $L \in \mathscr{R}^{n \times n}$, and unknown Carathéodory functions e: $\mathscr{R} \times \mathscr{R}^n \times \mathscr{R}^m \to \mathscr{R}^m$, f: $\mathscr{R} \times \mathscr{R}^n \times \mathscr{R}^m \to \mathscr{R}^n$ such that:

(i) $\bar{g} = \bar{B}e + f$
(ii) $B_2\Pi = 0$ where Π is the matrix of the orthogonal projection of $\mathscr{R}^m$ onto $\mathscr{S}$ and, for all $(t, x, u) \in \mathscr{R} \times \mathscr{R}^n \times \mathscr{R}^m$,
(iii) $\|f(t, x, u)\| \leqslant \alpha_1 \|x\| + \alpha_2$, $\alpha_1 < \alpha_1^* := 1/2[\sigma_{max}(K^2 Q^{-1}) \|Q^{-1}\|]^{-\frac{1}{2}}$
(iv) $\|(\mathrm{I}\text{-}\Pi)e(t, x, u)\| \leqslant \alpha_3 \|x\| + \alpha_4 + \beta_1 \|(\mathrm{I}\text{-}\Pi)u\|$, $\beta_1 < 1$
(v) $\|\Pi e(t, x, u)\| \leqslant \xi(t, Lx) + \beta_2 \|\Pi u\|$, $\beta_2 < 1$

In the familiar terminology, f models unmatched uncertainty in the reduced-order system and e models matched uncertainty. It is assumed that f is bounded above by a function which is affine in $\|x\|$ with sufficiently small coefficient α_1. The matched uncertainty e and control u admit a specific type of decomposition into orthogonal components $(\mathrm{I}\text{-}\Pi)e$, Πe and $(\mathrm{I}\text{-}\Pi)u$, Πu, respectively. The bound on $(\mathrm{I}\text{-}\Pi)e$ is affine in both $\|x\|$ and $\|(\mathrm{I}\text{-}\Pi)u\|$, and is independent of Πu; the bound on Πe is independent of $(\mathrm{I}\text{-}\Pi)u$ and is not required to be affine in $\|x\|$. If all uncertainty is affinely bounded (i.e. by a function of the form $(x, u) \mapsto c_1 + c_2\|x\| + c_3\|u\|$), then we simply set $\Pi = 0$ in A3.

Define $\nu \in \mathscr{R}$, A: $\mathscr{R} \to \mathscr{R}^{n \times n}$ and $\Gamma_1, \Gamma_2, \Gamma_3 \subset \mathscr{R}$ as follows:

$$\nu := 1 - \alpha_1/\alpha_1^* \ (>0 \text{ by A3 (iii)}) \tag{12.7a}$$

$$A(\gamma) := A_{21} - \gamma B_2 \bar{B}^T K \tag{12.7b}$$

$$\Gamma_1 := \begin{cases} [\underline{\gamma}, \infty); & \alpha_3 = 0 \\ (\underline{\gamma}, \infty); & \alpha_3 > 0; \ \underline{\gamma} := (1 - \beta_1)^{-1}[\gamma_0 + \tfrac{1}{2}\nu^{-1}\alpha_3^2 \| Q^{-1} \|] \end{cases} \tag{12.7c}$$

$$\Gamma_2 := \{\gamma: A(\gamma) B_1 \Pi = 0\} \tag{12.7d}$$

$$\begin{aligned} \Gamma_3 := \{\gamma: & |S - TA(\gamma)| \neq 0; \ \Pi B_1^T K[S - TA(\gamma)]^{-1} T = 0; \\ & L[S - TA(\gamma)]^{-1} T = 0; \ \kappa(\gamma) := \gamma \| B_2 \bar{B}^T K[S - TA(\gamma)]^{-1} T \| \\ & < \tfrac{1}{2}(1 - 2\kappa_c \|P\|)/(\|PC_0\| + \kappa_c \|P\|)\} \end{aligned} \tag{12.7e}$$

Then the following additional assumption is required.

Assumption A4: $\Gamma^* := \Gamma_1 \cap \Gamma_2 \cap \Gamma_3 \neq \varnothing$

12.4 Problem formulation

Suppose a (time-dependent) output feedback control function $(t, z) \mapsto q(t, z)$ is designed which guarantees that the feedback-controlled reduced-order system (viz. $u(t) = -q(t, z(t))$ in (12.4)) possesses some desired stability property $\mathscr{P}$; then the basic question to be addressed is that of robustness of $\mathscr{P}$ with respect to singular perturbation, where the singularly perturbed system is defined by (12.1) with $u(t) = -q(t, z(t))$; in particular, does there exist a (calculable) constant $\mu^* > 0$ such that the full-order system (12.1), under output feedback control $u(t) = -q(t, z(t))$, possesses property $\mathscr{P}$ for all values $\mu \in (0, \mu^*)$?

Here, we take the desired property $\mathscr{P}$ to be the existence of a compact set $\Sigma \subset \mathscr{R}^n$ (respectively $\Sigma \subset \mathscr{R}^{n+p}$) containing the origin which is a global uniform attractor for the reduced-order system (respectively, the full-order system) in the following sense:

Definition 1: A compact set $\Sigma \subset \mathscr{R}^q$ is a global uniform attractor for the system

$$\dot{w}(t) = \Xi(t, w(t)), \quad w(t) \in \mathscr{R}^q \tag{$*$}$$

if the following properties hold:

(i) *Existence and continuation of solutions:* For each pair $(t_0, w^0) \in \mathscr{R} \times \mathscr{R}^q$ there exists a solution $w: [t_0, t_1) \to \mathscr{R}^q$ (absolutely continuous function satisfying $(*)$ almost everywhere) with $w(t_0) = w^0$ and every such solution can be extended into a solution on $[t_0, \infty)$.

(ii) *Uniform boundedness of solutions:* For each $r > 0$ there exists $R(r) > 0$ such that $\|w(t)\| < R(r)$ for all t on every solution w: $[t_0, \infty) \to \mathscr{R}^q$ of $(*)$ with $\|w(t_0)\| < r$, where $t_0 \in \mathscr{R}$ is arbitrary.
(iii) *Uniform stability of* Σ: For each $d > 0$ there exists $D(d) > 0$ such that $w(t) \in \Sigma + d\mathscr{B}$ for all t on every solution w: $[t_0, \infty) \to \mathscr{R}^q$ of $(*)$ with $w(t_0) \in \Sigma + D(d)\mathscr{B}$, where t_0 is arbitrary (note, $\mathscr{B}$ denotes the open unit ball in $\mathscr{R}^q$ and, for $\delta > 0$, $\Sigma + \delta\mathscr{B}$ denotes the set $\{\sigma + \varrho\colon \sigma \in \Sigma;\ \|\varrho\| < \delta\}$).
(iv) *Global uniform attractivity of* Σ: For each $d > 0$ and $r > 0$ there exists $\tau(d, r) \geqslant 0$ such that $w(t) \in \Sigma + d\mathscr{B}$ for all $t > t_0 + \tau(d, r)$ on every solution w: $[t_0, \infty) \to \mathscr{R}^q$ of $(*)$ with $w(t_0) \in \Sigma + r\mathscr{B}$, where $t_0 \in \mathscr{R}$ is arbitrary.

In the next section, we construct a feedback strategy which ensures property $\mathscr{P}$ for the reduced-order system (12.4).

12.5 Nonlinear output feedback

Choose $\varepsilon_1, \varepsilon_2 > 0$; these are design parameters and can be chosen arbitrarily small. Define p: $\mathscr{R} \times \mathscr{R}^n \to \mathscr{R}^m$ as

$$p(t, x) := p_0(x) + (\mathrm{I}-\Pi)p_1(x) + \Pi p_2(t, x) \tag{12.8a}$$

The function p_0 is *linear* and is given by

$$p_0(x) := [\gamma_1(\mathrm{I}-\Pi) + \gamma_2\Pi]\bar{B}^T Kx \tag{12.8b}$$

where $\gamma_1, \gamma_2 \in \mathscr{R}^+$ satisfy

$$\gamma_1 \in \Gamma^*,\ \gamma_2 \geqslant (1-\beta_2)^{-1}\gamma_0 \tag{12.8c}$$

The function p_1 is *nonlinear* and bounded and is given by

$$p_1(x) := \begin{cases} \varrho_1\varphi_1(\varrho_1(\mathrm{I}-\Pi)\bar{B}^T Kx) & \text{if } B_2 = 0 \text{ or } \{T = 0 \text{ and } \Pi = 0\} \\ 0 & \text{otherwise} \end{cases} \tag{12.8d}$$

where $\varrho_1 \in \mathscr{R}^+$ satisfies

$$\varrho_1 \geqslant (1-\beta_1)^{-1}\,\alpha_4 \tag{12.8e}$$

and φ_1: $\mathscr{R}^m \to \mathscr{R}^m$ is any smooth (C^1) function which satisfies

$$\|\varphi_1(v)\| \leqslant 1, \quad \langle v, \varphi_1(v)\rangle \geqslant \|v\| - \varepsilon_1 \quad \forall v \in \mathscr{R}^m \tag{12.8f}$$

and which has bounded derivative $D\varphi_1$; i.e. there exists $\kappa_\varphi \in \mathscr{R}^+$ such that $\|D\varphi_1(v)\| \leqslant \kappa_\varphi$ for all $v \in \mathscr{R}^m$. The *nonlinear* function p_2 is given by

$$p_2(t, x) := \varrho_2(t, Lx)\varphi_2(\varrho_2(t, Lx)\Pi B_1^T Kx) \tag{12.8g}$$

where ϱ_2: $\mathscr{R} \times \mathscr{R}^n \to \mathscr{R}^+$ is any continuous function which satisfies

$$\varrho_2(t, Lx) \geqslant (1-\beta_2)^{-1}\xi(t, Lx) \tag{12.8h}$$

and φ_2: $\mathscr{R}^m \to \mathscr{R}^m$ is any continuous function which satisfies

$$\|\varphi_2(v)\| \leqslant 1, \quad \langle v, \varphi_2(v)\rangle \geqslant \|v\| - \varepsilon_2 \quad \forall\, v \in \mathscr{R}^m \tag{12.8i}$$

The proposed output feedback control function q: $\mathscr{R} \times \mathscr{R}^n \to \mathscr{R}^m$ is now defined by

$$q(t, z) := p(t, [S-TA(\gamma_1)]^{-1}z) \tag{12.9}$$

Loosely speaking, the linear component (12.8*b*) of the control stabilises (if necessary) the nominal linear system and counteracts part of the uncertainty $(I-\Pi)e$ while nonlinear component (12.8*d*) (when active) counteracts the remaining part of $(I-\Pi)e$; nonlinear control component (12.8*g*) counteracts the uncertainty Πe.

As an example of a function φ_1 satisfying the above requirements, consider the function

$$\varphi_1\colon v \mapsto [\|v\| + \varepsilon_1]^{-1}v$$

for which (12.8*f*) clearly holds and, moreover, φ_1 is C^1 with $\|D\varphi_1(v)\| \leqslant \varepsilon_1^{-1}$ for all $v \in \mathscr{R}^m$. As an example of a function φ_2 satisfying the above requirements, consider the function

$$\varphi_2\colon v \mapsto \begin{cases} \|v\|^{-1}v, & \|v\| > \varepsilon_2 \\ \varepsilon_2^{-1}v, & \|v\| \leqslant \varepsilon_2 \end{cases}$$

12.6 A compact attractor for the output feedback controlled reduced-order system

For the reduced-order system (12.4), it may be verified that $q(t, z(t)) = p(t, x(t))$. Hence, setting $u(t) = -q(t, z(t))$ in (12.4*a*) yields the system

$$\dot{x}(t) = F_r(t, x(t)), \quad x(t) \in \mathscr{R}^n \tag{12.10a}$$

with

$$F_r(t, x) := \bar{A}x - \bar{B}p(t, x) + \bar{g}(t, x, -p(t, x)) \tag{12.10b}$$

We will now establish that system (12.10) possesses stability property $\mathscr{P}$. To this end, we define V: $\mathscr{R}^n \to \mathscr{R}^+$ (a Lyapunov function candidate) by

$$V(x) := \langle x, Kx\rangle \tag{12.11}$$

Theorem 1: The closed ellipsoid

$$\Sigma_{r_0} := \{x \in \mathscr{R}^n\colon V(x) \leqslant r_0^2\}, \quad r_0 := \tfrac{1}{2}\,\tilde{\alpha}_1^{-1}[\tilde{\alpha}_2 + (8\varepsilon\tilde{\alpha}_1 + \tilde{\alpha}_2^2)^{1/2}]$$

where

$$\tilde{\alpha}_1 := \tilde{\nu}\,\sigma_{min}(QK^{-1}), \quad \tilde{\nu} := \nu - \tfrac{1}{2}[(1-\beta_1)\gamma_1 - \gamma_0]^{-1}\alpha_3^2\|Q^{-1}\|$$

$$\tilde{\alpha}_2 := \begin{cases} 2\alpha_2\|K\|^{1/2}; & p_1 \not\equiv 0 \\ 2[\alpha_2 + \alpha_4\|\bar{B}\|]\|K\|^{1/2}; & p_1 \equiv 0 \end{cases}$$

$$\varepsilon := \begin{cases} 0; & \Pi = 0, \ B_2 \neq 0, \quad T \neq 0 \\ \varepsilon_1; & \Pi = 0, \ B_2 = 0 \text{ or } T = 0 \\ \varepsilon_1 + \varepsilon_2; & \Pi \neq 0, \ B_2 = 0 \\ \varepsilon_2; & \Pi \neq 0, \ B_2 \neq 0 \end{cases}$$

is a global uniform attractor for system (12.10).

Proof: Note initially that, for all $(t, x) \in \mathscr{R} \times \mathscr{R}^n$, we have

$$\begin{aligned}
\tfrac{1}{2}\langle\nabla V(x), F_r(t,x)\rangle &= \langle Kx, \bar{A}x\rangle - \langle Kx, \bar{B}p(t,x)\rangle \\
&+ \langle Kx, \bar{g}(t,x,-p(t,x))\rangle \\
&= \langle Kx, \bar{A}x\rangle + \langle Kx, f(t,x,-p(t,x))\rangle \\
&+ \langle \bar{B}^T Kx, -p(t,x) + e(t,x,-p(t,x))\rangle \\
&\leqslant -\tfrac{1}{2}\langle x, Qx\rangle \\
&+ \gamma_0\|\bar{B}^T Kx\|^2 + \tfrac{1}{2}(\alpha_1/\alpha_1^*)\langle x, Qx\rangle + \alpha_2\|K\|^{1/2}V^{1/2}(x) \\
&+ \langle (I\text{-}\Pi)\bar{B}^T Kx, (I\text{-}\Pi)[-p_0(x) - p_1(x) + e(t,x,-p(t,x))]\rangle \\
&+ \langle \Pi\bar{B}^T Kx, \Pi[-p_0(x) - p_2(t,x) + e(t,x,-p(t,x))]\rangle \\
&\leqslant -\tfrac{1}{2}\nu\langle x, Qx\rangle - [(1-\beta_1)\gamma_1 - \gamma_0]\|(I\text{-}\Pi)\bar{B}^T Kx\|^2 \\
&+ \alpha_3\|Q^{-1}\|^{1/2}\|(I\text{-}\Pi)\bar{B}^T Kx\|\langle x, Qx\rangle^{1/2} \\
&- [(1-\beta_2)\gamma_2 - \gamma_0]\|\Pi\bar{B}^T Kx\|^2 + \tfrac{1}{2}\tilde{\alpha}_2 V^{1/2}(x) + \varepsilon \\
&\leqslant -\tfrac{1}{2}[\nu - \tfrac{1}{2}[(1-\beta_1)\gamma_1 - \gamma_0]^{-1}\alpha_3^2\|Q^{-1}\|]\langle x, Qx\rangle \\
&+ \tfrac{1}{2}\tilde{\alpha}_2 V^{1/2}(x) + \varepsilon
\end{aligned}$$

whence

$$\langle\nabla V(x), F_r(t,x)\rangle \leqslant -\tilde{\nu}\langle x, Qx\rangle + \tilde{\alpha}_2 V^{1/2}(x) + 2\varepsilon \tag{12.12}$$

Thus

$$\langle\nabla V(x), F_r(t,x)\rangle \leqslant -\tilde{\alpha}_1 V(x) + \tilde{\alpha}_2 V^{1/2}(x) + 2\varepsilon \tag{12.13}$$

and hence

$$\langle\nabla V(x), F_r(t,x)\rangle < 0 \ \forall\ (t,x) \notin \mathscr{R} \times \Sigma_{r_0} \tag{12.14}$$

Since, by assumption, g_1 is a Carathéodory function, it follows that $\bar{g}$ and hence F_r is also Carathéodory. Thus, by the standard theory of ordinary differential equations (Hartman, 1982), we conclude that, for each $(t_0, x^0) \in \mathscr{R} \times \mathscr{R}^n$, there exists at least one local solution $x: [t_0, t_1) \to \mathscr{R}^n$ with $x(t_0) = x^0$. Moreover, since $\dot{V}(x(t)) = \langle \nabla V(x(t)), F_r(t, x(t)) \rangle$ for almost all t on every such solution, it follows from (12.14) that every such solution evolves within the compact set

$$\{x \in \mathscr{R}^n : V(x) \leqslant V(x^0)\} \cup \Sigma_{r_0}$$

(i.e. the 'larger' of the two component nested ellipsoids) and hence can be continued indefinitely. This establishes property (i) of Definition 1.

Property (ii) of Definition 1 clearly holds with

$$R(r) = \begin{cases} [\|K\| \|K^{-1}\|]^{1/2} r & \text{if } r > \|K\|^{-1/2} r_0 \\ \|K^{-1}\|^{1/2} r_0 & \text{if } r \leqslant \|K\|^{-1/2} r_0 \end{cases}$$

Furthermore, property (iii) of Definition 1 holds with $D(d) = [\|K\| \|K^{-1}\|]^{-1/2} d$. Finally, in view of (12.13), for each pair r_1, r_2 with $r_2 \geqslant r_1 > r_0$ we have

$$0 \leqslant I(r_1, r_2) := \int_{r_1^2}^{r_2^2} [\tilde{\alpha}_1 V - \tilde{\alpha}_2 V^{1/2} - 2\varepsilon]^{-1} \, dV$$

Hence, it is readily verified that property (iv) of Definition 1 holds with

$$\tau(d, r) = \begin{cases} I(r_0 + \|K^{-1}\|^{-1/2} d, \; r_0 + \|K\|^{1/2} r) & \text{if } d < [\|K\| \|K^{-1}\|]^{1/2} r \\ 0 & \text{otherwise} \end{cases}$$

This completes the proof.

Our next objective is to show that property $\mathscr{P}$ is not destroyed by the re-introduction of the fast dynamics.

12.7 A compact attractor for the output feedback controlled full-order system

Define $W: \mathscr{R}^n \times \mathscr{R}^p \to \mathscr{R}^+$ by

$$W(x, y) := \langle y - h(x), P[y - h(x)] \rangle, \quad h(x) := H(x, -p(t, x))$$
$$= -A(\gamma_1)x + B_2 p_1(x)$$

Our final assumption is now made.

Assumption A5:

(i) For all (t, x),

$$\|g_1(t, x, y_1, -q(t, Sx + Ty_1)) - g_1(t, x, y_2 - q(t, Sx + Ty_2))\|$$
$$\leqslant \lambda \|y_1 - y_2\| \quad \forall \, y_1, y_2$$

where $\lambda \geqslant 0$ is a known constant;

(ii) for all (t, x, y) and $\mu \geqslant 0$,

$$\|g_2(t, x, y, -q(t, Sx + Ty), \mu)\| \leqslant \mu\,[\kappa_1\|y - h(x)\| + \kappa_2\|x\| + \kappa_3]$$

where $\kappa_1, \kappa_2, \kappa_3 \geqslant 0$ are known constants.

While Assumptions 1 – 5 might appear somewhat esoteric, it is stressed that the class of systems which satisfy these hypotheses is far from trivial; for example, the assumptions hold for a class of uncertain systems with parasitic actuator and sensor dynamics considered by Leitmann, Ryan & Steinberg (1986).

Let functions $F\colon \mathscr{R} \times \mathscr{R}^n \times \mathscr{R}^p \rightarrow \mathscr{R}^n$ and $G\colon \mathscr{R} \times \mathscr{R}^n \times \mathscr{R}^p \times \mathscr{R}^+ \rightarrow \mathscr{R}^p$ be given by

$$\begin{aligned} F(t, x, y) &:= A_{11}x + A_{12}y - B_1 q(t, Sx + Ty) \\ &\quad + g_1(t, x, y, -q(t, Sx + Ty)) \\ &= F_r(t, x) + A_{12}[y - h(x)] + B_1[p(t, x) - q(t, Sx + Ty)] \\ &\quad + g_1(t, x, y, -q(t, Sx + Ty)) - g_1(t, x, h(x), -p(t, x)) \\ G(t, x, y, \mu) &:= C(t)[A_{21}x + y - B_2 q(t, Sx + Ty)] \\ &\quad + g_2(t, x, y, -q(t, Sx + Ty), \mu) \\ &= C(t)[y - h(x)] + C(t)B_2[p(t, x) - q(t, Sx + Ty)] \\ &\quad + g_2(t, x, y, -q(t, Sx + Ty), \mu) \end{aligned}$$

Then the problem under consideration reduces to that of determining a threshold value $\mu^* > 0$ (if such exists) such that the system (two coupled subsystems)

$$\dot{x}(t) = F(t, x(t), y(t)) \tag{12.15a}$$

$$\mu\dot{y}(t) = G(t, x(t), y(t), \mu) \tag{12.15b}$$

possesses stability property $\mathscr{P}$ for all $\mu \in (0, \mu^*)$. We resolve this question via an analysis akin to that of Saberi & Khalil (1984). First, the following lemmas are required:

Lemma 1: $\langle \nabla V(x), F_r(t, x)\rangle \leqslant -\tilde{\nu}\theta_1^2(x) + \tilde{\alpha}_2 V^{1/2}(x) + 2\varepsilon \quad \forall\,(t, x)$
where

$$\theta_1(x) := \langle x, Qx\rangle^{1/2}$$

Proof: This is implicit in the proof of Theorem 1 (see Section 12.6).

Lemma 2: $\langle \nabla_y W(x, y), G(t, x, y, 0)\rangle \leqslant -\omega\theta_2^2(x, y)\ \forall\,(t, x, y)$
where

$$\omega := (1 - 2\|PC_0\|\kappa(\gamma_1)) - 2\kappa_c\|P\|(1 + \kappa(\gamma_1)) > 0$$

and $\theta_2(x, y) := \|y - h(x)\|$

Proof:

$$\langle \nabla_y W(x,y), G(t,x,y,0)\rangle = 2\langle P[y - h(x)], C(t)[y - h(x)]\rangle$$
$$+ 2\langle P[y - h(x)], C(t)B_2[p(t,x) - q(t, Sx + Ty)]\rangle$$
$$\leqslant -[1 - 2\kappa_c\|P\|]\|y - h(x)\|^2 + 2\kappa(\gamma_1)[\|PC_0\| + \kappa_c\|P\|]\|y - h(x)\|^2$$
$$= -\omega\theta_2^2(x,y)$$

Lemma 3: There exist calculable constants $\psi_1, \psi_2, \psi_3 \in \mathscr{R}^+$ such that

$$\langle \nabla_x W(x,y), F(t,x,y)\rangle \leqslant \psi_1\theta_2^2(x,y) + \psi_2\theta_1(x)\theta_2(x,y)$$
$$+ \psi_3 W^{1/2}(x,y) \quad \forall\,(t,x,y)$$

Proof: For all $(t,x,y) \in \mathscr{R} \times \mathscr{R}^n \times \mathscr{R}^p$ we have

$$\langle \nabla_x W(x,y), F(t,x,y)\rangle = -2\langle y - h(x), PDh(x)[F_r(t,x)$$
$$+ A_{12}(y - h(x)) + B_1[p(t,x) - q(t, Sx + Ty)]]\rangle$$
$$- 2\langle y - h(x), PDh(x)[g_1(t,x,y,-q(t, Sx + Ty))$$
$$- g_1(t,x,h(x),-p(t,x))]\rangle$$

Now

$$PDh(x) = -PA(\gamma_1) + PB_2Dp_1(x)$$

and, since the C^1 function φ_1 in the definition of p_1 has a uniformly bounded derivative, we conclude that PDh is uniformly bounded. Moreover, a straightforward calculation reveals that

$$B_1[p(t,x) - q(t, Sx + Ty)] = -B_1p_0([S - TA(\gamma_1)]^{-1}T[y - h(x)])$$
$$- B_1I - \Pi[p_1(x + [S - TA(\gamma_1)]^{-1}T[y - h(x)]) - p_1(x)] \tag{12.16}$$

and hence, noting that p_1 is globally Lipschitz, we may conclude that

$$\|PDh(x)[F_r(t,x) + A_{12}(y - h(x)) + B_1[p(t,x) - q(t, Sx + Ty)]]\|$$

can be bounded above affinely in $\theta_1(x)$, $\theta_2(x,y)$.

Finally, Assumption A5(i) ensures that

$$\|PDh(x)[g_1(t,x,y,-q(t, Sx + Ty)) - g_1(t,x,h(x),-p(t,x))]\|$$

is bounded above linearly in $\theta_2(x, y)$. The required result now follows.

Lemma 4: There exists calculable constant $\eta_0 \in \mathscr{R}^+$ such that

$$\langle \nabla V(x),\ F(t,x,y) - F_r(t,x) \rangle \leqslant \eta_0 \theta_1(x) \theta_2(x,y) \quad \forall\ (t,x,y)$$

Proof:

$$\begin{aligned}\langle \nabla V(x),\ F(t,x,y) - F_r(t,x) \rangle &= 2\,\langle Kx,\ A_{12}[y - h(x)] \\ &\quad + B_1[p(t,x) - q(t, Sx + Ty)] \rangle \\ &\quad + 2\,\langle Kx,\ g_1(t,x,y, -q(t, Sx + Ty)) - g_1(t,x,h(x), -p(t,x)) \rangle\end{aligned}$$

from which, together with Assumption A5(i), (12.16) and noting that p_1 is globally Lipschitz, the required result follows.

Lemma 5:

$$\begin{aligned}\langle \nabla_y W(x,y),\ g_2(t,x,y, -q(t, Sx + Ty), \mu) \rangle &\leqslant \mu\, [\tilde{\kappa}_1 \theta_2^2(x,y) \\ &\quad + \tilde{\kappa}_2 \theta_1(x) \theta_2(x,y) + \tilde{\kappa}_3 W^{1/2}(x,y)] \ \ \forall\ (t,x,y,\mu)\end{aligned}$$

where

$$\tilde{\kappa}_1 := 2\kappa_1 \|P\|,\ \tilde{\kappa}_2 := 2\kappa_2 \|P\| \|Q^{-1}\|^{1/2},\ \tilde{\kappa}_3 := 2\kappa_3 \|P\|^{1/2}$$

Proof: The result is a direct consequence of Assumption A5(ii).

For notational convenience, define

$$\eta_1 := \psi_1 + \tilde{\kappa}_1, \quad \eta_2 := \psi_2 + \tilde{\kappa}_2, \quad \eta_3 := \psi_3 + \tilde{\kappa}_3$$

Choose any $\delta > 0$ (arbitrarily small) and define

$$\mu^* := \begin{cases} \tilde{v}\omega(\tilde{v}\eta_1 + \eta_0\eta_2)^{-1}; & \eta_0\eta_2 \neq 0 \\ \tilde{v}\omega(\tilde{v}\eta_1 + \delta)^{-1}; & \eta_0\eta_2 = 0 \end{cases} \tag{12.17}$$

$$\xi_\mu := k(\mu/\mu^*)^{1/2}, \quad k := \begin{cases} \eta_0/\eta_2; & \eta_0\eta_2 \neq 0 \\ \eta_0^2/4\delta; & \eta_0 \neq 0,\ \eta_2 = 0 \\ 4\delta/\eta_2^2; & \eta_0 = 0,\ \eta_2 \neq 0 \\ \delta; & \eta_0 = 0 = \eta_2 \end{cases}$$

Let U_μ: $\mathscr{R}^n \times \mathscr{R}^p \to \mathscr{R}^+$, parameterised by $\mu > 0$, be given by

$$U_\mu(x,y) := V(x) + \xi_\mu W(x,y)$$

Then the following theorem establishes property $\mathscr{P}$ for the full order system under output feedback control.

Theorem 2: For each $\mu \in (0, \mu^*)$, the closed ellipsoid

$$\{(x,y) \in \mathscr{R}^n \times \mathscr{R}^p \colon U_\mu(x,y) \leqslant r_\mu^2\},$$

$$r_\mu := \tfrac{1}{2}\sigma_\mu^{-1}[(\tilde{\alpha}_2^2 + \xi_\mu \eta_3^2)^{1/2} + (8\varepsilon\sigma_\mu + \tilde{\alpha}_2^2 + \xi_\mu \eta_3^2)^{1/2}]$$

where

$$\sigma_\mu := \min\{\sigma_{\min}(QK^{-1}), \xi_\mu^{-1}\|P\|^{-1}\}\|M_\mu^{-1}\|^{-1},$$

$$M_\mu := \begin{bmatrix} \tilde{v} & -\tfrac{1}{2}(\eta_0 + \xi_\mu\eta_2) \\ -\tfrac{1}{2}(\eta_0 + \xi_\mu\eta_2) & \xi_\mu(\mu^{-1}\omega - \eta_1) \end{bmatrix} > 0$$

is a global uniform attractor for system (12.15).

Proof: For all $(t, x, y) \in \mathscr{R} \times \mathscr{R}^n \times \mathscr{R}^p$, define

$$\mathscr{U}_\mu(t,x,y) := \langle \nabla_x U_\mu(x,y), F(t,x,y)\rangle + \mu^{-1}\langle \nabla_y U_\mu(x,y), G(t,x,y,\mu)\rangle$$

Then, using Lemmas 1 – 5, we have

$$\begin{aligned}
\mathscr{U}_\mu(t,x,y) &= \langle \nabla V(x), F(t,x,y)\rangle + \xi_\mu\langle \nabla_x W(x,y), F(t,x,y)\rangle \\
&\quad + \mu^{-1}\xi_\mu\langle \nabla_y W(x,y), G(t,x,y,\mu)\rangle \\
&= \langle \nabla V(x), F_r(t,x)\rangle + \langle \nabla V(x), F(t,x,y) - F_r(t,x)\rangle \\
&\quad + \xi_\mu\langle \nabla_x W(x,y), F(t,x,y)\rangle \\
&\quad + \mu^{-1}\xi_\mu\langle \nabla_y W(x,y), G(t,x,y,0)\rangle \\
&\quad + \mu^{-1}\xi_\mu\langle \nabla_y W(x,y), g_2(t,x,y,-q(t,Sx+Ty),\mu)\rangle \\
&\leqslant -\tilde{v}\theta_1^2(x) + \tilde{\alpha}_2 V^{1/2}(x) + 2\varepsilon + \eta_0\theta_1(x)\theta_2(x,y) \\
&\quad + \xi_\mu[\psi_1\theta_2^2(x,y) + \psi_2\theta_1(x)\theta_2(x,y) + \psi_3 W^{1/2}(x,y)] \\
&\quad - \mu^{-1}\xi_\mu\omega\theta_2^2(x,y) \\
&\quad + \xi_\mu[\tilde{\kappa}_1\theta_2^2(x,y) + \tilde{\kappa}_2\theta_1(x)\theta_2(x,y) + \tilde{\kappa}_3 W^{1/2}(x,y)]
\end{aligned}$$

Equivalently,

$$\begin{aligned}
\mathscr{U}_\mu(t,x,y) \leqslant &- \left\langle \begin{bmatrix}\theta_1(x) \\ \theta_2(x,y)\end{bmatrix}, M_\mu \begin{bmatrix}\theta_1(x) \\ \theta_2(x,y)\end{bmatrix}\right\rangle \\
&+ \left\langle \begin{bmatrix}\tilde{\alpha}_2 \\ \xi_\mu^{1/2}\eta_3\end{bmatrix}, \begin{bmatrix}V^{1/2}(x) \\ \xi_\mu^{1/2} W^{1/2}(x,y)\end{bmatrix}\right\rangle + 2\varepsilon
\end{aligned}$$

Now it is readily verified that M_μ is strictly positive definite for all $\mu \in (0, \mu^*)$. Hence

$$\begin{aligned}\mathscr{U}_\mu(t,x,y) &\leqslant -\|M_\mu^{-1}\|^{-1}[\theta_1^2(x) + \theta_2^2(x,y)] \\ &\quad + (\tilde{\alpha}_2^2 + \xi_\mu \eta_3^2)^{1/2} U_\mu^{1/2}(x,y) + 2\varepsilon \\ &\leqslant -\sigma_\mu U_\mu(x,y) \\ &\quad + (\tilde{\alpha}_2^2 + \xi_\mu \eta_3^2)^{1/2} U_\mu^{1/2}(x,y) + 2\varepsilon\end{aligned}$$

Finally, noting that

$$\dot{U}_\mu(x(t), y(t)) = \mathscr{U}_\mu(t, x(t), y(t))$$

almost everywhere along every solution $(x(\cdot), y(\cdot))$ of (12.15), the proof now proceeds in a manner similar to that of Theorem 1.

Corollary: For each $\mu \in (0, \mu^*)$, on every trajectory $(x(\cdot), y(\cdot))$ of (12.15), the (projected) motion $x(\cdot)$ is ultimately bounded within every compact set (in $\mathscr{R}^n$) containing an open set which in turn contains the closed ellipsoid

$$\Sigma_{r_\mu} := \{x \in \mathscr{R}^n\colon V(x) \leqslant r_\mu^2\}.$$

Moreover, $\lim_{\mu \to 0} d(\Sigma_{r_\mu}, \Sigma_{r_0}) = 0$, where d is the Hausdorff metric. In this sense, the reduced-order dynamical behaviour is recovered in the limit as $\mu \downarrow 0$.

12.8 Example: Uncertain system with actuator and sensor dynamics

Consider the uncertain system

$$\dot{x}(t) = Ax(t) + [B + \Delta B(t)]y_1(t) + d(t, x(t)), \quad x(t) \in \mathscr{R}^n \tag{12.18a}$$

with actuator dynamics

$$\mu \dot{y}_1(t) = [C_1 + \Delta C_1(t)](y_1(t) - u(t)), \quad y_1(t), u(t) \in \mathscr{R}^m \tag{12.18b}$$

and sensor dynamics

$$\mu \dot{y}_2(t) = [C_2 + \Delta C_2(t)](y_2(t) - x(t)), \quad y_2(t) \in \mathscr{R}^n \tag{12.18c}$$

in which case

$$z(t) = y_2(t) \tag{12.18d}$$

and where the known nominal system matrices A, B, C_1, C_2 satisfy the following:

H1:
(i) (A, B) is a stabilisable pair
(ii) $\sigma(C_1) \subset \mathscr{C}^-$
(iii) $\sigma(C_2) \subset \mathscr{C}^-$

The uncertain functions $\Delta B(\cdot)$ and $d(\cdot,\cdot)$ are assumed to satisfy

H2:
(i) $\Delta B(\cdot) = BE(\cdot)$, where $E(\cdot)$ (unknown) is measurable with $\|E(t)\| \leqslant \beta < 1 \ \forall t$
(ii) $d(\cdot,\cdot) = f(\cdot,\cdot) + Bg(\cdot,\cdot)$, where $f(\cdot,\cdot)$ and $g(\cdot,\cdot)$ are Carathéodory functions with
$\|f(t,x)\| \leqslant \alpha_1\|x\| + \alpha_2$, $\|g(t,x)\| \leqslant \alpha_3\|x\| + \alpha_4 \ \forall \ (t,x)$
and where $\alpha_1, \alpha_2, \alpha_3, \alpha_4, \beta$ are known constants.

Let P (symmetric and positive definite) denote the unique solution of

$$P\begin{bmatrix} C_1 & 0 \\ 0 & C_2 \end{bmatrix} + \begin{bmatrix} C_1^T & 0 \\ 0 & C_2^T \end{bmatrix} P + I = 0$$

Then the uncertain functions $\Delta C_1(\cdot)$ and $\Delta C_2(\cdot)$ are assumed to satisfy

H3: $\|\mathrm{diag}\{\Delta C_1(t), \Delta C_2(t)\}\| \leqslant \kappa_c < \frac{1}{2}\|P\|^{-1} \ \forall t$, where κ_c is a known constant.

The above can be interpreted in the context of system (12.1) – (12.2) by making the following identifications:

$$y = \begin{bmatrix} y_1 \\ y_2 \end{bmatrix} \in \mathscr{R}^p, \quad p := m + n$$

$$A_{11} = A, \quad A_{12} = [B \ \vdots \ 0], \quad A_{21} = \begin{bmatrix} 0 \\ -I \end{bmatrix}$$

$$B_1 = 0, \quad B_2 = \begin{bmatrix} -I \\ 0 \end{bmatrix}, \quad S = 0, \quad T = [0 \ \vdots \ I]$$

$$C(t) = C_0 + \Delta C(t), \quad C_0 = \mathrm{diag}\{C_1, C_2\}$$

$$\Delta C(t) = \mathrm{diag}\{\Delta C_1(t), \Delta C_2(t)\}$$

$$g_1(t,x,y,u) = d(t,x) + BE(t)[I \ \vdots \ 0]y$$

$$g_2 \equiv 0$$

In view of H1(ii), (iii) and H3, it is clear that Assumption A1 holds for this system. Now

$$\bar{A} = A_{11} - A_{12}A_{21} = A_{11} = A$$

$$\bar{B} = B_1 - A_{12}B_2 = -A_{12}B_2 = B$$

and hence, in view of H1(i), it follows that Assumption A2 holds. Also,

$$H(x,u) = -[A_{21}x + B_2u] = \begin{bmatrix} u \\ x \end{bmatrix}$$

and

$$\bar{g}(t,x,u) = g_1(t,x,H(t,x),u) = f(t,x) + Be(t,x,u)$$

where

$$e(t,x,u) = g(t,x) + E(t)u$$

Thus, in view of H2, it is clear that Assumption A3 holds with $\mathscr{S} = \{0\}, \Pi = 0$ and $\beta_1 = \beta$, provided that $\alpha_1 < \alpha_1^* = \frac{1}{2}[\sigma_{max}(K^2Q^{-1})\|Q^{-1}\|]^{-1/2}$.

Proceeding,

$$A(\gamma) = A_{21} - \gamma B_2 \bar{B}^T K = \begin{bmatrix} \gamma B^T K \\ -I \end{bmatrix}$$

$$S - TA(\gamma) = I, \quad \kappa(\gamma) = \gamma\|B^T K\|$$

$$\Gamma_2 = \mathscr{R},$$

$$\Gamma_3 = (-\infty, \tfrac{1}{2}(1 - 2\kappa_c\|P\|)(\|PC_0\| + \kappa_c\|P\|)^{-1}\|B^T K\|^{-1}) \subset \mathscr{R}$$

Assumption A4 now reduces to the following:

$$\text{A4*: } \underline{\gamma} < \tfrac{1}{2}(1 - 2\kappa_c\|P\|)(\|PC_0\| + \kappa_c\|P\|)^{-1}\|B^T K\|^{-1}$$

Finally, it is readily verified that Assumption A5(ii) holds trivially (since $g_2 \equiv 0$) and A5(i) holds with $\lambda = \beta\|B\|$.

For the purposes of numerical illustration, suppose $n = 1 = m$ and $A = 0$, $B = 1$, $C_1 = -\frac{1}{2} = C_2$, $\alpha_1 = \alpha_2 = \alpha_4 = \kappa_c = 0, \alpha_3 = \frac{1}{4} = \beta$, i.e. consider the uncertain system

$$\dot{x}(t) = (1 + E(t))y_1(t) + g(t, x(t)), \quad x(t) \in \mathscr{R},$$

$$|E(t)| \leqslant \tfrac{1}{4},$$

$$|g(t, x(t))| \leqslant \tfrac{1}{4}|x(t)|$$

$$\mu\dot{y}_1(t) = -\tfrac{1}{2}(y_1(t) - u(t)), \quad y_1(t) \in \mathscr{R}$$

$$\mu\dot{y}_2(t) = -\tfrac{1}{2}(y_2(t) - x(t)), \quad y_2(t) \in \mathscr{R}$$

Selecting $Q = 1$ and $\gamma_0 = 9/32$ in (12.6) yields $K = 4/3$ and a straightforward calculation gives

$$\Gamma^* = (17/24, 18/24) \subset \mathscr{R}$$

so that A4 (equivalently A4*) is seen to hold.

The feedback control law for this example is now given by

$$u(t) = -q(y_2(t)) = -(4\gamma_1/3)\, y_2(t), \quad \gamma_1 \in \Gamma^*$$

With a view to calculating (via (12.17), a threshold value $\mu^* > 0$ for the singular perturbation parameter μ, first observe that

$$\tilde{v} = \frac{24\gamma_1 - 10}{24\gamma_1 - 9}, \quad \tilde{\alpha}_1 = \frac{3}{4}\tilde{v}, \quad \omega = 1 - (4\gamma_1/3), \quad \tilde{\alpha}_2 = 0$$

With reference to Lemma 3, the following choices suffice:

$$\psi_1 = \frac{5}{6}\sqrt{(16\gamma_1^2 + 9)}, \quad \psi_2 = \frac{2}{3}\left[\frac{4\gamma_1}{3} + \frac{1}{4}\right]\sqrt{(16\gamma_1^2 + 9)}, \quad \psi_3 = 0$$

and, with reference to Lemma 4,

$$\eta_0 = \frac{10}{3}$$

Finally, with reference to Lemma 5,

$$\tilde{\kappa}_1 = \tilde{\kappa}_2 = \tilde{\kappa}_3 = 0$$

Selecting $\gamma_1 = 35/48 \in \Gamma^*$, (12.17) yields

$$\mu^* \approx 0.00146$$

From Theorem 2, it may be concluded that, for each $\mu \in (0, \mu^*)$, the output feedback controlled system is globally uniformly ultimately bounded within every neighbourhood of the origin, or, equivalently, the origin is a globally uniformly asymptotically stable equilibrium.

12.9 References

BARMISH, B. R., CORLESS, M., and LEITMANN, G. (1983): 'A new class of stabilizing controllers for uncertain dynamical systems', *SIAM J. Control & Optimization*, **21**, pp. 246–255

BARMISH, B. R., and LEITMANN, G. (1982): 'On ultimate boundedness control of uncertain systems in the absence of matching conditions', *IEEE Trans.*, **AC-27**, 153–158

CORLESS, M., and LEITMANN, G. (1981): 'Continuous state feedback guaranteeing uniform ultimate boundedness for uncertain dynamic systems', *IEEE Trans.*, **AC-26**, pp. 1139–1144

GUTMAN, S. (1979): 'Uncertain dynamical systems: Lyapunov min–max approach', *IEEE Trans.*, **AC-24**, pp. 437–443

HARTMAN, P., (1982): 'Ordinary differential equations' (Birkhäuser; Boston, Basel, Stuttgart)

LEITMANN, G. (1980): 'Deterministic control of uncertain systems', *Astronautica Acta*, **7**, pp. 1457–1461

LEITMANN, G. (1981): 'On the efficacy of nonlinear control in uncertain linear systems', *J. Dynamic Systems Meas. Control*, **103**, 95–102

LEITMANN, G., RYAN, E. P., and STEINBERG, A. (1986): 'Feedback control of uncertain systems: robustness with respect to neglected actuator and sensor dynamics', *Int. J. Control*, **43**, pp. 1243–1256

RYAN, E. P., and CORLESS, M. (1984): 'Ultimate boundedness and asymptotic stability of a class of uncertain dynamical systems via continuous and discontinuous feedback control', *IMA J. Math. Control & Info.*, **1**, pp. 223–243

SABERI, A., and KHALIL, H. K. (1984): 'Quadratic-type Lyapunov functions for singularly perturbed systems', *IEEE Trans.*, **AC-29**, pp. 542–550

Chapter 13

Nonlinear composite control of a class of nominally linear singularly perturbed uncertain systems[†]

F. Garofalo[‡]

University of Naples

G. Leitmann

University of California

13.1 Introduction

Almost every process to be controlled is subject to parameter and disturbance uncertainty. If the 'uncertain' quantities are completely unknown and unpredictable, the design of the controller is carried out neglecting these quantities and giving an *a posteriori* evaluation of the robustness of the designed controller against canonical disturbances and/or typical parameter variations (the gain and phase margin in classical control, for instance).

Whenever a description, even very poor, of the uncertainties is available, it is convenient to incorporate this information in the design procedure, trying to get at least acceptable *a priori* guaranteed performance of the closed loop system in the presence of the hypothesised uncertainties.

One of the most common pieces of information that is available to the designer is the knowledge of the bounds of variation of the uncertain quantities during the time in which the control action takes place. This modest *a priori* information, together with some structural conditions regarding the manner in which the uncertainties enter into the control loop (the so-called 'matching'

[†] Work supported by the NSF and AFOSR under Grant ECS-8602524, and by FORMEZ, Italy
[‡] This work was performed during a stay at the University of California, Berkeley

assumptions about uncertainties), enables the designer to construct controllers which guarantee at least boundedness and ultimate boundedness of the controlled variables. We will refer to this behaviour as practical stability of the closed loop system; e.g. see Barmish *et al.* (1983).

The design methodology for practical stability controllers is based essentially on the constructive use of Lyapunov functions; e.g., see Kalman and Bertram (1960). Controllers have been designed for linear and nonlinear plants, and they turn out to be linear or nonlinear combinations of the state variables (e.g., see Barmish *et al.* (1983), Corless and Leitmann (1981), Leitmann (1978), Ambrosino *et al.* (1985)). Besides being robust, by their very nature, against bounded uncertainties, an *a posteriori* analysis reveals that they possess other robustness characteristics. In Barmish and Leitmann (1982) and Chen and Leitmann (1987) it was proved, for instance, that these controllers can tolerate a sufficiently small amount of unmatched uncertainty without loss of practical stability of the closed loop system. Recently attention has been focused on the analysis of the robustness of these controllers against neglected parasitics. In Leitmann *et al.* (1986), Ryan and Corless (1988), Leitmann and Ryan (1987), Corless (1987), it has been proved that a controller designed on the basis of a reduced order model retains the practical stability capabilities if the neglected parasitics do not introduce further uncertainties, are asymptotically stable and sufficiently fast.

In this paper we develop a two-time-scale controller design for a singularly perturbed uncertain system. A two-time-scale design was proposed in Garofalo (1988) for the case when the fast dynamics of the system have no uncertainties. In this chapter, the system is assumed to be nominally linear with both the slow and the fast components subject to parameter and disturbance uncertainties. The design procedure requires a sequential construction of (i) a controller that guarantees the practical stability of the slow dynamics of the system under the influence of the slowly varying uncertainties (the so-called 'slow' controller) and (ii) a controller that guarantees the practical stability of the fast dynamics of the system when the slow controller is active. The overall controller is then obtained as the sum of the two partial controllers, and the conditions under which this 'composite controller', e.g. Kokotovic (1984), guarantees the practical stability of the complete system are analysed. The composite control turns out to be nonlinear in the slow component of the state, even though the plant was considered nominally linear.

13.2 The singularly perturbed uncertain dynamical system

The two-time-scale uncertain dynamical system considered here has the form of two coupled subsystems described by the following equations:

$$\begin{aligned}\dot{x}(t) &= A_{11}(q(t))(t) + A_{12}(q(t))x(t) + B_1(q(t))u(t) + \omega_1(q(t)) \\ &\triangleq f(x(t), z(t), u(t), q(t), \omega_1(q(t))) \qquad (13.1a)\end{aligned}$$

$$\varepsilon \dot{z}(t) = A_{21}(q(t))x(t) + A_{22}(q(t))z(t) + B_2(q(t))u(t) + \omega_2(q(t))$$
$$\triangleq g(x(t), z(t), u(t), q(t), \omega_2(q(t))) \qquad (13.1b)$$

where $(x^T(t)\ z^T(t))^T \in R^{n+m}$ represents the state of the system, $u(t) \in R^p$ is the control input, $q(t) \in R^q$ is the uncertainty, $\omega_1(q(t))$ and $\omega_2(q(t))$ are the external disturbances, and $\varepsilon \in [0, \infty)$ is the 'small' singular perturbation parameter. Matrices $A_{ij}(\cdot)$ and $B_j(\cdot)$, $i,j = 1,2$, and vectors $\omega_i(\cdot)$, $i = 1,2$, are continuous in their argument. Furthermore it is assumed that the uncertainty, $q(\cdot)$: $R \rightarrow R^q$, is Lebesgue measurable and its value $q(t)$ lies in a prespecified bounding set $\mathcal{Q} \subset R^q$ for all $t \in R$.

The 'matching' assumptions, stated below, concern the manner in which the uncertainty enters structurally in the state equations and reflect our knowledge of the system.

Assumption 1: There exist known constant matrices $\bar{A}_{ij}$, $i, j = 1,2$, and full rank matrices $\bar{B}_j$, $j = 1,2$, such that the following decomposition holds:

$$A_{ij}(q) = \bar{A}_{ij} + \bar{B}_i D_{ij}(q), \quad D_{ij}(0) = 0, \quad i,j, = 1,2$$
$$B_i(q) = \bar{B}_i + \bar{B}_i E_i(q), \quad E_i(0) = 0, \quad i = 1,2$$
$$\omega_i(q) = \bar{B}_i d_i(q), \quad d_i(0) = 0, \quad i = 1,2$$

where $D_{ij}(\cdot)$, $E_i(\cdot)$ and $d_i(\cdot)$ are matrices, respectively a vector, of appropriate dimensions.

Another assumption concerns the range of variation of the uncertainty.

Assumption 2: The set $\mathcal{Q} \subset R^q$ is compact.

Under the above hypothesis the singularly perturbed uncertain system can be regarded as a two-time-scale linear time-invariant system subject to unknown bounded disturbances which lie in $\mathcal{R}(\bar{B}_1)$ and $\mathcal{R}(\bar{B}_2)$. The control objective is that of conferring a desired behaviour to the nominal part of the system, rejecting, as much as possible, the effect of the unknown disturbances; i.e. guaranteeing what we call 'practical' stability of the closed loop system. Of course, in order to assign arbitrary behaviour (or at least stable one), we need the following assumption:

Assumption 3: The pairs $(\bar{A}_{ii}, \bar{B}_i)$, i = 1,2, are controllable.

In the next Sections we shall construct a controller made up of the composition of the practical stability controller of the slow and the fast dynamics of the system. We shall investigate the additional conditions that guarantee the practical stability of the original system and the modifications of the bounding sets of the solutions.

In order to make possible the two-time-scale design we require that the singularly perturbed system be in *standard form*.

Assumption 4: Matrix $\bar{A}_{22}$ is nonsingular.

To carry out the design of the composite controller we shall need some more technical assumptions; they will be stated when they are needed.

13.3. Design objectives and mathematical preliminaries

In this section, with reference to a general uncertain system of the form

$$\dot{x}(t) = F(x(t), q(t)) + G(x(t), q(t))\, u(t), \quad x \in R^n,\ q \in R^q,\ u \in R^p \tag{13.2}$$

we give the formal definition of practical stability (Barmish *et al.* (1983)) and a criterion on how to state this property with the aid of a Lyapunov function.

Definition 1: A feedback controller $p(\cdot)$: $R^n \to R^p$ renders the uncertain dynamical system (13.2) practically stable iff there exist $\underline{d} \in [0, \infty)$ and $r_0 > 0$ such that the following properties hold:

(i) *Existence of solutions*: Given an initial condition x_0 at time t_0, the closed loop system possesses a solution on some interval $[t_0, t_1)$, $t_1 > t_0$.

(ii) *Uniform boundedness*: Given $r \in [0, r_0)$ there exists a positive $d(r) < \infty$ such that, for all solutions, $x(\cdot)$: $[t_0, t_1) \to R^n$, $x(t_0) = x_0$, of the closed loop system

$$\|x_0\| \leqslant r \Rightarrow \|x(t)\| \leqslant d(r) \quad \forall\, t \in [t_0, t_1)$$

(iii) *Extension of solutions*: Every solution of the closed loop system with initial condition x_0 at time t_0 and with $\|x_0\| < r_0$ can be extended over $[t_0, \infty)$.

(iv) *Uniform ultimate boundedness*: Given any $\bar{d} > \underline{d}$ and any $r \in [0, r_0)$, there exists a $T(\bar{d}, r) \in [0, \infty)$ such that, for every solution, $x(\cdot)$: $[t_0, \infty) \to R^n$, $x(t_0) = x_0$, of the closed loop system

$$\|x_0\| \leqslant r \Rightarrow \|x(t)\| \leqslant \bar{d} \quad \forall\, t \geqslant t_0 + T(\bar{d}, r)$$

(v) *Uniform stability*: Given any $\bar{d} > \underline{d}$ there is a positive $\eta(\bar{d}) < \infty$ such that, for every solution, $x(\cdot)$: $[t_0, \infty) \to R^n$, $x(t_0) = x_0$, of the closed loop system

$$\|x_0\| \leqslant \eta(\bar{d}) \Rightarrow \|x(t)\| \leqslant \bar{d} \quad \forall\, t \geqslant t_0$$

In Barmish *et al.* (1983), Corless and Leitmann (1982) and Chen and Leitmann (1987), it has been proved that all the conditions for practical stability can be met if functions $F(\cdot,\cdot)$ and $G(\cdot,\cdot)$ in (13.2) are continuous, $G(\cdot,\cdot)$ is full rank for every x and q, the feedback controller is continuous†, and there exists a Lyapunov candidate $V(\cdot)$: $R^n \to R^+$ for system (13.2) such that

$$\psi_1(\|x\|) \leqslant V(x) \leqslant \psi_2(\|x\|) \tag{13.3}$$

† This assumption can be weakened so as to allow mappings $F(\cdot,\cdot)$ and $G(\cdot,\cdot)$ and feedback controllers $p(\cdot)$ which are Caratheodory; see, for instance, Corless and Leitmann (1981). For the definition of class K functions see Hahn (1967).

where $\psi_1(\cdot)$, $\psi_2(\cdot)$ are class K functions, and for any realisation of the uncertainty $q(\cdot)$

$$\dot{V}(x,t) \triangleq \nabla_x V^{\mathrm{T}}(x)(F(x),q(t)) + G(x,q(t))p(x) \leqslant - l(\|x\|) \tag{13.4}$$

where $l(\cdot)$: $R^+ \rightarrow R^+$ is any function which is non-negative and strictly increasing in an interval (s_1, s_2) with $s_2 > (\psi_1^{-1} \circ \psi_2)(s_1)$ (for instance, see Chen and Leitmann (1987)).

Under these hypotheses all the trajectories of the closed loop system, starting from the ball of radius

$$r_0 = (\psi_2^{-1} \circ \psi_1)\,(s_2), \tag{13.5}$$

are attracted asymptotically toward a ball of ultimate boundedness of radius

$$\underline{d} = (\psi_1^{-1} \circ \psi_2)\,(s_1) \tag{13.6}$$

Moreover, every solution starting from a ball of radius $r < r_0$ and outside a ball of radius $\bar{d} > \underline{d}$, reaches a subset of that ball in a finite time whose upper bound is

$$T(\bar{d}, r) = \frac{\psi_2(r) - (\psi_1 \circ \psi_2^{-1} \circ \psi_1)(\bar{d})}{l \circ \psi_2^{-1} \circ \psi_1(\bar{d})} \tag{13.7}$$

and thereafter remains within it.

Notation: In this chapter the vector norm is taken to be Euclidean and the matrix norm is the corresponding induced one; thus, $\|M\| = \lambda_M^{1/2}(M^T M)$ where $\lambda_{M(m)}(\cdot)$ denotes the operation of taking the maximum (minimum) eigenvalue. A ball of radius r in the space R^n is denoted by $\mathscr{B}(r)$ and defined as

$$\mathscr{B}(r) \triangleq \{x \in R^n\colon \|x\| \leqslant r\}$$

13.4 The 'slow controller' for practical stability

The design of the slow time scale practical stability controller requires a model of the slow dynamics of system (13.1). To this end we define as 'nominal' behaviour of the variable z the one occurring in the absence of uncertainty (i.e., $q(t) \equiv 0$)

$$\varepsilon \dot{z}_n = \bar{A}_{21} x + \bar{A}_{22}\, z_n + \bar{B}_2 u \qquad (13.8)^\dagger$$

Correspondingly, the approximate model of the slow dynamics is taken as the one in which the nominal fast transients are neglected, namely

† When no confusion arises, we delete the argument t.

$$\dot{x}_s = A_{11}(q)x_s + A_{12}(q)\bar{z}_n + B_1(q)u_s + \omega_1(q) \tag{13.9a}$$

$$0 = \bar{A}_{21}\, x_s + \bar{A}_{22}\, \bar{z}_n + \bar{B}_2\, u_s \tag{13.9b}$$

where the subscript s has been used to denote slow variables.

By virtue of Assumption 4 eq. (13.9*b*) can be solved for $\bar{z}_n$ giving

$$\bar{z}_n = -\bar{A}_{22}^{-1}(\bar{A}_{21}\, x_s + \bar{B}_2 u_2) \triangleq h_s(x_s, u_s) \tag{13.10}$$

and, eliminating $\bar{z}_n$ from (13.9*a*), the 'slow control' problem becomes that of controlling the system

$$\dot{x}_s = A_0(q)\, x_s + B_0(q)\, u_s + \omega_1(q) = f(x_s, h(x_s, u_s), u_s, q, \omega_1) \tag{13.11}$$

with

$$A_0(q) \triangleq A_{11}(q) - A_{12}(q)\, \bar{A}_{22}^{-1}\, \bar{A}_{21} \tag{13.12a}$$

$$B_0(q) \triangleq B_1(q) - A_{12}(q)\, \bar{A}_{22}^{-1}\, \bar{B}_2 \tag{13.12b}$$

The matching properties contained in Assumption 1 can be utilised to obtain the following decomposition of matrices $A_0(q)$ and $B_0(q)$:

$$A_0(q) = \bar{A}_0 + \bar{B}_1\, D_0(q) + \bar{F}_0 \tag{13.13a}$$

$$B_0(q) = \bar{B}_1 + \bar{B}_1\, E_0(q) + \bar{G}_0 \tag*{(13.13a)}$$

with

$$\bar{A}_0 \triangleq \bar{A}_{11} - \bar{B}_1(\bar{B}_1^{\mathrm{T}}\bar{B}_1)^{-1}\bar{B}_1^{\mathrm{T}}\, \bar{A}_{12}\bar{A}_{22}^{-1}\bar{A}_{21} \tag{13.14a}$$

$$D_0(q) \triangleq D_{11}(q) - D_{12}(q)\, \bar{A}_{22}^{-1}\bar{A}_{21} \tag{13.14b}$$

$$E_0(q) \triangleq E_1(q) - D_{12}(q)\, \bar{A}_{22}^{-1}\bar{B}_2 - (B_1^{\mathrm{T}}\bar{B}_1)^{-1}\, \bar{B}_1^{\mathrm{T}}\, \bar{A}_{12}\bar{A}_{22}^{-1}\bar{B}_2 \tag{13.14c}$$

$$\bar{F}_0 \triangleq -T(T^T T)^{-1}T^T\, \bar{A}_{12}\bar{A}_{22}^{-1}\bar{A}_{21}$$

$$\bar{G}_0 \triangleq -T(T^T T)^{-1}\, T^T\, \bar{A}_{12}\bar{A}_{22}\bar{B}_2 \tag{13.14d}$$

where $T \in R^{n\times(n-p)}$ is a matrix whose columns span the subspace $\mathscr{R}(\bar{B}_1)^{\perp}$.

With the above decomposition the model of the slow dynamics appears as an uncertain system subject to matched uncertainties and unmatched (known) disturbances.

Define now the following constants related to the norm of matrices in (13.14):

$$k_{D_0} \triangleq \max_{q\in\mathscr{Q}} \|D_{11}(q) - D_{12}(q)\, \bar{A}_{22}^{-1}\bar{A}_{21}\| \tag{13.15a}$$

$$k_{E_0} \triangleq \max_{q\in\mathscr{Q}} \|E_1(q) - D_{12}(q)\, \bar{A}_{22}^{-1}\bar{B}_2 - (\bar{B}_1^T\bar{B}_1)^{-1}\, \bar{B}_1^{\mathrm{T}}\bar{A}_{12}\bar{A}_{22}^{-1}\bar{B}_2\| \tag{13.15b}$$

$$k_{G_0} \triangleq \|T(T^T T)^{-1}\, T^T\bar{A}_{12}\bar{A}_{22}^{-1}\bar{B}_2\| \tag{13.15c}$$

$$k_{\bar{F}_0} \triangleq \|T(T^T T)^{-1} T^T \bar{A}_{12} \bar{A}_{22}^{-1} \bar{A}_{21}\| \tag{13.15d}$$

$$k_{d_1} \triangleq \max_{q \in \mathcal{Q}} \|d_1(q)\|, \tag{13.15e}$$

whose existence is guaranteed by the continuity of matrices $A_{ij}(\cdot)$, $B_i(\cdot)$, $i,j = 1,2$, and of vectors $\omega_i(\cdot)$, $i = 1,2$, and by Assumption 2.

To carry out the design of the practical stability controller we need the following assumption.

Assumption 5: The constant $\alpha_s > 0$ where

$$\alpha_s \triangleq \lambda_m(Q_s) - 2\, k_{\bar{F}_0} \|P_s\|$$

where P_s is the positive definite solution of the Riccati equation

$$\bar{A}_0^T P_s + P_s \bar{A}_0 + Q_s - 2\varrho_s P_s \bar{B}_1 \bar{B}_1^T P_s = 0 \tag{13.16}$$ †

with $\varrho_s > 0$ and Q_s symmetric, positive definite. Moreover, $k_{E_0} < 1$.

With this assumption satisfied we can proceed to the construction of the practical stability 'slow controller' as suggested in the following theorem:

Theorem 1: Consider system (13.11) with the system matrices decomposed as in (13.13) and (13.14), and Assumptions 1,2 and 5 satisfied. A linear feedback controller of the form

$$u_s = -\frac{\gamma_s + \varrho_s}{(1 - k_{E_0})} \bar{B}_1^T P_s x_s \triangleq -\tilde{\gamma}_s \bar{B}_1^T P_s x_s \tag{13.17}$$

guarantees the practical stability of the closed loop system (13.11) for any $\gamma_s \in (\gamma_1, \gamma_2)$ where

$$\gamma_1 \triangleq \max_{r \geqslant 0} \frac{(k_{D_0} r + k_{d_1})^2}{4(c_1 \alpha_s r^2/2 + c_2)} = \frac{1}{4}\left[\frac{2k_{D_0}^2}{c_1 \alpha_s} + \frac{k_{d_1}^2}{c_2}\right] \tag{13.18a}$$

$$\gamma_2 \triangleq \frac{(1 - k_{E_0})}{4} \frac{(1 - c_1)}{\|\bar{B}_1\|^2 \|\bar{G}_0\|} \frac{\alpha_s}{\|P_s\|^2} - \varrho_s \tag{13.18b}$$

and the nonnegative constants c_1 and c_2 are such that

(i) $c_1 < 1$
(ii) $c_1 \neq 0$ whenever $k_{D_0} \neq 0$
(iii) $c_2 \neq 0$ whenever $k_{d_1} \neq 0$

and $\gamma_1 < \gamma_2$

Proof: Consider as Lyapunov function candidate for system (13.11) the function

† Note that in view of Assumption 3, $(\bar{A}_0, \bar{B}_1)$ is a controllable pair. Furthermore, if A_0 is asymptotically stable, we may choose $\varrho_s = 0$ so that (13.16) becomes a Lyapunov equation.

$$V(x_s) = \tfrac{1}{2} x_s^T P_s x_s \tag{13.19}$$

Evaluating its derivative along the solutions of system (13.11) with the closed loop controller (13.17), recalling (13.16), Assumption 1 and decomposition (13.13), we have

$$\begin{aligned} \dot{V}(x_s, t) &\triangleq \nabla_x^T V(x_s) f(x_s, h_s(x_s, u_s), u_s, q, \omega_1) \\ &= -\tfrac{1}{2} x_s^T Q_s x - (\tilde{\gamma}_s - \varrho_s) x_s^T P_s \bar{B}_1 \bar{B}_1^T P_s x_s \\ &+ x_s^T P_s \bar{B}_1 D_0(q) x_s + x_s^T P_s \bar{F}_0 x_s - \tilde{\gamma}_s x_s^T P_s \bar{B}_1 E_0(q) \bar{B}_1^T P_s x_s \\ &- \tilde{\gamma}_s x_s^T P_s \bar{G}_0 \bar{B}_1^T P_s x_s + x_s^T P_s \bar{B}_1 d_1(q) \end{aligned} \tag{13.20}$$

For every realisation of the uncertainty $q(t) \in \mathcal{Q}$, recalling (13.15) and (13.18), we have

$$\begin{aligned} \dot{V}(x_s, t) &\leqslant -\tfrac{1}{2}\alpha_s \|x_s\|^2 - \gamma_s \|\bar{B}_1^T P_s x_s\|^2 \\ &+ k_{D_0} \|\bar{B}_1^T P_s x_s\| \, \|x_s\| + k_{d_1} \|\bar{B}_1^T P_s x_s\| \\ &+ \tilde{\gamma}_s \|B_1\| \, \|G_0\| \, \|P_s\|^2 \|x_s\|^2 \\ &\leqslant -\tfrac{1}{2}\alpha_s \|x_s\|^2 + \frac{(k_{D_0}\|x_s\| + k_{d_1})^2}{4\gamma_s} \\ &+ \tilde{\gamma}_s \|\bar{B}_1\| \, \|G_0\| \, \|P_s\|^2 \|x_s\|^2 \\ &\leqslant -\tfrac{1}{2}\alpha_s \left(1 - \frac{\gamma_1}{\gamma_s} c_1 - \frac{\gamma_s + \varrho_s}{\gamma_2 + \varrho_s}(1 - c_1)\right) \|x_s\|^2 + \frac{\gamma_1}{\gamma_s} c_2 \end{aligned} \tag{13.21}$$

Choosing a particular value $\gamma_s \in (\gamma_1, \gamma_2)$ we have a Lyapunov candidate such that

$$\lambda_m(P_s)\|x_s\|^2 \leqslant V(x_s) \leqslant \lambda_M(P_s)\|x_s\|^2 \tag{13.22}$$

and whose derivative satisfies

$$V(x_s, t) \leqslant -v_1 \|x_s\|^2 + v_2 \tag{13.23}$$

where

$$v_1 \triangleq \tfrac{1}{2}\alpha_s\left(1 - \frac{\gamma_1}{\gamma_s} c_1 - \frac{\gamma_s + \rho_s}{\gamma_2 + \rho_s}(1 - c_1)\right) > 0 \tag{13.24a}$$

$$v_2 \triangleq \frac{\gamma_1}{\gamma_s} c_2 > 0 \tag{13.24b}$$

Since $\dot{V}(x_s, t)$ is negative and strictly decreasing for any $\|x_s\| > (v_2/v_1)^{1/2}$ and $t \in R$, based on the results in Section 13.2, we can conclude that system (13.11) with the feedback controller (13.17) is practically stable with a ball of ultimate boundedness whose radius is given by

$$\underline{d}_s = \left[\frac{\lambda_M(P_s)}{\lambda_m(P_s)}\frac{v_2}{v_1}\right]^{1/2} \tag{13.25}$$

This ball is attractive everywhere in $R^n \setminus \mathscr{B}(\underline{d}_s)$. Moreover, with initial condition x_s^0 such that $\|x_s^0\| > \underline{d}_s$, under the action of the slow controller, a subset of every ball of radius $\bar{d}_s > \underline{d}_s$ is reached in a finite time not greater than

$$T_s = \frac{\lambda_M(P_s)\ \|x_s^0\|^2 - [\lambda_m^2(P_s)/\lambda_M(P_s)]\ \bar{d}_s^2}{v_1(\lambda_m(P_s)/\lambda_M(P_s))\ \bar{d}_s^2 - v_2} \tag{13.26}$$

with v_1 and v_2 given by (13.24), and thereafter remains within it.

13.5 The 'fast controller' for practical stabilisation

An appropriate model of the fast dynamics of system (13.1) can be obtained as follows. Substituting the expression (13.17) for the 'slow controller' in eq. (13.1*b*), we obtain

$$\varepsilon\dot{z} = (A_{21}(q) - \tilde{\gamma}_s B_2(q)\bar{B}_1^T P_s)\ x + A_{22}(q)\ z + B_2(q)u_f + \omega_2(q) \tag{13.27}$$

where variable $x_s(t)$ has been approximated by $x(t)$, and u_f denotes the control of the fast system (13.27).

Let us now define

$$z_f \triangleq z - h(x) \tag{13.28}$$

where

$$h(x) \triangleq h_s(x, u_s(x)) = -\bar{A}_{22}^{-1}(\bar{A}_{21} - \tilde{\gamma}_s\bar{B}_2\bar{B}_1^T P_s)\ x \tag{13.29}$$

In the fast time scale, the variable $x(t)$ is treated as a fixed parameter so that, from eqn. (13.27), the boundary layer model of system (13.1) can be expressed by

$$\begin{aligned}\frac{dz_f}{d\tau} &= A_{22}(q)z_f + B_2(q)u_f + F_0(q)x + \omega_2(q)\\ &= g(x, z_f + h(x), \tilde{\gamma}_s\bar{B}_1^T P_s x + u_f, q, \omega_2)\end{aligned} \tag{13.30}$$

with $\tau \triangleq t/\varepsilon$, and

$$\begin{aligned}F_0(q) \triangleq\ &(A_{21}(q) - \tilde{\gamma}_s B_2(q)\bar{B}_1^T P_s)\\ &- A_{22}(q)\bar{A}_{22}^{-1}(\bar{A}_{21} - \tilde{\gamma}_s\bar{B}_2\bar{B}_1^T P_s)\end{aligned} \tag{13.31}$$

In view of Assumption 1 matrix $F_0(q)$ can be rewritten as

$$F_0(q) \triangleq \bar{B}_2 H(q) \tag{13.32a}$$

$$H(q) \triangleq (D_{21}(q) - \tilde{\gamma}_s E_2(q)\bar{B}_1^T P_s) - D_{22}(q)\bar{A}_{22}^{-1}(\bar{A}_{21} - \tilde{\gamma}_s \bar{B}_2 \bar{B}_1^T P_s \tag{13.32b}$$

With this decomposition the boundary layer system (13.30) appears as an uncertain system subject to matched uncertainties and a matched constant (in the fast time scale) disturbance signal (i.e. signal $x(t)$). Let us now define the following quantities:

$$k_H \triangleq \max_{q \in \mathcal{Q}} \|H(q)\|, \quad k_{E_2} \triangleq \max_{q \in \mathcal{Q}} \|E_2(q)\|,$$

$$k_{D_{22}} \triangleq \max_{q \in \mathcal{Q}} \|D_{22}(q)\|, \quad k_{d_2} \triangleq \max_{q \in \mathcal{Q}} \|d_2(q)\| \tag{13.33}$$

where the existence of the maxima is guaranteed by Assumption 2 and the continuity with respect to q of the matrices involved. In order to construct a practical stability controller for the boundary layer system we need the following assumption.

Assumption 6: $k_{E_2} < 1$.

The feedback controller utilised to counteract the effect of the uncertainties and to eliminate the offset that would be produced by the constant disturbance, making the boundary layer model practically stable, has the following structure (see Ambrosino *et al.*, 1985):

$$u_f(z_f) = -K_f z_f - \gamma_f(z_f, x)\, \bar{B}_2^T P_f z_f \tag{13.34}$$

where K_f is such that $\bar{A}_{22} - \bar{B}_2 K_f$ is asymptotically stable, and

$$\gamma_f(z_f, x) \triangleq \frac{\gamma_{f_1}\|z_f\| + \gamma_{f_2}\|x\| + \gamma_{f_3}}{\|\bar{B}_2^T P_f z_f\| + \delta}, \quad \delta > 0 \tag{13.35}$$

with P_f the solution of the Lyapunov equation

$$(\bar{A}_{22} - \bar{B}_2 K_f)^T P_f + P_f(\bar{A}_{22} - \bar{B}_2 K_f) = -Q_f, \quad Q_f > 0 \tag{13.36}$$

Remark: It is worthwhile to note that this controller depends on x only through the nonlinear scalar gain $\gamma_f(z_f, x)$. This ensures that, when used in the composite controller, it will not modify the nominal dynamics of the x variable (see Suzuki (1981)), allowing the separate construction of the 'slow controller' $u_s(x)$ and the fast controller $u_f(z_f)$.

The stabilising capability of this controller is stated in the following theorem.

Theorem 2: Consider system (13.30) with (13.32) and Assumptions 1–3 and 6 satisfied. The feedback controller (13.34) renders system (13.30) practically stable provided that the gains γ_{f_i}, $i = 1,2,3$, are chosen so that

$$\gamma_{f_1} \geqslant \frac{k_{D_{22}} + k_{E_2}\|K_f\|}{1 - k_{E_2}} \tag{13.37a}$$

$$\gamma_{f_2} \geqslant \frac{k_{\rm H}}{1 - k_{E_2}} \tag{13.37b}$$

$$\gamma_{f_3} \geqslant \frac{k_{d_2}}{1 - k_{E_2}} \tag{13.37c}$$

Proof: Consider as Lyapunov candidate for system (13.30) the function

$$W(z_f) = \tfrac{1}{2} z_f^T P_f z_f \tag{13.38}$$

with P_f the solution of (13.36). Evaluating the time derivative of $W(z_f)$ along the solutions of (13.30), with u_f defined as in (13.34), we have

$$\begin{aligned}\frac{\mathrm{d}}{d\tau} W(z_f) &\triangleq \nabla_{z_f}^T W(z_f)\, g(x, z_f + h(x), \tilde{\gamma}\, \bar{B}_1^T P_s x + u_f, q, \omega_2)\\
&= -\tfrac{1}{2} z_f^T Q_f z_f - \gamma_f(z_f, x)\, z_f^T P_f \bar{B}_2 \bar{B}_2^T P_f z_f\\
&\quad + z_f^T P_f \bar{B}_2 D_{22}(q) z_f - z_f^T P_f \bar{B}_2 E_2(q) K_f z_f\\
&\quad - \gamma_f(z_f, x)\, z_f^T P_f \bar{B}_2 E_2(q)\, \bar{B}_2^T P_f z_f + z_f^T P_f \bar{B}_2 H(q) x\\
&\quad + z_f^T P_f \bar{B}_2 d_2(q)\end{aligned} \tag{13.39}$$

For every realisation of the uncertainty, recalling that $\|v\|^2/(\|v\| + \delta) \geqslant \|v\| - \delta$, $\delta > 0$, and in view of (13.35), we have

$$\begin{aligned}\frac{d}{d\tau} W(z_f) &\leqslant -\tfrac{1}{2} z_f^{\rm T} Q_f z_f - \gamma_f(z_f, x)(1 - k_{E_2})\, \|\bar{B}_2^T P_f z_f\|^2\\
&\quad + (k_{D_2} + k_{E_2}\|K_f\|)\, \|\bar{B}_2^T P_f z_f\|\, \|z_f\| + k_H \|\bar{B}_2^T P_f z_f\|\, \|x\|\\
&\quad + k_{d_2} \|\bar{B}_2^T P_f z_f\|\\
&\leqslant -\tfrac{1}{2} \lambda_m(Q_f)\, \|z_f\|^2 - (\gamma_{f_1} \|z_f\| + \gamma_{f_2} \|x\| + \gamma_{f_3})(\|\bar{B}_2^T P_f z_f\| - \delta)(1 - k_{E_2})\\
&\quad + (k_{D_{22}} + k_{E_2}\|K_f\|)\|\bar{B}_2^T P_f z_f\|\, \|z_f\|\\
&\quad + k_H \|\bar{B}_2^T P_f z_f\|\, \|x\| + k_{d_2} \|\bar{B}_2^T P_f z_f\|\end{aligned}$$

$$\leqslant -\tfrac{1}{2}\lambda_m(Q_f)\,\|z_f\|^2 + \delta\,\gamma_{f_1}\|z_f\| + \delta(\gamma_{f_2}\|x\| + \gamma_{f_3}) \tag{13.40}$$

Hence we have a Lyapunov candidate satisfying

$$\lambda_m(P_f)\|z_f\|^2 \leqslant W(z_f) \leqslant \lambda_M(P_f)\|z_f\|^2 \tag{13.41}$$

and

$$\frac{d}{d\tau}W(z_f) \leqslant -w_1\|z_f\|^2 + w_2\|z_f\| + w_3' \tag{13.42}$$

where

$$w_1 \triangleq \tfrac{1}{2}\lambda_m(Q_f),\; w_2 \triangleq \delta\gamma_{f_1},\; w_3' \triangleq w_3 + w_4\|x\|,\; w_3 \triangleq \delta\gamma_{f_3},\; w_4 \triangleq \delta\gamma_{f_2}\,. \tag{13.43}$$

The derivative (13.42) is negative and strictly decreasing for

$$\|z_f\| > R_f \triangleq (w_2 + \sqrt{w_2^2 + 4w_1w_3'})\,/\,2w_1 \tag{13.44}$$

so that, based on the results reported in Section 13.2, we can conclude that system (13.30) with the feedback controller (13.34) is practically stable with a ball of ultimate boundedness whose radius is given by

$$\underline{d}_f = \left[\frac{\lambda_M(P_f)}{\lambda_m(P_f)}\right]^{1/2} R_f \tag{13.45}$$

This ball is attractive everywhere in $R^m \setminus \mathscr{B}(\underline{d}_f)$. With initial condition z_f^0 such that $\|z_f^0\| > \underline{d}_f$, under the action of the 'fast controller' (13.34), a subset of every ball of radius $\bar{d}_f > \underline{d}_f$ is reached in a finite time not greater than

$$T_F = \frac{\lambda_M(P_f)\|z_f^0\|^2 - [\lambda_m^2(P_f)/\lambda_M(P_f)]\,\bar{d}_f^2}{w_1(\lambda_m(P_f)/\lambda_M(P_f))\bar{d}_f^2 - w_2(\lambda_m(P_f)/\lambda_M(P_f))^{1/2}\bar{d}_f - w_3'}$$

with w_1, w_2 and w_3' given by (13.43), and thereafter remains within it.

Note from (13.40)–(13.42) the effect of the (designer chosen) constant δ in (13.35) on the radius of the ball of ultimate boundedness.

13.6 Composite control and practical stability of the complete system

The composite controller for system (13.1) is formed as the sum of the 'slow' and the 'fast controller' with variables x_s and z_f replaced by their approximations x and $z - h(x)$, respectively; that is,

$$u = u_c(x,z) \triangleq u_s(x) + u_f(z - h(x)) = -\tilde{\gamma}_s\bar{B}_1^T P_s x$$
$$-\gamma_f(z - h(x), x)\bar{B}_2^T P_f(z - h(x)) - K_f(z - h(x)) \tag{13.46}$$

This controller has the practical stability capabilities pointed out in the following theorem.

Theorem 3: Consider system (13.1) with Assumptions 1 – 6 satisfied. Suppose that the 'slow' and the 'fast controllers' (13.17) and (13.34) guarantee practical stabilisation of the slow and the fast time scale approximations of system (13.1). Moreover suppose that constant γ_{f_2} in (13.29), selected satisfying (13.37), is such that

$$\beta\, \gamma_{f_2} < \frac{v_1}{\|P_s\|}, \quad \beta \triangleq \max_{q \in \mathcal{Q}} \|B_1(q)\| \tag{13.47}$$

where v_1 is defined in (13.24*a*). Then it is possible to find an ε^* (given by (13.67)) such that for every $\varepsilon \in (0, \varepsilon^*)$ the composite control (13.46) is a practical stability controller for system (13.1) with a ball of ultimate boundedness whose radius (given by (13.70)) can be evaluated *a priori* on the basis of the selected value for ε and the bounds of the uncertainties affecting the system.

Proof: As suggested in Saberi and Khalil (1984), the proof of the theorem is based on the combined use of Lyapunov functions for the two subsystems forming the complete system; see also the proofs in Ryan and Corless (1989), Leitmann and Ryan (1987), Corless (1987), Garofalo (1988). Let us start by noting that the dynamics of the x component of system (13.1), under the action of the composite feedback controller, can be rewritten as

$$\begin{aligned} \dot{x} &= f(x, z, u_s + u_f, q, \omega_1) = f(x, h(x), u_s, q, \omega_1) \\ &\quad + (f(x, z, u_s + u_f, q, 0) - f(x, h(x), u_s, q, 0)) \\ &= (A_0(q) - \tilde{\gamma}_s B_0(q) \bar{B}_1^T P_s)x + F_{12}(x, z, q)(z - h(x)) + \omega_1(q) \end{aligned} \tag{13.48}$$

with

$$F(x, z, q) \triangleq A_{12}(q) - B_1(q)(K_f + \gamma_f(z - h(x), x)\bar{B}_2^T P_f) \tag{13.49}$$

Let us observe that

$$\begin{aligned} \max_{q \in \mathcal{Q}} \|F_{12}(x, z, q)(z - h(x))\| &\leqslant \max_{q \in \mathcal{Q}} \|(A_{12}(q) - B_1(q)K_f)\| \, \|z - h(x)\| \\ &\quad + \max_{q \in \mathcal{Q}} \|B_1(q)\| \, | \gamma_f(z - h(x), x | \, \|B_2^T P_f(z - h(x))\| \end{aligned} \tag{13.50}$$

But, from (13.35), we have

$$\begin{aligned} | \gamma_f(z - h(x), x) | \, \|B_2^T P_f(z - h(x))\| &\leqslant \gamma_{f_1} \|z - h(x)\| \\ &\quad + \gamma_{f_2} \|x\| + \gamma_{f_3} \end{aligned} \tag{13.51}$$

We can conclude that

$$\max_{q \in \mathcal{Q}} \|F_{12}(x, z, q)(z - h(x))\| \leqslant \beta\gamma_{f_2}\|x\| + a_2'\|z - h(x)\| + \beta\gamma_{f_3} \quad (13.52)$$

which, in turn, implies

$$\max_{q \in \mathcal{Q}} \|f(x, z, u_s + u_f, q, \omega_1)\| \leqslant a_1'\varepsilon x\| + a_2'\|z - h(x)\| + a_3' \quad (13.52b)$$

where

$$a_1' \triangleq \beta\gamma_{f_2} + \max_{q \in \mathcal{Q}} \|A_0(q) - \tilde{\gamma}_s B_0(q)\bar{B}_1^T P_s\| \quad (13.53a)$$

$$a_2' \triangleq \beta\gamma_{f_1} + \max_{q \in \mathcal{Q}} \|A_{12}(q) - B_1(q)K_f\| \quad (13.53b)$$

$$a_3' \triangleq \beta\gamma_{f_3} + \max_{q \in \mathcal{Q}} \|\omega_1(q)\| \quad (13.53c)$$

The closed loop dynamics of the z component of system (13.1) under the action of the composite controller (13.46), using (13.35), can be rewritten as

$$\begin{aligned}\varepsilon\dot{z} &= g(x, z, u_s + u_f, q, \omega_2) = (A_{22}(q) - B_2(q)K_f \\ &\quad - \gamma_f(z - h(x), x)B_2(q)\bar{B}_2^T P_f)(z - h(x)) + F_0(q)x + \omega_2(q)\end{aligned} \quad (13.54)$$

Consider now the function

$$W(x, z) = \frac{1}{2}(z - h(x))^T P_f (z - h(x)) \quad (13.55)$$

with P_f the solution of (13.36). Evaluating its time derivative along the solutions of system (13.1) with the composite controller (13.46), we have

$$\begin{aligned}\dot{W}(x, z, t) &\triangleq \nabla_x^T W(x, z) f(x, z, u_s + y_f, q, \omega_1) \\ &\quad + \frac{1}{\varepsilon}\nabla_z^T W(x, z)\, g(x, z, u_s + u_f, q, \omega_2) \\ &= -(z - h(x))^T P_f J_h f(x, z, u_s + u_f, q, \omega_1) \\ &\quad + \frac{1}{\varepsilon}(z - h(x))^T P_f g(x, z, u_s + u_f, q, \omega_2)\end{aligned} \quad (13.56)$$

where $J_h = \partial h(x)/\partial x$ is the Jacobian matrix associated with vector $h(x)$; it is constant and given by

$$J_h = -\bar{A}_{22}^{-1}(\bar{A}_{21} - \tilde{\gamma}_s \bar{B}_2 B_1^T P_s) \quad (13.57)$$

For every realisation of the uncertainty $q(\cdot)$, recalling (13.52), (13.54) and (13.43), we have

$$\dot{W}(x, z, t) \leqslant \|P_f J_h\| \, (a_1' \|x\| \, \|z - h(x)\| + a_2' \|x - h(x)\|^2$$
$$+ a_3' \|z - h(x)\|) - \frac{w_1}{\varepsilon} \|z - h(x)\|^2 + \frac{w_2}{\varepsilon} \|z - h(x)\|$$
$$+ \frac{w_3}{\varepsilon} + \frac{w_4}{\varepsilon} \|x\|$$
$$= -\left(\frac{w_1}{\varepsilon} - a_2\right) \|z - h(x)\|^2 + a_1 \|x\| \, \|z - h(x)\|$$
$$+ \left(a_3 + \frac{w_2}{\varepsilon}\right) \|z - h(x)\| + \frac{w_3}{\varepsilon} + \frac{w_4}{\varepsilon} \|x\| \qquad (13.58)$$

where

$$a_i \triangleq d_i \|P_f J_h\|, \quad i = 1,2,3$$

Consider now the function

$$V(x) = \frac{1}{2} x^T P_s x \qquad (13.59)$$

where P_s is the positive definite solution of Riccati equation (13.16). Along the solutions of the closed loop system (13.1) with control (13.46), recalling (13.48), we have

$$\dot{V}(x, t) \triangleq \nabla_x^T V(x) f(x, z, u_s + u_f, q, \omega_1)$$
$$= x^T P_s f(x, h(x), u_s, q, \omega_1) + x^T P_s F_{12}(x, z, q)(z - h(x)) \qquad (13.60)$$

For every realization of the uncertainty $q(\cdot)$, in view of (13.23) and (13.52*a*), we have

$$\dot{V}(x, t) \leqslant -v_1 \|x\|^2 + v_2 + \|x\| \, \|P_s\| \, (\beta\gamma_{f_2} \|x\| + a_2' \|z - h(x)\| + \beta\gamma_{f_3})$$
$$= -(v_1 - b_1)\|x\|^2 + b_2 \|x\| \, \|z - h(x)\| + b_3 \|x\| + v_2 \qquad (13.61)$$

with

$$b_1 \triangleq \|P_s\| \, \beta\gamma_{f_2} \qquad (13.62a)$$

$$b_2 \triangleq \|P_s\| \, a_2' \qquad (13.62b)$$

$$b_3 \triangleq \|P_s\| \, \beta\gamma_{f_3} \qquad (13.62c)$$

and where $v_1 > b_1$ by virtue of condition (13.47). With the scalar functions $W(x, z)$ and $V(x)$, as suggested in Khalil and Saberi (1984), we now form a Lyapunov function candidate $\mathscr{L}(x, z)$ for system (13.1):

$$\mathscr{L}(x,z) = \begin{bmatrix} x \\ z - h(x) \end{bmatrix}^T P(c) \begin{bmatrix} x \\ x - h(x) \end{bmatrix} \tag{13.63}$$

where

$$P(c) = \begin{bmatrix} (1-c)P_s & 0 \\ 0 & cP_f \end{bmatrix}, \quad c \in (0,1). \tag{13.64}$$

In view of (13.58) and (13.61), the time derivative of (13.63) along the solutions of (13.1) with composite control (13.46) satisfy, for any realisation of the uncertainty $q(\cdot)$, the following inequality:

$$\begin{aligned}\dot{\mathscr{L}}(x,z,t) &= \nabla_x \mathscr{L}(x,z) f(x,z,u_s + u_f, q, \omega_1) \\ &\quad + \frac{1}{\varepsilon} \nabla_z \mathscr{L}(x,z) g(x,z,u_s + u_f, q, \omega_2) \\ &\leqslant -\begin{bmatrix} \|x\| \\ \|z - h(x)\| \end{bmatrix}^T M(c) \begin{bmatrix} \|x\| \\ \|z - h(x)\| \end{bmatrix} \\ &\quad + m^T(c) \begin{bmatrix} \|x\| \\ \|z - h(x)\| \end{bmatrix} + l(c) \end{aligned} \tag{13.65}$$

where

$$M(c) \triangleq \begin{bmatrix} (1-c)(v_1 - b_1) & -1/2(ca_1 + (1-c)b_2) \\ -\frac{1}{2}(ca_1 + (1-c)b_2) & c\left(\frac{w_1}{\varepsilon} - a_2\right) \end{bmatrix} \tag{13.66a}$$

$$m(c) \triangleq \left[(1-c)\, b_3 + \frac{w_4}{\varepsilon} c \quad \left(a_3 + \frac{w_2}{\varepsilon}\right)\right]^T \tag{13.66b}$$

$$l(c) \triangleq (1-c)v_2 + c\,\frac{w_3}{\varepsilon} \tag{13.66c}$$

with $c \in (0,1)$.

The upper bound of parameter ε that guarantees the positivity of matrix $M(c)$ in (13.66a) can be evaluated as (see Khalil and Saberi, 1984)

$$\varepsilon^* \triangleq \frac{(v_1 - b_1)w_1}{(v_1 - b_1)a_2 + a_1 b_2} \tag{13.67}$$

and corresponds to $c = c^* = b_2/(a_1 + b_2)$. For any $\varepsilon < \varepsilon^*$ there is an interval (c_1, c_2) of values of c for which matric $M(c)$ is positive definite. Corresponding to a $c \in (c_1, c_2)$

$$\lambda_m(P(c))(\|x\|^2 + \|z - h(x)\|^2) \leqslant \mathscr{L}(x, z) \leqslant \lambda_M(P(c))(\|x\|^2 + \|z - h(x)\|^2) \tag{13.68}$$

and

$$\dot{\mathscr{L}}(x, z, t) \leqslant -\lambda_m(M(c))(\|x\|^2 + \|z - h(x)\|^2) + \|m(c)\|(\|x\|^2 + \|z - h(x)\|^2)^{1/2} + l(z) \tag{13.69}$$

In view of the results reported in Section 13.2, we can conclude that for every $\varepsilon < \varepsilon^*$ the complete system is practically stable with a ball of ultimate boundedness whose radius is given by

$$\underline{d} = \left[\frac{\lambda_M(P(c))}{\lambda_m(P(c))}\right]^{1/2} \mathrm{R} \tag{13.70}$$

where R is defined as

$$R \triangleq (\|m(c)\| + (\|m(c)\|^2 + 4\,\lambda_m(M(c))\,l(c))^{1/2}/2\,\lambda_m(M(c)) \tag{13.71}$$

The freedom in the choice of $c \in (c_1, c_2)$ can be employed to obtain smaller approximations of the ball of ultimate boundedness.

With initial conditions x_0, z_0 such that $(\|x_0^2 + \|z_0 + h(x_0)\|)^{1/2} > \underline{d}$, under the action of the composite controller, a subset of every ball of radius $\bar{d} > \underline{d}$ is reached in a finite time not greater than

$$T = \frac{\lambda_M(P(c))(\|x_0\|^2 + \|z - h(x_0)\|^2 - [\lambda_m(P(c))/\lambda_M(P(c))]\,\bar{d}^2}{\dfrac{\lambda_M(P(c))}{\lambda_m(P(c))}\lambda_m(M(c))\,\bar{d}^2 - \sqrt{\dfrac{\lambda_M(P(c))}{\lambda_m(P(c))}}\,\|m(c)\|\,\bar{d} - l(c)} \tag{13.72}$$

Remarks: Condition (13.47) of Theorem 3, together with condition (13.37*b*) of Theorem 2, constrains the gain γ_{f_2}(which appears in the expression for the composite controller through (13.35) to belong to an interval $[\underline{\gamma}_{f_2}, \bar{\gamma}_{f_2}]$ where with

$$\underline{\gamma}_{f_2} \triangleq \frac{k_H}{1 - k_{E_2}},\ \bar{\gamma}_{f_2} \triangleq \frac{1}{2}\frac{\lambda_m(Q_s)}{\|P_s\|}\left(1 - \frac{\gamma_1}{\bar{\gamma}_s}c_1 - \frac{\bar{\gamma}_s + \varrho_s}{\gamma_2 + \varrho_s}(1 - c_1)\right)\beta^{-1}$$

provided that $\bar{\gamma}_{f_2} > \underline{\gamma}_{f_2}$. This latter requirement depends on the choice of the gain of the 'slow controller' as well as on some structural conditions involving the uncertainties affecting the system. One way to separate as much as possible $\underline{\gamma}_{f_2}$ from $\bar{\gamma}_{f_2}$ is to maximise the ratio $\alpha_s/\|P_s\|$ by a suitable choice of Q_s in (13.16); this can be done easily following Patel and Toda (1980), when the choice $\varrho_s = 0$ in (13.16) is possible.

Given an $\varepsilon \in (0, \varepsilon^*)$, the radius $\underline{d}$ of the ball of ultimate boundedness can be adjusted by appropriate choice of design constants δ, γ_{f_3}, and c.

13.7 Example

The design procedure is now illustrated by a simple example. Consider the second order system

$$\dot{x} = -x + u + \omega(q)$$

$$\varepsilon\dot{z} = a(q)x + z + u$$

which trivially satisfies Assumptions 1,3 and 4.
The functions ω: $\mathcal{Q} \to R$, a: $\mathcal{Q} \to R$ are such that

$$\max_{q \in \mathcal{Q}} | \omega(q) | \leqslant \bar{\omega} = 1$$

$$\max_{q \in \mathcal{Q}} | a(q) | \leqslant \bar{a} = 0{\cdot}35$$

The approximation of the slow time scale system is easily obtained as

$$\dot{x}_s = -x_s + u_s + \omega(q)$$

The control law (13.17), noting that $k_{E_0} = 0$ and having selected $\varrho_s = 0$ and $Q_s = 2$ in (13.16), can be rewritten as

$$u_s = -\gamma_s x_s$$

From (13.18), with c_1 arbitrary, the interval in which γ_s must lie is easily evaluated as $(0, \infty)$. For simplicity we select $\gamma_s = 1$ and correspondingly, from (13.24), we have $v_1 = 1$ and $v_2 = 0{\cdot}03$.

The nominal equilibrium (i.e. the equilibrium in the absence of uncertainties) of variable z in the fast time scale is evaluated through (13.10) as $\bar{z}_n = -u_s = x_s$, so that, from (13.30), the boundary layer model can be written as

$$\frac{d_{z_f}}{d\tau} = z_f + u_f + a(q)x$$

where $z_f = z - x$. In order to counteract the effect of the external disturbance $a(q)x$, we adopt the control law (13.34). In this example $k_{f_0} = \bar{a}$, $k_{E_2} = k_{d_2} = 0$, and choosing $K_f = -2$ and $Q_f = 2$ in (13.36), the control (13.34) becomes

$$u_f = -2z_f - \frac{\bar{\gamma}_{f_2}|x|}{|z_f| + \delta} z_f, \quad \delta > 0$$

where $\bar{\gamma}_{f_2} \geqslant \bar{a}$ (from 13.37*b*). For simplicity we select $\delta = 1$ and $\gamma_{f_2} = 0.4$ and, correspondingly, from (13.43), we find $w_1 = 1$, $w_2 = w_3 = 0$, $w_4 = 0{\cdot}4$. Note that $\bar{\gamma}_{f_2} = 0{\cdot}4 < v_1/\|P_s\| = 1$, and hence the results of Theorem 3 are applicable.

The composite control takes the form

$$u_c = -x - 2(z - x) - \frac{|x|}{|z - x| + 1}(z - x)$$

From (13.52) we have $a_1' = 2{\cdot}4$, $a_2' = 2$, $a_3' = 1$, and $J_h = 1$ from (13.56), so that $a_i = d_i$, $i = 1,2,3$. Moreover, from (13.60), $b_1 = 0{\cdot}4$, $b_2 = 2$ and $b_3 = 0$. These values allow us to evaluate $\varepsilon^* = 0{\cdot}1$ corresponding to $c^* = 0{\cdot}45$. For instance, selecting $\varepsilon = 0{\cdot}09$ and $c = 0{\cdot}5$, the ball of ultimate boundedness has the radius $\underline{d} = 0.7$.

13.8 Conclusions

The conditions that allow the construction of a composite control for the practical stabilisation of a singularly perturbed uncertain system have been analysed. Even though the system has been considered nominally linear, the presence of the uncertainties couples the fast and the slow dynamics of the system and, in consequence, the controller must contain a nonlinear term that allows the separate design of the fast and the slow controllers. It has been found that the feedback gains of the slow variable, besides having a lower bound (as is usual in the control of uncertain systems), also possess an upper bound that depends mainly on the structural coupling between the slow and the fast parts of the system. This upper bound prevents the variables which have been considered as having slow dynamics from becoming fast under the action of a high gain feedback.

13.9 Acknowledgment

The authors wish to thank Professor M. Corless (Purdue University) for his critical reading of the manuscript.

13.10 References

AMBROSINO, G., CELENTANO, G., and GAROFALO, F. (1985): 'Robust model tracking control for a class of nonlinear plants,' *IEEE Trans.*, **AC-30**, pp. 275–279.

BARMISH, B. R., CORLESS, M., and LEITMANN, G. (1983): 'A new class of stabilizing controllers for uncertain dynamical systems,' *SIAM J. Control and Optimiz.*, **21**, pp. 246–255

BARMISH, B. R., and LEITMANN, G. (1982): 'On ultimate boundedness control of uncertain systems in the absence of matching assumptions,' *IEEE Trans.*, **AC-27**, 153–158

CHEN, Y. H., and LEITMANN, G. (1987): 'Robustness of uncertain systems in the absence of matching assumptions,' *Int. J. Cont.*, **45**, 1527–1542

CORLESS, M., (1987): 'Robustness of a class of feedback controlled uncertain nonlinear systems in the presence of singular perturbations.' *Proc. Am. Contr. Conf.*, Minneapolis, MN

CORLESS, M., and LEITMANN, G. (1981): 'Continuous state feedback guaranteeing uniform ultimate boundedness for uncertain dynamic systems,' *IEEE Trans.*, **AC-26**, 1139–1144

GAROFALO, F. (1988): 'Composite control of a singularly perturbed uncertain system with slow nonlinearities' *Int. J. Contr.*, **48**, 1979–1991

HAHN, W. (1967): 'Stability of motion,' Springer-Verlag, (Berlin & New York)

KALMAN, R. L., and BERTRAM, J. E. (1960) 'Control system analysis and design via the second method of Lyapunov I: Continuous-time systems, *J. Basic Enging.*, **82,** 371–393

KOKOTOVIC, P. V. (1984): 'Applications of singular perturbation techniques to control problems,' *SIAM Rev.*, **26,** 501–550

LEITMANN, G. (1978): 'Guaranteed ultimate boundedness for a class of uncertain linear dynamical systems,' *IEEE Trans.* **AC-23,** 1109–1110

LEITMANN, G., and RYAN, E. P. (1987): 'Output feedback control of a class of singularly perturbed uncertain dynamical systems' *Proc. Am. Contr. Conf.* Minneapolis, MN

LEITMANN, G., RYAN, E. P. and STEINBERG, A. (1986); 'Feedback control of uncertain systems: Robustness with respect to neglected actuator and sensor dynamics,' *Int. J. Cont.*, **43,** 1242–1256

PATEL, R. V., and TODA, M. (1980): 'Quantitative measures of robustness for multivariable systems,' *Proc. JAC,* San Francisco, p. 8-A

RYAN, E. P., and CORLESS, (1989): 'Robust feedback control of singularly perturbed uncertain dynamical systems,' (to be published)

SABERI, A., and KHALIL, H. (1984): 'Quadratic-type Lyapunov functions for singularly perturbed system,' *IEEE Trans.*, **AC-29,** 542–550

SUZUKI, M. (1981): 'Composite controls for singularly perturbed systems,' *IEEE Trans.* **Ac-26** pp. 505–507

Chapter 14

Some extensions of variable structure control theory for the control of nonlinear systems

G. Bartolini and T. Zolezzi†

University of Genoa

14.1 Introduction

Recent work by the authors in the area of variable structure control (VSC) for nonlinear systems is presented in this chapter. Most of the available results about variable structure control systems deal mainly with linear control systems (see Utkin, 1978; Utkin, 1984), or more generally with systems affine in the control variables. The theory and some applications of VSC of nonlinear systems of the form

$$\dot{x} = f(t, x, u) \quad \text{(state equation)},$$

$$u \in U \quad \text{(control constraint)}$$

$$s(x) = 0 \quad \text{(sliding manifold)}$$

will be considered. The need for such a theory has been mentioned in the early development of VSC (see Utkin, 1978), and more recently in Slotine and Sastry (1983).

The mathematical description of the motion of the given systems on the sliding manifold, or near it, will be studied. An appropriate definition of the solution to the relevant ordinary differential equations with discontinuous terms is that of Filippov (1964). An attempt to extend the VSC methodology to distributed systems will also be presented.

† Work partially supported by MPI (funding 40%)

14.2 VSC of nonlinear ordinary differential systems

Consider the following VSC systems:

$$\dot{x} = f(t, x, u), \quad t \in [0, T] \text{ (state equation)}, \tag{14.1}$$

$$u(t, x) \in U \quad \text{(control constraint)} \tag{14.2}$$

$$s(x) = 0 \quad \text{(sliding manifold } S\text{)}. \tag{14.3}$$

Here $x \in R^n$, $u \in R^m$, U is a given subset of R^m, Ω is an open subset of R^m (containing all instantaneous systems states), and

$$f: [0, T] \times \Omega \times U \rightarrow R^n$$

is a given Carathèodory function, i.e. $f(\cdot, x, u)$ is Lebesgue measurable for every x, u and $f(t, \cdot, \cdot)$ is continuous for every t. Moreover

$$s: R^n \rightarrow R^m$$

is a given continuously differentiable mapping.

Assume that it is possible for the state vector x to reach the sliding manifold S and to remain on S. How does one describe mathematically the dynamic behaviour of the VSC system under discontinuous feedback control laws in the sliding mode?

Roughly speaking, if some control law u allows us to keep the state vector x on the sliding manifold S, then for all t,

$$\frac{d}{dt} s(x) = 0 = \frac{\partial s}{\partial x} f(t, x, u).$$

We introduce the following assumption:

There exists a neighbourhood V of S such that for every $t \in [0, T]$ and

$$x \in V \text{ the map } f(t, x, \cdot) \text{ is one-to-one on } U \text{ and its range contains } 0. \tag{14.4}$$

The unique solution $u \in U$ (if any) of the equation

$$\frac{\partial s}{\partial x} f(t, x, u) = w$$

for a given $w \in R^m$ will be denoted by

$$u^* (t, x, w).$$

With the assumption (14.4) the equivalent control for the system (14.1) – (14.3) is the mapping

$$(t, x) \rightarrow u^* (t, x, 0)$$

for $t \in [0, T]$, $x \in V$. This definition generalises that given in Utkin (1978) when

$$U = R^m, f(t, x, u) = A(t, x) + B(t, x)u$$

since in this case (14.4) amounts to the nonsingularity condition

$$\det \frac{\partial s}{\partial x} B(t, x) \neq 0.$$

A suitable definition of solutions to (14.1) corresponding to a discontinuous feedback $u(t, x)$ was introduced and developed by Filippov (1964) (see also Filippov, 1985; 1961 for control theoretic motivations). Consider

$$g(t, x) = f[t, x, u(t, x)].$$

Then g is a R^n-valued Lebesgue measurable function on $[0, T] \times \Omega$. Denote by $B(z, \delta)$ the ball of radius δ around z. The y is a Filippov solution in $[0, T]$ to $\dot{y} = g(t, y)$ iff y is absolutely continuous there, and for a.e. $t \in [0, T]$

$$\dot{y}(t) \in \text{cl co } g[t, B(y(t), \delta) \setminus N]$$

for every $\delta > 0$ and every set N of zero Lebesgue n-dimensional measure (here 'cl' means closure, and 'co' the convex hull).

The next result shows that the state trajectories on the sliding manifold S corresponding to any discontinuous feedback control law may be obtained as a.e. solutions of (14.1) corresponding to the equivalent control and conversely. This justifies the above definition of equivalent control in the nonlinear setting, and shows that (under some assumptions on the systems) the dynamics of the state trajectory on the sliding manifold are completely specified by the constraint (14.3). Thus these dynamics are insensitive to parameter variations, disturbances and to the particular feedback control law employed.

To state this result formally, we shall assume that

$$m \leqslant n \text{ and rank } \frac{\partial s}{\partial x} = m \text{ for all } x \in S. \tag{14.5}$$

Then there exists a positive integer p such that for every $x_0 \in S$, a suitable neighbourhood of x_0 may be written as a disjoint union of subsets of the surfaces

$$S_j = \{x \in R^n : s_j(x) = 0\}, \quad j = 1, \ldots, m$$

and of p open connected regions $c_1, \ldots, c_p$. Then a discontinuous feedback u will be a Carathèodory function

$$u:[0, T] \times \left(\Omega \Big\backslash \bigcup_{j=1}^{n} S_j\right) \to U$$

such that $t \to f[t, x, u(t, x)]$ is locally integrably bounded, and for every $x_0 \in S$ there exists a finite limit as $x \to x_0$,

$$u^j(t, x_0) = \lim u^j(t, x), J = 1, \ldots, p$$

where u^j denotes the restriction of u to c_j.

Theorem 1.1: Assume that $f(t, x, U)$ is convex for every $t \in [0, T]$ and $x \in S$.

(*a*) If x is a Filippov solution to (14.1) in $[0, T]$ corresponding to some discontinuous feedback, and $x \in S$, then x is an a.e. solution to (14.1) corresponding to the equivalent control.
(*b*) Let U be closed. Let x be an a.e. solution to (14.1) in $[0, T]$ corresponding to the equivalent control, with $x(0) \in S$. Then x is a Filippov solution to (14.1) in $[0, T]$ corresponding to any discontinuous feedback such that $x(t) \in S$, $0 \leqslant t \leqslant T$, provided the following hold:

$s \in C^2(\Omega)$, $f(t, \cdot, \cdot)$ is continuously differentiable, $\partial s/\partial x \; \partial f/\partial u \; (t, x, u)$ is nonsingular when x is near S. $\partial f/\partial u$ is locally bounded and aj, $a \; \partial f_i/\partial x_j$ are locally integrably bounded near S for every i, j and every element a of $(\partial s/\partial x \; \partial f/\partial u)^{-1}$.
For the proof see Bartolini and Zolezzi (1986*a*).

Simple examples show that the convexity of the velocity set in Theorem 1 cannot be omitted.

Remark: A viability approach in the sense discussed in Aubin and Cellina (1984) is also possible. The control system (14.1), (14.2) may be written as a differential inclusion

$$\dot{x}(t) \in F(t, x(t)), \; F(t, x) = f(t, x, U)$$

with viability domain $s^{-1}(0)$. The notion of equivalent control reduces here to that of the feedback map considered in Aubin and Cellina (1984).

In practice, the sliding condition (14.3) will be fulfilled only approximately owing to imperfections, disturbances and perturbations acting on the system (such as delays, hysteresis, small time constants, errors in the control law etc.) A distinction then arises between real states x (such that $s[x(t)]$ is, in some sense, very small but possibly not exactly zero) and ideal states x (which fulfil exactly (14.3)). As discussed in Utkin (1978), the equivalent control has a physically acceptable meaning only if the real states of the control system converge to the ideal states generated by the equivalent control as the perturbations acting on the system disappear.

Not every nonlinear system satisfies such a property. For example (see Utkin, 1978, p. 64).

$$\dot{x}_1 = u_1, \; \dot{x}_2 = u_2, \; \dot{x}_3 = u_1 u_2 \qquad |u_1| \leqslant 1, \qquad |u_2| \leqslant 1$$

$$s_1(x) = x_1, \; s_2(x) = x_2.$$

By using the control law

$$\underset{1}{u^j}(t) = \underset{2}{u^j}(t) = \begin{cases} 1 & \text{if } \; k/j \leqslant t < (2k+1)/2j \\ -1 & \text{if } \; (2k+1)/j \leqslant t < (k+1)/j \end{cases}$$

$j = 1, 2, 3, \ldots, 0 \leqslant k \leqslant j - 1$, and noticing that

$$u^*(t, x, w) = w$$

we get the real states x^j with $x^j(0) = 0$, $x_3^j(t) = t$, while the real state with $x(0) = 0$ has $x_3(t) = 0$.

So for some nonlinear systems the concept of equivalent control may lack physical meaning. The following definition isolates the class of variable structure control systems for which the equivalent control has a physical meaning, according to the above discussion.

System (14.1) – (14.3) fulfils the approximability property iff (14.4) holds and the following is true:

There exist $p > 1$, $M \in L^p(0, T]$ such that for all sequences $a_\varepsilon \in L^p(0, T)$, $\varepsilon > 0$, with $|a_\varepsilon(t)| \leqslant M(t)$ and $\sup \{|\int_0^t a_\varepsilon ds| : 0 \leqslant t \leqslant T\} \to 0$ as $\varepsilon \to 0$, if $\dot{x}_\varepsilon = f[t, x_\varepsilon, u^*(t, x_\varepsilon, a_\varepsilon)]$ a.e. in $[0, T]$, $s[x_\varepsilon(0)] \to 0$ and

$$\dot{x} = f[t, x, u^*(t, x, 0)]; \quad s[x(0)] = 0$$

then as $\varepsilon \to 0$, $x_\varepsilon(0) \to x(0)$ implies $x_\varepsilon \to x$ uniformly in $[0, T]$.

The meaning of approximability is the following. The real parameter ε describes the imperfections causing approximate satisfaction of the sliding condition (14.3). The (essentially arbitrary) functions a_ε satisfy (a.e. in $[0, T]$)

$$a_\varepsilon(t) = \frac{d}{dt} s[x_\varepsilon(t)] = \frac{\partial s}{\partial x}(x_\varepsilon) f(t, x_\varepsilon, u_\varepsilon) \tag{14.6}$$

where x_ε are real states and

$$u_\varepsilon = u^*(t, x_\varepsilon, a_\varepsilon)$$

By integrating (14.6) between 0 and t we get

$$s[x_\varepsilon(t)] \to 0 \text{ uniformly on } [0, T] \text{ as } \varepsilon \to 0.$$

Therefore approximability holds iff we can (uniformly) approximate any ideal sliding state by real states, disregarding the particular nature of the disturbances causing the sliding error (which is measured by a_ε), and the real states of the systems converge towards a well-defined ideal state provided the initial values tend to the sliding manifold, as the disturbances disappear. In particular, according to the terminology of Utkin (1978) no ambiguous behaviour arises when approximability is present.

Theorem 1.2: System (14.1) – (14.3) fulfils the approximability property in each of the following cases:

$$(a) \quad f(t, x, u) = A(t, x) + B(t, x)\, h(u)$$

$$(b) \quad f(t, x, u) = (x_2, x_3, \ldots, x_n, g(t, x, u)).$$

under suitable smoothness assumptions (see Bartolini and Zolezzi, 1986*a*) for the precise technical details).

As a matter of fact, a generalised notion of approximability may be defined without any reference to the equivalent control. After all, existence of a (unique) equivalent control and uniqueness of the sliding state are (to a large extent) independent properties. We do not assume existence of the equivalent control. The system (14.1) – (14.3) verifies generalised approximability iff the following is true. For every sequence $(u_\varepsilon, x_\varepsilon)$ fulfilling a.e. (14.1), (14.2) such that as $\varepsilon \to 0$

$s(x_\varepsilon) \to 0$ uniformly on $[0, T]$ and for every a.e. solution (u, x) to (14.1) – (14.3), such that $x_\varepsilon(0) \to x(0)$, it follows that $x_\varepsilon \to x$ uniformly on $[0, T]$.

Approximability implies generalised approximability, which is a form of sliding state uniqueness, as seen from the following theorem.

Theorem 1.3: Generalised approximability implies that: if (u_1, x_1), (u_2, x_2) are solutions to (14.1) – (14.3) such that

$$x_1(0) = x_2(0), \text{ then } x_1 = x_2 \tag{14.7}$$

Conversely, (14.7) implies generalised approximability under the following assumption: for every $k > 0$ there exist $A, B \in L^1(0, T)$ such that

$$|f(t, x, u)| \leqslant A(t)|x| + B(t)$$

whenever $|s(x)| \leqslant k$.

Example: Linear-time invariant systems. Let

$$f(t, x, u) = Ax + Bu, \; s(x) = Cx + D$$

for suitable constant matrices A, B, C, D and U compact. Then generalised approximability holds whenever (C, A) is observable.

14.3 Output feedback with observers

In accordance with the evolution of linear control theory we consider the possible extension of the previous theory to the case in which the state is accessible for control purpose only through the outputs. Actually we consider the case in which there is a perfect knowledge of the plant structure (the state equation) as in the well known linear case (Luenberger observer). Though restricted to known systems the following analysis (see Bartolini and Zolezzi (1986*b*) may be applied to systems not necessarily in controllable canonical form.

Consider the following observed variable structure control system:

$$\dot{x} = f(t, x, u), \; u \in U \qquad \text{(the state equation)} \tag{14.8}$$

$$s(x) = 0 \qquad \text{(sliding manifold)} \tag{14.9}$$

$$y = h(x) \qquad \text{(the output)} \tag{14.10}$$

$$\dot{z} = f(t, z, u) + P[y(t) - h(z)] \qquad \text{(the full order observer)} \tag{14.11}$$

Here $x, z \in R^n$, $y \in R^k$, $u, s \in R^m$ and $m \leqslant n$

under the standard hypotheses of regularity of the data (s, f, h continuously differentiable in x, f measurable in t and continuous in (x, u)).

Denote by s_x, f_x, h_x the Jacobian matrices of s, f, h, respectively. Given a constant matrix $P \in R^{n \times k}$ consider the following assumption:

(*a*) for every $t \geqslant 0$, y, z there exists a unique solution $u^*(t, y, z) \in U$ of the equation

$$s_x(z)\,\{f(t, z, \cdot) + P(y - h(z)\} = 0.$$

Such a mapping u^* is by definition the obvserver's equivalent control corresponding to the output y. The previous condition is true if $s_x(z)f(t, z, \cdot)$ maps U onto R^n in a one-to-one fashion.

For example for

$$f(t, x, u) = A(x) + B(x)u; \quad U = R^m$$

condition (*a*) holds if $s_x B$ is everywhere nonsingular.

The following analysis is mainly devoted to formalising the estimation problem within the theory already developed for the available state case.

Let us consider, under condition analogous to those developed in Theorem 1.1, the overall state equation with a control law coincident with the observer equivalent control, i.e.

$$\dot{x} = f(t, x, u^*(t, h(x), z) \tag{14.12}$$

$$\dot{z} = f(t, z, u^*(t, h(x), z)] + P[h(x) - h(z)] \tag{14.13}$$

Suppose that the relevant conditions to guarantee the existence and uniqueness of the solution are satisfied. The existence of solutions in the large for such a system is assured, for example, in the case of bounded continuous h if U is compact, f independent of t and bounded, s, f, h continuous and $f(x, U)$ always convex (see Aubin and Cellina, 1984). Under the further hypothesis that

$$\{(f(t, x, u), f(t, z, u))' : u \in U\} \text{ is convex in } R^{2n} \tag{14.14}$$

it is possible to show (Bartolini and Zolezzi, 1986*b*) that the Filippov solution $(x, z)'$ in $[0, +\infty)$ to (14.8), (14.20), (14.11) with $s[z(0)] = 0$ corresponding to the set of admissible output feedback (defined analogously as before: Theorem 1.1) coincides with the a.e. solution to (14.12), (14.13). Hypothesis (14.14) is stronger than the convexity of $f(t, x, U)$.

We use results contained in Kou, Elliot and Tarn (1983). We assume that:

(*b*) There exist matrices $P \in R^{n \times k}$, $Q \in R^{n \times n}$ and positive numbers $\delta, \varepsilon, \omega$ such that the eigenvalues of the symmetric matrix Q are between δ and ω and those of the symmetric part of $Q(f_x - Ph_x)$ are $\leqslant -\varepsilon$ everywhere. It is then possible to show that

$$|S[x(t)]| \leqslant M(\omega/\delta)^{1/2}|x(0) - z(0)| \exp(ct), \ t \geqslant 0$$

where $\omega c = -\varepsilon$ for every a.e. solution $(x, z)'$ in $[0 + \infty]$ of (14.12), (14.13) starting from a time instant in which $s(z) = 0$. (see Bartolini and Zolezzi, 1986*b*).

Hypothesis (*b*) in the case of linear time invariant systems

$$f(t, x, u) = Ax + Bu, h(x) = C\mathbf{x}$$

holds if (C, A) is detectable.

Example 2.1

$$\dot{x}_1 = f_1(x, u)$$

$$\dot{x}_2 = f_2(x, u)$$

with f_1, f_2 continuously differentiable in x and $f_{1x} \leqslant F, f_{2x} \leqslant G$,

$$A \leqslant f_{1x} + f_{2x} \leqslant B, G + F < A$$

for some constants $A, B > 0, F, G$. Then (*b*) holds with Q = identity,

$$0 < 2\varepsilon < A - G - F, P = (P_1, P_2)'$$

and

$$P_2 - P_1 \leqslant G - F, P_2 \geqslant G + \varepsilon,$$

$$P_1 - P_2 \leqslant A - 2\varepsilon - 2G, P_1 + 3P_2 \geqslant B + 2\varepsilon + 2G.$$

Remark: When the system is a differential equation $y^{(n)} = f(y, y', \ldots, y^{(n-1)}, u)$, under the usual transformation condition (*b*) can be reduced to the stability of the following matrix:

$$\begin{bmatrix} 0 & 1 & \cdot & \cdot & \cdot & \cdot & P_1 \\ \cdot & 0 & \cdot & \cdot & \cdot & & \cdot \\ \cdot & & \cdot & & & & \cdot \\ \cdot & & & \cdot & & & \cdot \\ \cdot & & & & \cdot & & (1 + P_{n-1}) \\ \dfrac{\partial f}{\partial x_1} & & & & & & \dfrac{\partial f}{\partial x_n} + P_n \end{bmatrix}$$

which under suitable hypotheses may be accomplished even when only some bounds, such as

$$\delta_i \leqslant \frac{\partial f}{\partial x_i} \leqslant \beta_i$$

are known.

The extension of the previous analysis to the case of partially unknown systems is characterised by difficulties analogous to those encountered in the case of the adaptive estimation of partially unknown linear systems. The relevant stability theory for this case requires (Landau, 1979) some passivity conditions to be satisfied. For this reason suitable dynamical systems were introduced in the control scheme. The possibility of a generalisation of such an approach is under investigation.

14.4 Asymptotic linearisation by VSC for uncertain systems

Feedback linearisation of nonlinear control systems has been of considerable interest in recent years. Such results (see e.g. Su, 1982; Boothby, 1984; Hunt, Su and Meyer, 1983*a* and *b* and their references) obtain local feedback equivalence of nonlinear smooth control systems to linear ones via differential geometric methods, requiring a complete knowledge of their dynamics.

On the other hand, a more realistic approach takes into account uncertainty acting on the given system. For nonlinear control systems of special form with deterministically uncertain dynamics, VSC techniques are effective to get at least asymptotically a prescribed linear behaviour. This is accomplished by a suitable discontinuous feedback so that the equivalence in our approach is meant in a different sense than, for example, Su (1982). A few related results may be found in Ryan and Corless (1984), Marino (1985), Fernandez and Hedrick (1987).

We consider control systems described by a scalar differential inclusion of order n

$$\gamma^{(n)} \in G(t, \gamma, \gamma', \ldots, \gamma^{(n-1)}, u),\ t \geqslant 0 \tag{14.15}$$

$$u \in U \tag{14.16}$$

where $x = (\gamma, \gamma', \ldots, \gamma^{(n-1)})'$ is the available state vector. U is a known interval, u is the scalar control variable and G is a multifunction which describes the unknown systems dynamics through known upper and lower bounds, so that

$$G(t, x, u) = [g^-(t, x, u), g^+(t, x, u)]$$

where g^-, g^+ are known Carathèodory functions.

The initial state $x(0)$ is uncertain but bounded,

$$|x(0)| \leqslant C \tag{14.17}$$

for some known constant C.

We are given a fixed time–invariant known model

$$\dot{y}_1 = y_2, \dot{y}_2 = y_3, \ldots, \dot{y}_{n-1} = y_n, \dot{y}_n = \sum_{i=1}^{n} a_i y_i + bv \tag{14.18}$$

with state y and scalar control variable v.

Given $\delta > 0$ and any control state pair v, y of the model, we wish to construct a feedback u such that every possible system state corresponding to it will satisfy both

$$|x(t) - y(t)| \leqslant (\text{const.})\, e^{-\delta t} \tag{14.19}$$

and the model dynamics (14.18) up to an exponentially decaying error term

$$\left| \sum_{i=1}^{n} q_i[x_i(t) - y_i(t)] \right| \leqslant (\text{const.})\, e^{-\delta t}$$

for some $q_i \in R$, all $t \geqslant T$ and some T independent of x and explicitly estimated by known data.

Since in general such a feedback u will be discontinuous, states x corresponding to it through (14.15) will be a Filippov solution on $[0, +\infty]$ to

$$\dot{x}_i = x_{i+1},\ i = 1, \ldots, n-1,\ \dot{x}_n = g(t, x, u)$$

for some (uncertain) dynamics $g(t, x, u) \in G(t, x, u)$. Conditions yielding piecewise continuity of the linearising feedback u are considered in Bartolini and Zolezzi (1988*a*).

Having fixed any model control state pair v, y, we pick real numbers $c_1, \ldots, c_{n-1}$ such that all the roots of the polynomial (in h)

$$h^{n-1} + c_{n-1} h^{n-2} + \ldots + c_2 h + c_1$$

have a negative real part. Then consider the error vector

$$e = y - x$$

and set

$$s(e) = e_n + \sum_{i=1}^{n-1} c_i e_i,$$

$$p(x, y, v) = bv + \sum_{i=1}^{n} a_i y_i + \sum_{i=1}^{n-1} c_i(y_{i+1} - x_{i+1}).$$

We assume that there exists a constant $k \neq 0$ such that for all $t \geqslant 0$ and x we can find $u(t, x) \in U$ fulfilling

$$\begin{aligned} g^-(t, x, u) &\geqslant k^2 + p(x, y, v) \text{ if } s(y - x) > 0 \\ g^+(t, x, u) &\leqslant -k^2 + p(x, y, v) \text{ if } s(y - x) < 0 \end{aligned} \tag{14.20}$$

Of course (14.20) is true if

$$g^+(t, x, U) = g^-(t, x, U) = R.$$

Such a (measurable) u is the required feedback, as stated by the following theorem, proved in Bartolini and Zolezzi (1988*a*).

Theorem 3.1: If (14.20) holds, any system state z corresponding to u (for any system dynamics g) satisfies

$$\dot{z}_1 = z_{i+1}, i = 1, \ldots, n-1,$$

$$\dot{z}_n = \sum_{i=1}^{n} a_i z_i + bv + \sum_{i=1}^{n} q_i e_i, \; t \geqslant T$$

where $T \leqslant kL$ and $|s(0)| \leqslant$ L;

$$q_1 = a_1, q_i = a_i + c_{i-1} \text{ if } i = 2, \ldots, n$$

and (14.19) is fulfilled.

The error term decays (exponentially) arbitrarily fast (i.e. δ may be arbitrarily prescribed) according to the choice of $c_1, \ldots, c_{n-1}$.

Example 3.1: Let the uncertain system dynamics be

$$\dot{x}_1 = x_2, \dot{x}_2 = a(x_1, x_2) + u$$

where the continuous function a is uncertain, but for some known constant A, $|a| \leqslant A$. A (piecewise continuous) asymptotically linearising feedback is given by

$$u = A + k^2 + p \quad \text{if} \quad y_2 - x_2 + c(y_1 - x_1) > 0$$

$$u = A - k^2 + p \quad \text{if} \quad y_2 - x_2 + c(y_1 - x_1) < 0,$$

p as before, any $k \neq 0$.

The linear behaviour of the model is then obtained up to an error term involving e, whose dynamics is now

$$\dot{e}_1 = e_2, \dot{e}_2 + ce_2 = 0 \text{ for large } t.$$

Approximability holds here. (This example has been treated by continuous state feedback in the framework of uncertain dynamic systems (Barmish, Corless and Leitmann, 1983) by using Lyapunov functions. Here we do not require them).

14.5 VSC for model reference control of non–minimum phase linear systems

In Bartolini and Zolezzi (1988*b*) a problem derived from the theory of adaptive control systems has been studied. Often in the control of nonlinear plant about some equilibrium point it happens that the linear approximation adopted to synthesise the control law is nonminimum phase (i.e. the linearised system has no stable inverse).

This situation prevent the use of traditional continuous adaptive model reference control schemes which explicity tend to introduce in the regulator 'cancellations' of suitable components of the transfer function.

'Cancellation' means that by using feedback or series compensation the whole controlled system is characterised by unobservable modes coincident with those (in particular the unstable ones) of the inverse plant system. The only way to avoid this phenomenon consists in choosing a reference model having the same unstable modes in the inverse system.

We can clarify this well known procedure by a simple example involving transfer function representation.

If

$$Wp\,(s) \;=\; \frac{B^{+}(s)\;B^{-}(s)}{A^{-}(s)}$$

is the plant transfer function in which the signs + − indicate that the polynomials have roots, respectively, in the right and left complex halfplane, we want, by means of series compensation $R(s)\;Wp(s)$, to achieve an input–output behaviour described by a transfer function $\hat{B}(s)/\hat{A}(s)$.

This is possible if

$$R(s) \;=\; \frac{\hat{B}(s)}{\hat{A}(s)}\,\frac{A^{-}(s)}{B^{+}(s)\,B^{-}(s)}$$

which is unstable and the product $R(s)Wp(s)$ is not stabilisable. If $\hat{B}(s) = B^{+}(s)\,B^{*-}(s)$, then

$$R(s) \;=\; \frac{B^{*-}(s)\,A^{-}(s)}{\hat{A}(s)\,B^{-}(s)}$$

which is stable and results in an overall stabilisable system.

On the basis of these considerations let us consider the scheme of Fig. 14.1 in which $Wp = B^{+}(s)\,B^{-}(s)/A(s)$ is the plant transfer function.

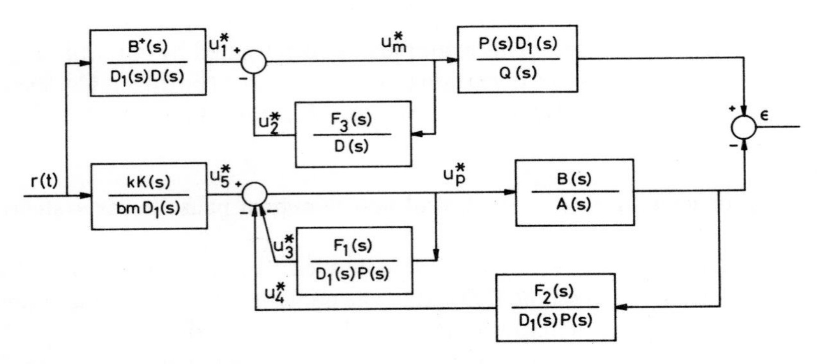

Fig. 14.1 *Plant transfer function*

There are two parallel branches. The upper represents the model and is equivalent by means of simple manipulation to a transfer function

$$W_m(s) \;=\; \frac{B^{+}(s)}{B^{+}(-s)}\,\frac{P(s)}{Q(s)}$$

if

$$F_3(s) + D(s) = B^+(-s).$$

If $B^+(s)$ is non Hurwitz, $B^+(-s)$ is Hurwitz (both of degree h). $P(s)$, $Q(s)$ are arbitrary Hurwitz polynomials of degree $l \leqslant m$ and $h < n$ where m, n are, respectively, the degree of B^+/B^- and A. D_1, D are arbitrary monic Hurwitz, of degree $n - 1$ and h.

The transfer function of the lower branch must be equal to the upper one. This may be accomplished if $F_1(s)$ and $F_2(s)$, the numerators of the feedback regulators, are chosen to satisfy the following Bezoutian equation:

$$A(s)\,(D(s)P(s) + F_1(s)) + F_2(s)\,B^-(s)\,B^+(-s) = B^+(-s)\,B^-(s)\,K(s)\,Q(s) \qquad (14.21)$$

The control signals u_m and u_p are obtained by summing the output of linear filters. It is well known that it is possible to represent such outputs as the scalar product of a parameter vector (with coefficients identical to those of the numerator) with a state vector of controllable canonical systems, whose characteristic polynomials coincide with the denominators, i.e.

$$\frac{N(s)}{D(s)} = \frac{\theta^i_m s^m + \theta^i_{m-1} s^{m-1} + \ldots + \theta^i_0}{s^n + d_{n-1} s^{n-1} + \ldots + d_0} \quad \text{is realised by}$$

$$\dot{x}_i = \begin{bmatrix} 0 & 1 & & & \cdot \\ \cdot & & & & \cdot \\ \cdot & & & & \cdot \\ \cdot & & & & \cdot \\ d^i_0 & \cdot & \cdots & \cdot & d^i_{n-1} \end{bmatrix} x_i + \begin{bmatrix} 0 \\ \cdot \\ \cdot \\ \cdot \\ \cdot \\ 1 \end{bmatrix} u_i$$

$$y_i = [\theta^i_0 \ldots\ldots \theta^i_m]\, x.$$

Moreover, if the starred symbols are used to indicate the known parameter case,

$$u^*_m = \theta^{T*}_m x_m = [\theta^{*T}_1, \theta^{*T}_2] \begin{bmatrix} x_1 \\ x_2 \end{bmatrix},$$

$$u^*_p = \theta^{T*}_p x_p = [\theta^{*T}_3, \theta^{*T}_4, \theta^{*T}_5] \begin{bmatrix} x_3 \\ x_4 \\ x_5 \end{bmatrix}.$$

When considering the unknown parameter case a control structure is chosen similar to that previously defined by Landau (1979), i.e.

$$u_m(t) = \theta_m(t)\, x_m,$$

$$u_p(t) = \theta_p(t)\, x_p.$$

This means that only the output coefficient vectors are considered variable or 'adaptable' while the regulators state equations are time invariant. Then

$$u_m(t) = [\theta_m^{*T} + \tilde{\theta}^T(t)]\, x_m = u_m^* + \tilde{u}_m$$

$$u_p(t) = [\theta_p^{*T} + \tilde{\theta}_p^T(t)]\, x_p = u_p^* + \tilde{u}_p$$

The zero state response of the system subject to a control law pair with time-varying parameter vectors is equal to that obtained in the known parameter vector case. In addition there is the effect of a pair of disturbances acting at the input of the reduced order model and the plant.

When dealing with the zero state response one can interchange the operator s with the time derivative operator d/dt. The output errors (in operational notation) are obtained as the difference of the output of the response of the control scheme of Fig. 14.1 by simple operations (see also Ambrosino, Celentano and Garofalo, 1984):

$$e(t) = k\,\frac{P(s)\,D_1(s)\,D(s)}{Q(s)\,B^+(-s)}\,u(t) + b_m\,\frac{D_1(s)\;P(s)\,B^+(s)}{K(s)\,B^+(-s)\;Q(s)}\,\tilde{u}_p(t). \quad (14.22)$$

If we consider $e_F(t)$ as the output of an all-pole filter with characteristic polynomial $P(s)\,D_1(s)\,D(s)$, the previous equation becomes

$$Q(s)\,B^+(-s)\;e_F(t) = k\tilde{u}_m(t) + b_m\,\frac{B^+(s)}{D(s)\,K(s)}\,\tilde{u}_p(t).$$

This differential equation is zero state equivalent to the following state equation:

$$\dot{e} = \begin{bmatrix} 0 & 1 & 0 & \cdot & \cdot & & \\ 0 & 0 & 1 & & & & \\ \cdot & & & & & & \\ \cdot & & & & & & \\ -\hat{a}_1 & -a & & \cdot & \cdot & \cdot\ \cdot & -\hat{a}_p \end{bmatrix} e + k \begin{bmatrix} 0 \\ \cdot \\ \cdot \\ \cdot \\ \tilde{u}_m(t) \end{bmatrix} + b_m \begin{bmatrix} 0 \\ \cdot \\ \cdot \\ \cdot \\ \eta(t) \end{bmatrix}$$

where $e = [e_F, se, \ldots .\, s^{p-1} e_F]^T$.

The coefficients $\{a_i\}$ are equal to the coefficients of the unknown polynomial $Q(s)\,B^+(-s)$, $p = \deg B^+ + \deg Q$, $\deg PD_1D = p - 1$.

The state equation in canonical form with accessible states (e_F and its derivatives) may be controlled by means of a variable structure control strategy once relevant bounds on the coefficients $\hat{a}_i$ and of the disturbance $\eta(t)$ are known.

The knowledge of bounds $\delta(t)$ ($\delta(t) > 0$) on the modulus of $\eta(t)$ may be ensured if suitable bounds on $\tilde{u}_P(t)$ are known. The boundedness of $\tilde{u}_P(t)$ is obtained by means of suitable saturation level on $u_P(t)$. In this analysis we simply suppose that $\delta(t)$ is known and bounded.

We consider

$$s(e) = e_P + \sum_{1}^{P-1} \gamma_i e_i.$$

The sliding and reaching condition become in this case

$$\text{sign}\left(-\sum_{0}^{P-1}\hat{a}_i e_{i+1} + k\tilde{u}_m(t) + b_m\eta(t) + \sum_{1}^{P-1}\gamma_i e_{i+1}\right) = -\text{sign}\, s(e) \tag{14.23}$$

It can be shown that the following control law satisfies (14.23):

$$u_m(t) = \mu_1'(t)\,\text{sign}\, s + w_1(t), \quad \text{where}$$

$$\mu_1' = 1/k\left\{b_m\,\delta(t) + \sum_{i=0}^{P-1}|\alpha_i e_{i+1}| + \sum_{i=1}^{P-1}|\gamma_i e_{i+1}|\right\},\ |\alpha_i| > \hat{a}_i$$

$$w_1(t) = \sum_{i=0}^{h-1}\theta_{1i}x_{1i} + \sum_{i=0}^{h-1}\theta_{2i}x_{2i},$$

$$\theta_{1i}(t) = -\theta_{1i}^M(t)\,\text{sign}\, x_{1i}\,\text{sign}\, s,\ \theta_{2i}(t) = -\theta_{2i}^M\,\text{sign}\, x\,\text{sign}\, s,$$

$$\theta_{1i}^M > |\theta_{1i}^*|,\ \theta_{2i}^M > |\theta_{2i}^*|$$

where α_i, θ_{1i}^M, θ_{2i}^M are easily evaluated if an upper bound on the maximum modulus of the root of $B^+(s)$ is known.

As far as the plant control law $u_P(t)$ is concerned we can choose a control law discontinuous on $s(e) = 0$ (eventually bounded by known bounded function). If $\dot{x}_P = f_P(t, x_P, u_P)$ (with $f(t, \cdot)$ linear) is the plant equation and we choose a control law $u_P(t) = M_P(t)\,\text{sgn}\,(s)$, $M_P(t) > 0$ discontinuous on $s(e) = 0$, then from Lemma 3 of Filippov (1964), we get

$$\dot{x}_P \in f_P(t, x_P, (2\alpha - 1)M_P(t)), \quad 0 < \alpha < 1. \tag{14.24}$$

Owing to the linearity of f_P, this set contains the control law u_P^* if $M_P(t) > |u_P^*|$.

It must be explicitly remarked that in the case actually considered an upper bound of the time parameters θ_P^*, depending on plant parameters uncertainties, is obtained by means of an approximate solution of the relevant Bezoutian equation (14.21).

This operation requires the approximate inversion of a suitable coefficient matrix. The following theorem proved in Bartolini and Zolezzi(1986*c*) establishes a connection between upper bounds on system parameters and upper bounds on θ_P^*.

Theorem 4.1: Let $\{a_i\}$, $\{b_i\}$ denote the plant parameters and $\lambda_k(P)$ be the kth root of a polynomial P. Assume

(i) $a_i^m \leqslant a_i \leqslant a_i^H$, $i = 1 - n$, $b_i^m \leqslant b_i \leqslant b_i^H$, $i = 1, \ldots m$,
(ii) $\lambda^+ \geqslant \max \lambda_j(B)$,

(iii) $\Delta < \min_{i,j} \prod |\lambda_i(A) - \lambda_j(B)|$.

Suppose that a_i^m, a_i^H, b_i^m, b_i^H, Δ are known. Then a number H_P can be found such that $\|\theta_P\| \leqslant H_P$.

Obviously a control law

$$u_P(t) = H_P \|x_P\| \operatorname{sgn}(s)$$

gives rise to a convex set containing $u_P^*(t)$.

At this point of the analysis we have ensured the existence of a control law pair $(u_m(t), u_P(t))$ discontinuous on $s(e) = 0$ such that

(i) a sliding motion on $s(e) = 0$ is attained in finite time;
(ii) the convex set on the RHS of the model and plant equation contains, respectively, $u_m^*(t)$ and $u_P^*(t)$, the ideal known parameter case control strategies.

This fact does not prevent the possibility of equivalence, in the sense of Filippov, of the discontinuous control law with control laws different from the ideal one. It is easy to show that the equivalence of the various discontinuous controls, even with unstable overall transfer function, is possible on $s(e) = 0$.

We have to prove, by applying Filippov's theory, that the only equivalent a.e. solution of our discontinuous control system is, at least asymptotically, the one, previously derived, which corresponds to the perfect knowledge of plant parameters. This may be accomplished by the following theorem:

Theorem 4.2: Under the hypotheses of the Theorem 4.1, assume moreover that:

(iv) $r(t)$ is bounded
(v) there is a known bound $g(t)$ on the modulus of the ideal control $u_m^*(t)$
(vi) the manifold $s(e) = 0$ has been reached and a sliding regime has been established.

Denote by $\hat{u}_m(t) = \mu_1'(t) \operatorname{sgn}(s) + w_1(t)$ the model control law previously defined. Choose

$$w_1(t) = \sum_0^{n-1} |\theta_{1i}^H (x_{1i} + (-1)^i x_{2i}| \operatorname{sign}(s) + \sum_0^{n-1} d_i^* x_{2i}$$

where d_i^* are the coefficients of the polynomial $D(s)$. Then a control law

$$u_m(t) = \begin{cases} \hat{u}_m(t) & \text{if } |\hat{u}_m(t)| \leqslant g(t), \\ -g(t) \operatorname{sgn}(s) & \text{if } |\hat{u}_m(t)| > g(t) \end{cases},$$

$$u_P(t) = H_P \|x_P\| \operatorname{sgn}(s)$$

causes the a.e. equivalent linear solution (in the sense of Filippov) of the whole control scheme to be equal to the solution of the known parameter case.

Details of the proof may be found in Bartolini and Zolezzi (1988*b*). Unsolved problems, related to this topic, include the elimination of the knowledge of the degrees of $B^+(s)$; the relaxation of hypotheses concerning the upper bound on

$\eta(t)$ and $u_m^*(t)$; the chattering phenomenon reduction and the extension of this approach to MIMO systems.

14.6 VSC of semilinear distributed systems

Variable structure control theory may be extended to distributed systems described by semilinear evolution equations in Banach spaces. As is well known, some control systems described by, for example, parabolic partial differential equations may be cast in such an abstract framework (e.g. the control of diffusion).

We consider the control system

$$\begin{aligned} \dot{y} + A(y) &= Bu \qquad \text{on } [0, T], \\ y(0) &= y_0 \end{aligned} \tag{14.25}$$

$$s(y(t)) = 0, \qquad 0 \leqslant t \leqslant T. \tag{14.26}$$

Here the state $y(t)$ takes values in a Banach space V and the (not necessarily linear) mapping

$$A: V \rightarrow V^*$$

acts between V and its dual space. Moreover

$$B: W \rightarrow V^* \quad \text{is bounded linear,}$$

$$s: D \rightarrow Z \quad \text{is continuously Frechet differentiable,}$$

with W a further Banach space and Z a Hilbert space. D is either V or a Hilbert space H such that

$$V \subset H \subset V^*$$

with continuous and dense imbedding. All spaces are assumed separable.

Admissible controls are all functions $u: [0, T] \rightarrow U$, a given subset of W, representing pointwise control constraints such that

$$\int_0^T \|u\|^q \, dt \leqslant C$$

for given constants $q > 1$ and C.

A solution y to (14.25) is meant in the (suitably weak) sense of Lions (1969). The mapping A is assumed monotone and satisfies

$$\langle Av, v \rangle \geqslant \alpha \|v\|^p, \ \|Av\| \leqslant C_1 \|v\|^{p-1} + c_2$$

for suitable constants $\alpha > 0$, c_1 and c_2, and $1/p + 1/q = 1$. Finally mild continuity conditions are imposed on A. An example of a mapping A which meets all these requirements is obtained by taking $V = H_0^1(\Omega)$, $H = L^2(\Omega)$ and

$$A(y) = -\sum_{i,j=1}^{N} \partial/\partial x_i [a_{ij}(x, y)\, \partial y/\partial x_j + b_i(x, y)]$$

where (a_{ij}) is a uniformly positive definite matrix with a_{ij}, b_i Carathèodory functions on $\Omega \times R^N$, Ω open bounded in R^N, and

$$\left| \sum_{j=1}^{N} a_{i_j} (x, y) \lambda_j \right| \leqslant h(x) + \omega (|y| + |\lambda|),$$

$$|b_i (x, y)| \leqslant k(x) + \omega |y|.$$

for every $i = 1, \ldots, N$, $x \in \Omega$, $\lambda \in R^N$, some h and k in $L^2(\Omega)$ and $\omega > 0$.

A suitable procedure to solve (14.25) and (14.26) is the following. We fix bases $\{v_i\}$ in V, $\{w_i\}$ in W, and $\{Z_i\}$ in Z. Then we project the infinite-dimensional control system (14.25), (14.26) on finite-dimensional subspaces in V, W, Z, spanned by elements of the bases there. At each step we obtain a Faedo–Galerkin approximate version of the original control system, to which we apply the existing finite-dimensional (ordinary differential equation) theory. Then we pass to the limit, letting the approximate system dimension tend to $+\infty$. Convergence theorems have been obtained (see Zolezzi, 1989). The following very simple example (to which the above theory applies) is a particular case of linear temperature control problem treated in Orlov and Utkin (1982).

We want to control to zero, near the final time T, the state $y = y(t, x)$ given by

$$\begin{cases} y_t = y_{xx} + u; & 0 \leqslant t \leqslant T, \quad 0 \leqslant x \leqslant 1; \\ y(0, x) = y_0(x), & \\ y(t, 0) = y(t, 1) = 0. & \end{cases}$$

Therefore we take the (trivial) sliding manifold $s(y) = y = 0$, and

$$V = H_0^1 (0, 1) = D = Z, H = L^2 (0, 1) = W.$$

By taking the eigenfunction basis

$$v_j(x) = \sin (\pi_{j_x}) = w_j(x), i = 1, 2, \ldots,$$

we get the following Faedo–Galerkin approximations:

$$\dot{x}_{inm} = -(i\pi) x_{inm} + u_{inm}, 0 \leqslant t \leqslant T,$$

$$x_{inm}(0) = a_i; i = 1, \ldots n-1, n \geqslant m,$$

where $y_0 \in L^2 (0, 1)$ and

$$y_0(x) \sim \sum_1^{\infty} a_j \sin (\pi_{j_x})$$

Each $y_{nm} = \Sigma_1^n x_{inm} v_i$ may be sent to the sliding manifold using a feedback

$$u_{inm} = -k^2 \operatorname{sgn}(x), k \neq 0$$

which yields

$$\dot{x}_{inm} x_{inm} \leqslant -k^2 |x_{inm}|.$$

Hence

$$x_{inm}(t) = 0 \text{ near } T.$$

provided $\sup |a_i| < k^2 T$

A convergence theorem may be applied and this yields convergence to a solution of the given control problem.

Interesting (still partially unsolved) related problems centre around a reasonable definition of approximability and a more general convergence theory (allowing distributional limits of finite-dimensional discontinuous feedback). We feel that the applications of variable structure control techniques to distributed systems are still at an early stage. The potential applications are undoubtedly interesting and very promising. Much work remains to be done in this area.

14.7 Conclusions

The main goal of the chapter has been to describe new tools in the analysis of discontinuous control systems and to show, with some significant control problems, how the VSC approach yields efficient and robust control, compared with other techniques used to deal with the same problems.

With the reduction of the chattering phenomenon using suitable smoothing control (Burton and Zinober, 1986), feasible, simple, low cost control strategies are available to control nonlinear systems in a global fashion. This possibility greatly improves the traditional technique based on approximate linearisation around some equilibrium point. If uncertainty and disturbance rejection are considered for an important class of nonlinear systems, the effectiveness of this approach is further increased.

Nevertheless, owing to limitations intrinsic in every theory applied to real problems, we cannot exclude the possibility of adopting approaches in which the VSC is active only outside a neighbourhood of the sliding manifold. Inside this neighbourhood other control techniques may be used. By varying the dimension of such a neighbourhood we can take into account real phenomena such as resonance, chattering etc.

The direction of future research in this field might be oriented towards the following objectives: extension of approximability theory to the case in which the equivalent control is not uniquely defined; linearisation of systems not necessarily in 'canonical' form; observers for unknown but bounded nonlinear plants; VSC of a wide class of distributed systems.

14.8 References

AMBROSINO, G., CELENTANO, G., and GAROFALO, F. (1984): *Int. J. Control*, **39**, pp. 1339–1349

AUBIN, J. P., and CELLINA, A. (1984): 'Differential Inclusions' (Springer)

BARMISH, B., CORLESS, M., and LEITMANN, G. (1983): *SIAM J. Control Optimiz.* **21**, pp. 246–255

BARTOLINI, G., and ZOLEZZI, T. (1985): *IEEE Trans.*, **AC-30**, pp. 681–684; (1986*a*): *J. Math. Anal. Appl.*, **118**, pp. 42–62; (1986*b*), *Systems Control Lett.*, **7**, pp. 19–193; (1986*c*), *Lecture Notes Control Inf. Sci*, **84**, pp. 54–62; (1988*a*): *System Control Lett.*, **10**, pp. 111–117; (1988*b*): *IEEE Trans.*, **AC** 33 (1988), pp. 859–63

BOOTHBY, W. (1984): *Systems Control Lett.*, **4**, pp. 143–147

BURTON, J. A., and ZINOBER, A. S. I. (1986): *Int. J. Systems Sci.*, **17**, pp. 875–885

FERNANDEZ, B., and HEDRICK, J. K. (1987): *Int. J. Control*, **46**, pp. 1019–1040

FILIPPOV, A. F. (1961): Proc. First IFAC Congress, Vol. 1, pp. 923–927; (1964): *Amer. Math. Soc. Transl.*, **42**, pp. 199–231; (1985): 'Differential equations with discontinuous right-hand side (Nauka, Moscow) (in Russian)

HUNT, L., SU, R., and N. MEYER, G. (1983*a*): *IEEE Trans.* **AC-28**, pp. 24–31; (1983*b*): 'Differential geometric control theory' (Birkhäuser) pp. 268–98

KOU, S. R., ELLIOTT, D. L., and TARN, T. J. (1975): '*Inform. Control*, **29**, pp. 204–216

LANDAU, I. D. (1979): 'Adaptive control. The model reference approach' (M. Dekker)

LIONS, J. L. (1969): 'Quelques methodes de resolution des problems aux limites non lineaires' (Dunod & Gauthier-Villars)

MARINO, R. (1985): *Int. J. Control*, **42**, pp. 1369–1385

VORLOV, V. J., and UTKIN, V. I. (1982): *Autom. Remote Control*, **43**, pp. 1127–1135

RYAN, E., and CORLESS, M. (1984): *IMA J. Math. Control Inform.*, **1**, pp. 223–242

SLOTINE, J., and SASTRY, S. S. (1983): '*Int. J. Control*, **38**, pp. 465–492

UTKIN, V. I. (1978): 'Sliding modes and their application in variable structure systems (MIR, Moscow); (1984): *Autom. Remote Control*, **44**, pp. 1105–1117

SU, R. (1982): *Systems Control Lett.*, **2**, pp. 48–52

ZOLEZZI, T. (1989): 'Partial differential equations and the calculus of variations, Vol. II' (Birkhäuser), pp. 997–1018

Chapter 15

Continuous self-adaptive control using variable structure design techniques

J. A. Burton

Simulation Modelling and Control Ltd.

A. S. I. Zinober

University of Sheffield

15.1 Introduction

Variable structure control theory (VSC) is well established in the control of time-varying uncertain systems and the basic techniques are well documented (Utkin, 1978; Itkis, 1976; Zinober, 1981). The main feature of a scalar VSC system is the sliding mode; this occurs when the discontinuous control action keeps the direction of the state trajectory towards or along the switching surface. In the sliding mode the system is invariant to parameter variations and rejects a certain class of disturbance (Draženović, 1969).

In the sliding mode infinitely fast chattering control maintains the state on the switching surfaces. In the application of VSC this produces an undesirable chatter motion; to obviate this a continuous approximation to VSC has been developed (Ambrosino *et al.*, 1984; Burton and Zinober, 1986) which gives a good approximation to ideal sliding motion and retains the basic VSC system design procedures and the associated properties.

Discontinuous self-adaptive control (SAC), developed from VSC, incorporates a time-varying switching surface. The self-adaptive scheme results in responses that are close to the time-optimal for a wide range of system parameter values without the need for the direct identification of the system parameters (Zinober, 1975, 1977, 1986).

Here a continuous counterpart of this self-adaptive scheme, continuous self-adaptive control (CSAC), is developed. The CSAC is implemented more practically than SAC (Burton, 1987), but possesses similar self-adaptive properties. In the continuous self-adaptive scheme the position of the switching surface is altered by a small amount when some continuous function reaches zero.

CSAC is demonstrated for simple scalar, higher order scalar and multivariable systems. A practical application is also considered.

15.2 The variable structure control system

The VSC system considered here is the linear time-invariant regulator system

$$\dot{\boldsymbol{x}} = A\boldsymbol{x} + B\boldsymbol{u} \tag{15.1}$$

where $\boldsymbol{x}$ and $\boldsymbol{u}$ are of dimension n and m, respectively, and the VSC law is of the form

$$u_i = \begin{cases} u^+(\boldsymbol{x}) & s_i(\boldsymbol{x}) > 0 \\ u^-(\boldsymbol{x}) & s_i(\boldsymbol{x}) < 0 \end{cases} \qquad i = 1, \ldots, m. \tag{15.2}$$

The fixed switching hyperplane is defined by

$$\boldsymbol{s}(\boldsymbol{x}) = G\boldsymbol{x} = 0 \tag{15.3}$$

When sliding motion occurs in the VSC system, the state slides on $N(\boldsymbol{s})$ the null space of the switching function. The system is designed to ensure attainment of a stable sliding mode (Utkin, 1978; El-Ghezawi *et al.*, 1983*a*).

In the sliding mode

$$s(\boldsymbol{x}) = G\boldsymbol{x} = 0 \quad \text{and} \quad \dot{s}(\boldsymbol{x}) = G\dot{\boldsymbol{x}} = 0 \tag{15.4}$$

which are the equations that define the system dynamics. These dynamics also arise when the linear equivalent control $\boldsymbol{u}_{eq}$ is applied. If GB is nonsingular, then from (15.4), during sliding

$$G(A\boldsymbol{x} + B\boldsymbol{u}_{eq}) = 0 \tag{15.5}$$

and therefore

$$\boldsymbol{u}_{eq} = -(GB)^{-1}GA\boldsymbol{x} \tag{15.6}$$

The resulting system dynamics are

$$\dot{\boldsymbol{x}} = [I - B(GB)^{-1}G]A\boldsymbol{x} \tag{15.7}$$

$$= A_{eq}\boldsymbol{x} \tag{15.8}$$

During sliding the system behaves as an unforced system and, for asymptotic stability, system (15.8) should have $n - m$ stable eigenvalues; the remaining m eigenvalues being zero (Utkin, 1978; El-Ghezawi *et al.*, 1983*b*).

15.3 The smoothed variable structure control system

One of the inherent difficulties of VSC in its application to real plant is 'chatter' motion that results about the switching hyperplane. Recently a continuous nonlinear control approximation (Ambrosino *et al.*, 1984; Burton and Zinober, 1986) has been developed. Termed smoothed variable structure control (SmVSC), analysis has been carried out that gives a bound on the deviation from perfect sliding, predetermines the profile of the resulting motion and proves that chatter motion is indeed eliminated (Burton and Zinober, 1986).

The nonlinear approximation is generated by noting that, for scalar control, the discontinuous relay control is

$$u = -\varrho \operatorname{sgn}(s) \tag{15.9}$$

where ρ is some positive scalar, and can be written in the form

$$u = -\varrho \frac{s}{|s|} \tag{15.10}$$

For δ a small positive constant

$$u = -\varrho \frac{s}{|s| + \delta} \tag{15.11}$$

is a continuous control with upper and lower bounds $\pm \varrho$.

This control function can be approximated (Burton and Zinober, 1986), in a small enough neighbourhood of the switching line, to first order by

$$u = ks \tag{15.12}$$

where

$$k = \varrho / \delta \tag{15.13}$$

The approximation is essentially a high gain feedback control in the neighbourhood of $N(\mathbf{s})$ and the system response therefore possesses properties associated with such systems (Young *et al.*, 1977). The design procedure, however, remains similar to the VSC system design approach.

15.3.1 Analysis of the system response in a SmVSC system

Since SmVSC is nonlinear, exact analysis of the resulting response is extremely complex. The key to the analysis is the inspection of the null spaces that correspond to zero values of the successive derivatives of the switching function.

Consider the configuration of these subspaces in n-dimensional space. For scalar SmVSC the closed loop system is given by

$$\dot{\boldsymbol{x}} = (A - kBG)\boldsymbol{x} = F\boldsymbol{x} \tag{15.14}$$

and the derivatives of the switching function by

$$s = G\boldsymbol{x}, \quad \dot{s} = GF\boldsymbol{x}, \ldots, \frac{d^p s}{dt^p} = GF^p\boldsymbol{x} \tag{15.15}$$

Since the columns of F do not necessarily lie in $N(G)$, if the state lies in $N(s)$, it does not lie in $N(\dot{s})$. By considering a similar expression for a system in the ideal sliding mode (15.8) it is seen that during ideal sliding motion these null spaces coincide (Burton and Zinober, 1986). In SmVSC the null spaces are fanned out and the condition for sliding motion is only satisfied on one side of the line. This is best illustrated in the second-order scalar control case in a phase plane diagram (Fig. 15.1).

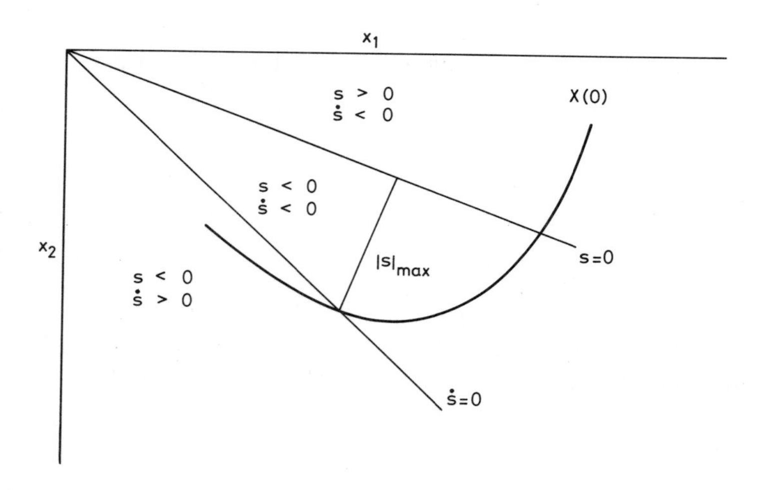

Fig. 15.1 *Sliding conditions in a SmVSC system*

Using a standard numerical analysis tool, the power method (Wilkinson, 1965), the following results have been proved (Burton and Zinober, 1986):

(i) In SmVSC the full space given by $N(\boldsymbol{w}_1^T)$, namely the null space of the dominant left eigenvector of the closed loop system matrix F, represents the limiting null space of the derivatives of the switching function.
(ii) The state approaches $N(\boldsymbol{w}_1^T)$ asymptotically.
(iii) Chatter motion is eliminated since the switching hyperplane can be crossed at most $n - 1$ times, where n is the order of the system.

15.4 Continuous self-adaptive control of scalar control systems

15.4.1 Self-adaptive control

Self-adaptive control (SAC) based on VSC theory has been presented by Zinober (1975, 1977, 1986), where a time-varying 'linear' switching surface, $G(t)\boldsymbol{x} = 0$, is adjusted by altering a single design parameter $g(t)$ whenever sliding motion occurs. The system is designed so that the design parameter has direct influence on the eigenvalues during the sliding mode, which are successively shifted further into the left half plane of the complex plane state at each adaptation step, resulting in faster system dynamics. The system is analysed by studying certain nonlinear differential equations and the following properties have been shown to result:

(i) The time taken to reach the state origin is finite (unlike perfect sliding with G constant)
(ii) The state trajectory approaches the origin on a singular sliding trajectory (for definition see Zinober, 1977)

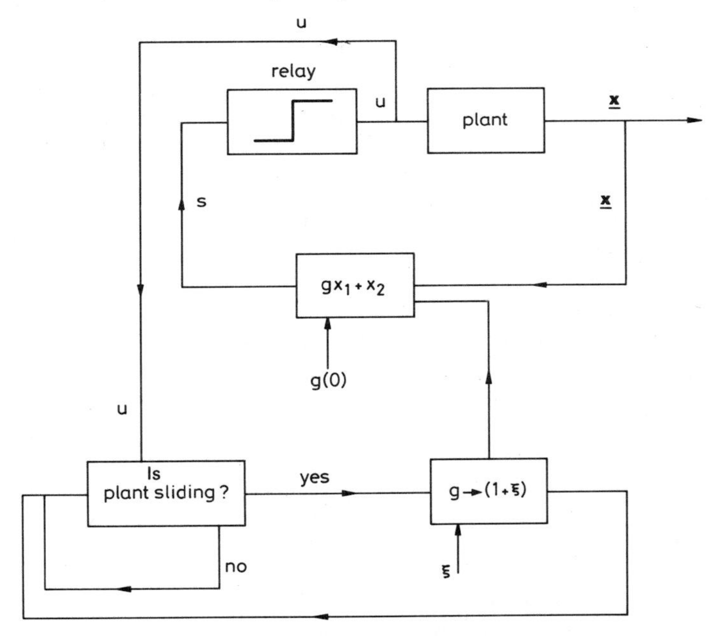

Fig. 15.2 *The SAC system for a second-order scalar system*

(iii) At all points of adaptation $g(t) > 0$
(iv) A dynamic response close to that of time-optimal control is achieved without the need for the identification of certain plant parameters.

The SAC system is described in Fig. 15.4.2 for second-order systems. For the system (15.1) the relay control is given by

$$u = \begin{cases} +\varrho & s^{(m)} < 0 \\ -\varrho & s^{(m)} > 0 \end{cases} \tag{15.16}$$

where $s^{(m)}$ represents the mth adaptation of the original switching line $s^{(0)}$. When sliding motion occurs the switching line is adapted according to the following law:

$$s^{(m)} = g^{(m)}x_1 + x_2 \qquad t_{m-1} < t < t_m \tag{15.17}$$

$$\begin{aligned} s^{(m+1)} &= g^{(m)}(1 + \xi)x_1 + x_2 \qquad t > t_m \\ &= g^{(m+1)}x_1 + x_2 \end{aligned} \tag{15.18}$$

where ξ is some positive design parameter. Here t_m represents the time of attainment of $N(s^{(m)})$, the null space of the mth adaptation of the switching line.

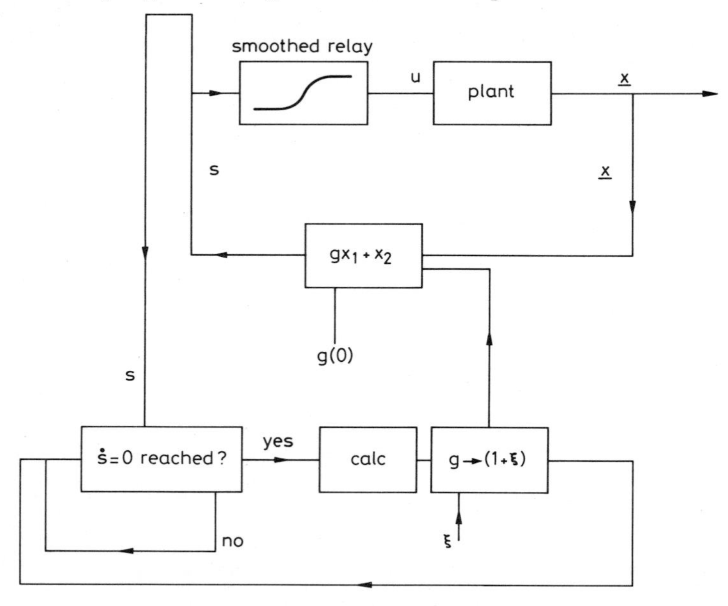

Fig. 15.3 *The CSAC system for a second-order scalar system*

15.4.2 CSAC of second-order scalar systems

CSAC is derived from SmVSC; in particular, use is made of the resulting 'fan' of derivatives of the switching function (Fig. 15.1). The CSAC system defers adaptation of the design parameter $g(t)$ until the line $\dot{s} = 0$, the first derivative of the switching function, is reached. This point is considered analogously to sliding motion as the sliding condition is again satisfied (Fig. 15.1). The design parameter is then adjusted so that the new switching line lies in $N(\dot{s})$. CSAC is described pictorially in Fig. 15.3.

The CSAC function is the SmVSC function

$$u = -\varrho \frac{s^{(m)}}{|s^{(m)}| + \delta} \tag{15.19}$$

with the continually adapting switching line $s^{(m)}$. CSAC performs adaption when both

$$s^{(m)} = g^{(m)} x_1 + x_2 = 0 \qquad t_m < t < t_{m+1} \tag{15.20}$$

and $N(s^{(m)})$ have been crossed. The new switching function is then

$$s^{(m+1)} = \dot{s}^{(m)} = g^{(m)} \dot{x}_1 + \dot{x}_2 \qquad t > t_{m+1} \tag{15.21}$$

The time t_{m+1} represents the time of attainment of the state trajectory of $N(s^{(m+1)})$. This equation can be expressed in the form

$$s^{(m+1)} = g^{(m+1)} x_1 + x_2 \tag{15.22}$$

The adaptive parameter in (15.18) is given implicitly in (15.22) by a predetermined amount that is governed by the parameters of the SmVSC function and the parameters of the system itself.

15.4.2.1 Justification of the adaptive control scheme: Firstly, it is necessary to show that the time taken to reach $N(\dot{s})$ is indeed short; if this is not so adaptation will take place at intervals that would have no significant advantage over SmSVC or ideal sliding motion. The adaptation steps should regularly improve the speed of response of the system by continually adapting the switching line, increasing the magnitude of the absolute value of the eigenvalue of the closed loop system.

It is required that

$$\tau = t_{m+1} - t_m \simeq 0 \tag{15.23}$$

Firstly the system eigenvalue of (15.21) is calculated. For scalar control, has G the form $G = [g, 1]$ and the system is in phase canonical form. The eigenvalues $\{\lambda_1, \lambda_2\}$ of the closed loop system are given by

$$2\lambda_i = -a_2 - k \pm k[1 + (2a_2 - 4g)/k]^{1/2} \tag{15.24}$$

where k is given in (15.13) and a_i $(i = 1, 2)$ are the elements of the system matrix A. This can be expanded to give

$$2\lambda_i = -a_2 - k \pm k(1 + \xi) \tag{15.25}$$

provided that k is sufficiently large where

$$\xi = (a_2 - 2g)/k \tag{15.26}$$

This implies that δ in (15.11) has been chosen suitably small; so from (15.25),

$$\lambda_1 \simeq -g, \qquad \lambda_2 \simeq -k \tag{15.27}$$

if a_i are small in comparison with k. Let $\{\boldsymbol{u}_1, \boldsymbol{u}_2\}$ be the eigenvectors of A. Using a modal expression the switching function can be written in the form

$$s(t) = \alpha_1 e^{\lambda_1 t} G\boldsymbol{u}_1 + \alpha_2 e^{\lambda_2 t} G\boldsymbol{u}_2 \tag{15.28}$$

where $\{\alpha_1, \alpha_2\}$ are constants dependent upon the initial condition of the system. If $s(t_m) = 0$ then

$$\alpha_1 e^{\lambda_1 t_m} G\boldsymbol{u}_1 + \alpha_2 e^{\lambda_2 t_m} G\boldsymbol{u}_2 = 0 \tag{15.29}$$

and if $\dot{s}(t_{m+1}) = 0$

$$\alpha_1 \lambda_1 e^{\lambda_1 t_{m+1}} G\boldsymbol{u}_1 + \alpha_2 \lambda_2 e^{\lambda_2 t_{m+1}} G\boldsymbol{u}_2 = 0 \tag{15.30}$$

These two equations can be combined to give

$$\lambda_1 e^{\lambda_1 (t_{m+1} - t_m)} = \lambda_2 e^{\lambda_2 (t_{m+1} - t_m)} \tag{15.31}$$

so that

$$\tau = [\ln (\lambda_2/\lambda_1)]/(\lambda_1 - \lambda_2) \tag{15.32}$$

from which it is immediately apparent from (15.27) that

$$\tau \to 0 \quad \text{as} \quad k \to \infty \tag{15.33}$$

The case $\lambda_1 = \lambda_2$ will be discussed at a later stage. A CSAC system has then small time intervals between adaptations provided that k is large compared to the other parameters. This assumption is that required for the approximation of VSC by SmVSC (Burton and Zinober, 1986). The condition for short time intervals between adaptation steps is therefore automatically satisfied by choice of a valid SmVSC function.

15.4.2.2 Method of adaptation: Adaptation in a CSAC system is now investigated and comparisons made with the SAC scheme. It is required for a switching line given by

$$s^{(m)} = G^{(m)}\boldsymbol{x} = g^{(m)}x_1 + x_2 = 0 \tag{15.34}$$

that adaptation results in $s^{(m+1)}$ being formed when the state reaches the null space of the derivative of $s^{(m)}$,

$$s^{(m+1)} = G^{(m+1)}\boldsymbol{x} = g^{(m+1)}x_1 + x_2 = 0 \tag{15.35}$$

Also

$$\dot{s}^{(m)} = G^{(m)}G\boldsymbol{x} = 0 \tag{15.36}$$

Combining (15.35) and (15.36) it is apparent that

$$G^{(m+1)} = G^{(m)}F \tag{15.37}$$

This iterative formula allows derivation of the amount of adaptation required in the parameter $g^{(m)}$. If A is in canonical form it can be shown that

$$G^{(m+1)} = [(kg^{(m)} + a_1)/(k + a_2 - g^{(m)}), 1] \tag{15.38}$$

which can be written as

$$G^{(m+1)} = [(kg^{(m)} + a_1)/\{k(1 - (g^{(m)} - a_2)/k)\}, 1] \tag{15.39}$$

Using the fundamental assumption that k is large in comparison with other parameters, expanding and neglecting terms of order $1/k^2$ and higher yields

$$G^{(m+1)} = [g^{(m)}(1 + \xi), 1] \tag{15.40}$$

where

$$\xi = (g^{(m)} - a_2)/k \tag{15.41}$$

The resulting adaptation of the switching line is

$$g^{(m+1)} = (1 + \xi)g^{(m)} \tag{15.42}$$

In CSAC the successive values of $g^{(m)}$ are given by discrete values predetermined by the design of the SmVSC function and the system parameters. This is in direct contrast to the SAC scheme where the control is discontinuous and the adaptation is presented in a continuous manner.

15.4.2.3 Stability of the adaptive control scheme: For any initial stable switching line, the use of valid SmVSC results in a stable response (Burton and Zinober, 1986); stability of a CSAC system therefore depends on the new switching line generated being stable. It is not always the case that a new switching line will be stable. There is in fact an upper limit to the adaptation procedure resulting in what is termed 'fine tuning' of the switching line to yield a maximum possible speed of response.

Fine tuning is most easily illustrated for the double integrator system. The line

$$s^{(m)} = g^{(m)}x_1 + x_2 \tag{15.43}$$

after adaptation becomes

$$s^{(m+1)} = g^{(m+1)}x_1 + x_2 \tag{15.44}$$

where, from (5.39)

$$g^{(m+1)} = kg^{(m)}/(k - g^{(m)}) \tag{15.45}$$

It is apparent that if $g^{(m)} > k$ the adaptation step would generate $g^{(m+1)} < 0$ and hence, as the eigenvalue of the second-order system in the sliding mode is given by the negative of the adaptive parameter, an unstable switching line results. Thus there exists the necessary condition that

$$g^{(m)} > k \tag{15.46}$$

The adaptation technique is halted when $g^{(m+1)}$ violates this condition. The system then continues as a SmVSC system with a response that is faster than the initial design response (since $g^{(m)} > g^{(0)}$). Closer inspection of the time interval between adaptation steps via (15.27) and (15.32) shows that at this value of $g^{(m)}$ CSAC breaks down as τ becomes undefined.

15.4.3 CSAC of higher order scalar systems

For $n > 2$ the switching hyperplane in SAC is also adapted by the alteration of a single parameter; to achieve this a suitable switching hyperplane has coefficients which are the binomial coefficients to allow easy factorisation. A similar approach is employed for the higher order CSAC scheme.

In CSAC no straightforward analysis is available to justify the scheme by generating a rigorous expression for the time interval between adaptation steps. It can be inferred that the time interval is small by comparing SmVSC with a high gain feedback control and examining the fast modes of such systems towards the switching hyperplane (Young *et al.*, 1977; Burton and Zinober, 1986). It has been shown to be a valid assumption via numerous computer simulation studies.

In higher order CSAC the n-dimensional switching function is expressed in the adaptive form

$$s = \sum_{i=1}^{n-1} c_i g^{n-i} x_i + x_n \tag{15.47}$$

where g is the parameter to be adapted and c_i, $i = 1, \ldots n - 1$, are positive constants dependent on the desired initial properties of the system. The iteration superscript of g has been dropped to ease the notation. Any initial switching function can be expressed in this form. The CSAC iteration procedure is again given by (15.37) and it can be shown for a system in canonical form that

$$G^{(m)} = \left[\frac{kc_1 g^{n-1} + a_1}{k + a_n - c_{n-1} g}, \frac{kc_2 g^{n-2} - c_1 g^{n-1} + a_2}{k + a_n - c_{n-1} g} \ldots \right. \tag{15.48}$$

$$\left. \ldots \frac{kc_{n-1} g - c_{n-2} g^2 + a_{n-1}}{k + a_n - c_{n-1} g}, 1 \right] \tag{15.48}$$

where g is the mth adaptation of the initial design parameter, and by similar manipulation, as in the scalar CSAC, it can be shown that the generated switching hyperplane can be written as

$$G^{(m+1)} = [c_1 g^{n-1}(1 + \xi), c_2 g^{n-2}(1 + \xi), \ldots, c_{n-1} g(1 + \xi), 1] \tag{15.49}$$

with

$$\xi = (c_{n-1} g - a_n)/k \tag{15.50}$$

The resulting adaptation is then given approximately as (15.42) and a similar response is obtained as in the second-order CSAC.

The stability of CSAC depends only on the necessity that a stable switching hyperplane is generated. If this is not possible, the adaptation must be halted. Inspection of the eigenvalues generated in ideal sliding motion by any new switching hyperplane can be carried out using the theory of equivalent systems (Utkin, 1978). For CSAC of an nth-order integrator system the equivalent system matrix (15.8) is given by

$$A_{eq} = \begin{bmatrix} 0 & 1 & 0 \ldots 0 \\ 0 & 0 & 1 \ldots 0 \\ 0 & 0 & 0 \ldots 0\ 1 \\ \dfrac{kc_1 g^{n-1}}{k - c_{n-1}g} & \dfrac{kc_2 g^{n-2} - c_1 g^{n-1}}{k - c_{n-1}g} & \ldots\ 1 \end{bmatrix} \qquad (15.51)$$

A necessary condition for the stability of this system is that the coefficients in the bottom row are positive; therefore the necessary condition

$$k - c_{n-1}g > 0$$

$$g < \frac{k}{c_{n-1}} \qquad (15.52)$$

results, giving an upper limit on the allowable fine tuning within the switching hyperplane.

Owing to the complexity of deriving an analytical expression for the time interval between adaptation steps an expression for the derivative of the adaptive parameter is not found. It is apparent from simulation that it is possible for the derivative of the adaptive parameter to become negative and the new hyperplane to be a stable hyperplane for the system. This phenomenon can be explained by examining the properties of SmVSC.

In SmVSC it is known that the switching function can possess at most $n - 1$ zeros (Burton and Zinober, 1986). For an initial condition on the switching hyperplane the time portrait of the initial switching function versus time may look typically as shown in Fig. 15.4, where it is assumed that the system is of order 4 and uses the three available zeros. As $g^{(m)}$, representing the 'slope', is progressively increased, the SmVSC response changes with the zero at point A, moving to the left until it eventually coincides with O as shown in Fig. 15.5. Initial adaptation is found to result in increasing 'slope' until the critical case is reached where adaptation will tend to decrease the speed of response. In general, adaptation can take place in any system if the distance OA is not zero.

When adaptation is stopped the system must wait until the second zero of the critical switching function (Fig. 15.5) is reached (point B), when adaptation can resume and (in this example) is then only constrained by the upper limit on the procedure. Fine tuning of the system response is therefore achieved in distinct phases.

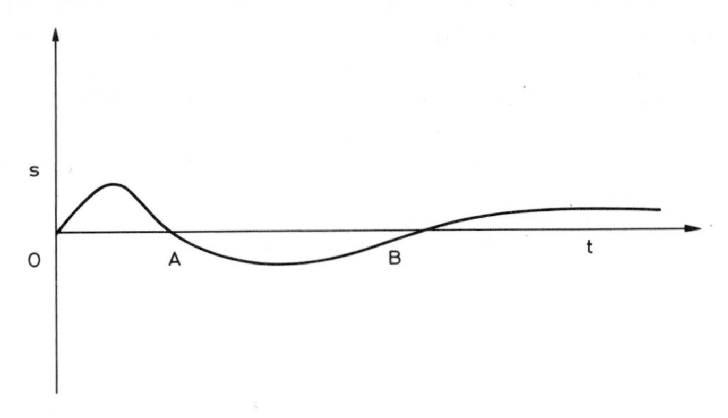

Fig. 15.4 *A switching function time portrait*

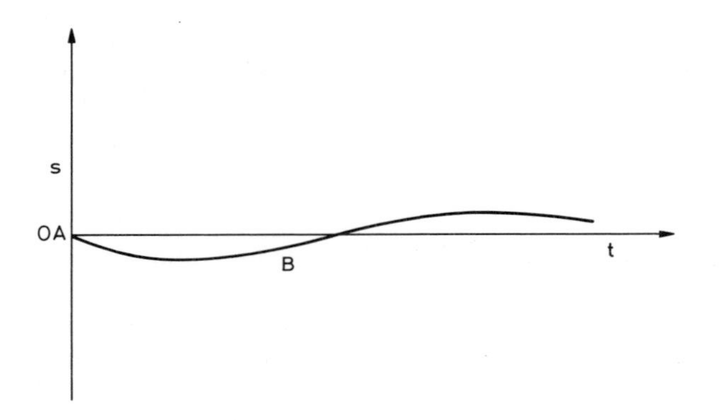

Fig. 15.5 *A switching function time portrait at a critical slope*

15.5 Continuous self-adaptive control of multivariable control systems

Having derived a scalar system its application to the potentially more interesting multivariable control system is now investigated. In SAC a rigid structure is assumed for the whole switching hyperplane ($n - 1$)th degree polynomial in the adaptive parameter); such a structure proves unmanageable in the multivariable case. In multivariable CSAC a structure is imposed only on the parameters within the switching hyperplane matrix which have direct influence upon the eigenvalues of the sliding mode. This structure is non-restrictive and is in fact useful in making generation of the initial controller an easier task, since the system has a form that is amenable to initial multivariable VSC design.

Multivariable CSAC is implemented in a different manner; each switching hyperplane is adapted separately (and some not at all), rather than developing a methodology for adaptation of all the switching hyperplanes concurrently. Particular elements (the position of which is governed by the structure of the system) which have direct influence on the eigenvalues of the system during sliding motion, are singled out and adapted. The point in time at which an adaptation step takes place is again when the state trajectory reaches the null space of the derivative of a switching function, and adaptation again brings a switching hyperplane into correspondence with the null space of its derivative.

An important feature of a multivariable CSAC system is that it is possible for the increase in speed of response to be varied by adapting one or more switching hyperplanes. It is also possible that one, or more, switching hyperplanes may not be adapted as the null space of that switching function may not be reached. This corresponds to multivariable SmVSC where a particular hyperplane is approached asymptotically and therefore the null space of the first derivative of the switching hyperplane is not reached. Even if the null space of a switching hyperplane is reached, adaptation does not take place if that hyperplane has no direct influence on an eigenvalue in the sliding mode.

The iteration procedure for scalar CSAC (15.37) can be carried over from the analysis with the subscript j denoting any of the m switching hyperplanes that are to be adapted. In the multivariable case, however, not all the coefficients of a switching hyperplane need possess the adaptive design parameter.

At some instant the nth adaptation of the jth hyperplane is given for a system in canonical form as

$$s_j^{(m)} = \sum_{i=1}^{n_a} c_{ij} g_{a_j}^{n_a+1-i} x_i + \sum_{i=n_a}^{n} c_{ij} x_i \qquad (15.53)$$

The states have been re-ordered in the jth switching function so that the first summation represents a polynomial, the power of which is dependent upon the number n_a of key elements of that switching function in which adaptation will take place. The second summation represents those elements that have no direct influence on the eigenvalue of ideal sliding motion and hence in which no adaptation will take place. The iterative superscript has been dropped for ease of notation. Hyperplanes which are not adapted are expressed as simple linear combination of states.

The control function used throughout is the multivariable SmVSC

$$u_i = -\varrho \frac{s_i^{(m)}}{|s_i^{(m)}| + \delta} \qquad i = 1, \ldots m \qquad (15.54)$$

and a subspace of the state space is considered that allows the approximation

$$u_i = -ks_i \qquad i = 1, \ldots m \quad \text{where} \quad k = \varrho/\delta \qquad (15.55)$$

to be valid when required (Burton and Zinober, 1986).

15.5.1 Justification of the adaptive control scheme

Precise analytical expressions to justify the scheme prove difficult to derive since any expression of the time interval between adaptation steps is a linear combination of exponential functions and exact solution is unobtainable. It is sufficient to note that the controller approximates to a high gain feedback controller in the subspace considered, and in such systems it is a reasonable assumption that the time interval from the switching function will be small owing to the particular separation of the eigenvalues (Young *et al.*, 1977) into fast eigenvalues in the range space of the switching hyperplane and slow within the null space. This assumption is also borne out by actual simulation.

15.5.2 Method of adaptation

The CSAC system requires that the switching hyperplane matrix is expressed in a suitable form such that any adaptation of the design parameter directly increases the eigenvalue that would result if sliding motion were to take place at that instance. For convenience this is termed increasing the 'slope' of the set of hyperplanes.

The system matrix is assumed to be in canonical form and the block nature of the input matrix is given below; if a structure is not in this form it is a simple task to transform it to such a structure

$$\dot{\boldsymbol{x}} = A\boldsymbol{x} + B\boldsymbol{u} \tag{15.54}$$

$$\boldsymbol{s} = G\boldsymbol{x} \tag{15.55}$$

where

$$B = \begin{bmatrix} 0 \\ B_2 \end{bmatrix}, \quad G = [G_1, G_2] \tag{15.56}$$

and the submatrices have appropriate dimension. The response of the system in the sliding mode is that of the reduced order system (Utkin, 1978)

$$\dot{z}_1 = (A_{11} - A_{12}G_2^{-1}G_1)\dot{z}_1 \tag{15.57}$$

The multivariable CSAC system is most easily analysed by considering first the case $n - m = 1$ and then extending the approach to the case $n - m > 1$.

15.5.2.1 Multivariable CSAC: $n - m = 1$: For $n - m = 1$ the sub-matrices of A have particular form and (15.57) reduces to the scalar equation

$$\dot{z}_1 = (-A_{12}G_2^{-1}G_1)z_1 \tag{15.58}$$

where the eigenvalue of ideal sliding motion is now given by

$$\sigma = -A_{12}G_2^{-1}G_1 \tag{15.59}$$

In order to adapt the eigenvalue it is necessary therefore to inspect the component $G_2^{-1}G_1$.

Useful results are derived if, for the choice of switching function G, the additional constraint $GB = I_m$ is added. This can always be done for any initial choice of eigenvalue, and is indeed beneficial in multivariable VSC owing to the decoupling of the control actions. It immediately follows that

$$G_2^{-1} = B_2 \tag{15.60}$$

and the eigenvalue of sliding motion (15.59) is now given by

$$\sigma = -A_{12}B_2G_1 \tag{15.61}$$

where the only degree of freedom for adaptation is now in the submatrix G_1. Expressing the matrices G_1 and B_2 by their elements and multiplying by A_{12} this can be written as

$$\sigma = -\sum_{i=1}^{m} b_{2i}g_{1i} \tag{15.62}$$

It is a fundamental assumption of VSC theory that the inverse of GB exists; therefore it can be guaranteed that at least one of the coefficients of the second row of the input matrix is non-zero. Adaptation of the eigenvalue is clearly possible by adapting in some pre-described way the parameters in the first column of the switching hyperplane matrix that correspond to non-zero elements in the second row of the system input matrix.

Those hyperplanes that correspond to zero elements in the second row of the input matrix need not be adapted if the null space of their derivatives is attained, as they have no direct influence on the eigenvalue during sliding motion. The switching hyperplanes are partitioned into 'adaptable' and 'non-adaptable' hyperplanes. Note also that not all adaptable hyperplanes need be adapted to improve system response; the adaptation of just one parameter would prove sufficient. Those elements g_{ij} of the switching matrix G that are to be adapted are written as follows. For the ith element of the jth hyperplane

$$g_{ij} = c_{ij}g_{a_j} \tag{15.63}$$

where g_{aj} represent the adaptive design parameter for the jth hyperplane and c_{ij} constants dependent on the initial design criteria.

In order to investigate the amount of adaptation that takes place at any adaptation step consider the expression for the null space of the derivative of the switching functions. Closed loop SmVSC is given by

$$\dot{\boldsymbol{x}} = \begin{bmatrix} 0 & A_{12} \\ A_{21} & A_{22} \end{bmatrix} \boldsymbol{x} - k \begin{bmatrix} 0 \\ B_2 \end{bmatrix} [G_1, B_2^{-1}]\boldsymbol{x} \tag{15.64}$$

Hence

$$\dot{\boldsymbol{x}} = \begin{bmatrix} 0 & A_{12} \\ A_{12} - kB_2G_1 & A_{22} - kI \end{bmatrix} \boldsymbol{x} \tag{15.65}$$

The derivative of the switching function is given by the vector

$$\dot{\boldsymbol{s}} = G\dot{\boldsymbol{x}} = [B_2^{-1}A_{21} - kG_1, \;\; G_1A_{12} + B_2^{-1}A_{22} - kB_2^{-1}]\boldsymbol{x} \tag{15.66}$$

If adaptation takes place for the jth hyperplane, then

$$s_j^{(m+1)} = ([B_2^{-1}]_j A_{21} - k[G_1^{(m)}]_j)x_1 + ([G_1^{(m)}]_j A_{12} + [B_2^{-1} A_{22}]_j - k[B_2^{-1}]_j)\begin{bmatrix} x_2 \\ \cdot \\ \cdot \\ x_n \end{bmatrix} \quad (15.67)$$

where $[\cdot]_j$ represents the jth row of a particular matrix. From (15.53) and (15.63), if $n - m = 1$ then

$$[G_1^{(m)}]_j = c_{1j}g_{a_j}, \qquad [B_2^{-1}]_j = [G_2]_j = (c_{2j}, \ldots, c_{nj}) \quad (15.68)$$

and from (15.67)

$$\begin{aligned} s_j^{(m+1)} &= (c_{2j} - a_1 c_{nj} - kc_{1j}g_{a_j})x_1 + (c_{1j}g_{a_j} - c_{nj}a_2 - kc_{2j})x_2 \\ &\quad + (c_{2j} - c_{nj}a_3 - kc_{3j})x_2 \\ &\quad + \ldots (c_{n-1,j} - c_{nj}a_n - kc_{nj})x_n. \end{aligned} \quad (15.69)$$

Comparison can be **made** with the previous expression for the switching function by dividing throughout by one of the elements. This, by similar techniques as in scalar CSAC, results in

$$\begin{aligned} s_j^{(m+1)} &= c_{1j}g_{a_j}(1 + \xi)x_1 + c_{2j}(1 + \xi)x_2 \\ &\quad + \ldots c_{qj}x_q + \ldots c_{nj}(1 + \xi)x_n \end{aligned} \quad (15.70)$$

where q is the index of the coefficient chosen for division and

$$\xi = (c_{qj}a_n - c_{q-1,j})/k \quad (15.71)$$

Adaptation cannot be allowed in the coefficients c_{qj} $(i = 2, \ldots n)$ as these must at all times be the rows of the matrix B_2^{-1}. In any case these have no effect on the magnitude of the desired eigenvalue during sliding motion. The new switching hyperplane is made to lie as close as possible to the required null space by adapting the first parameter, that is

$$s_j^{(m+1)} = c_{1j}g_{a_j}(1 + \xi)x_1 + \sum_{i=2}^{n} c_{ij}x_i \quad (15.72)$$

Analysis cannot be carried out using a fixed value for the adaptation parameter as the choice of coefficient for division (15.70) can be varied. It is sufficient to state that the value is bounded by the value resulting from use of the largest coefficient and is sufficiently small, if k is large in comparison with parameters in the system.

It is immediately apparent, from (15.59), that adaptation of different parameters will have varying degrees of effect on the system eigenvalue. If for two non-zero elements of the input matrix the respective element associated with one hyperplane is of an order of magnitude greater than the other, then the effect of adaptation on the eigenvalue during sliding motion will be of that same order of magnitude difference. Choice of rate of adaptation can be made through parameter inspection.

15.5.2.2 Multivariable CSAC: $n - m > 1$: In the general multivariable CSAC system the eigenvalues during ideal sliding motion are given from

$$\{\sigma_1, \sigma_2, \ldots, \sigma_{n-m}\} = \sigma(A_{11} - A_{12}G_2^{-1}G_1) \tag{15.73}$$

where the submatrices have dimensions as previously defined. Under the imposed constraint the eigenvalues are the roots of the polynomial

$$|\lambda I - A_{11} + A_{12}B_2G_1| = 0 \tag{15.74}$$

It can be shown that a polynomial of the following form results:

$$\lambda^{n-m} + h_{n-m}\lambda^{n-m-1} + \ldots + h_2\lambda + h_1 = 0 \tag{15.75}$$

where

$$h_k = f(g_{a_1}, \ldots, g_{a_m}) \tag{15.76}$$

(Burton, 1987).

Adaptation of the eigenvalues in the sliding mode will depend upon adaptation of the coefficients of this polynomial being possible. Foresight into the manner and magnitude of the resulting adaptation, however, is more complex. There exist many more degrees of freedom and many more alternative strategies. Adaptation may be carried out with respect to many criteria: eigenvalues may be adapted such that all are increased simultaneously; some remain constant whilst others are increased, or the product or sum of eigenvalues is increased. The system responses derived in all of these schemes deserve closer attention and interesting results may well be forthcoming. Here attention is focused on forming an exact polynomial in the adaptive parameter and so adapting the eigenvalues simultaneously.

To illustrate, the case $n - m = 2$ is considered. The quadratic equation for the eigenvalues is written in a slightly amended form:

$$\lambda^2 + \Phi_2(g_a)h_2\lambda + \Phi_1(g_a)h_1 = 0 \tag{15.77}$$

where the functions $\Phi_1(g_a)$ and $\Phi_2(g_a)$ represent the adaptive element in the coefficients of the polynomial achieved by adapting coefficients in the hyperplane matrix. The eigenvalues themselves are given by

$$\{\sigma_1, \sigma_2\} = -\frac{\Phi_2(g_a)h_2}{2} \pm \frac{\Phi_2(g_a)h_2}{2}\left(1 - \frac{4\Phi_1(g_a)h_1}{(\Phi_2(g_a)h_2)^2}\right)^{1/2} \tag{15.78}$$

If these functions possess the correct powers of the design parameter, then exact factorisation is possible. In order for both eigenvalues to be directly increased, then the same adaptive parameter must be used in the switching hyperplanes, resulting in increase of the eigenvalues simultaneously. If separate parameters are used, the degree of increase of each eigenvalue will depend on the relative value of the functions in (15.78). Detailed expressions for the magnitude of adaptation are not derived as only bounds will result. The scheme is demonstrated by example.

15.5.3 Stability of the adaptive control scheme

The stability of the scheme, in both cases, in dependent only on the generation of a stable switching hyperplane matrix, as, when adaptation is not occurring, the system behaves as a multivariable SmVSC system. It is obvious, in the $n - m = 1$ case, that, whilst the first parameter of any switching hyperplane (for adaptation on one hyperplane) or the combination of the first parameters are positive (for adaptation on more than one hyperplane) the system will be stable. It is a requirement of the implementation of CSAC that any adaptation step is checked, and if the condition is not satisfied then the adaptation step should not take place.

The fine tuning achieved compares with that of scalar CSAC system where the process may be stopped if the state reference point lies in a region of the state space in which adaptation is undesirable. If adaptation is stopped, the system is allowed to continue under SmVSC (improved on the initial requirement) until the state re-enters a region where adaptation can again take place. In general this will be at the next attainment of the null space of the derivative of the switching function.

15.6 Application of the CSAC

One of the advantages of CSAC is its extreme simplicity of application. A complicated test for the occurrence of sliding motion does not need to be implemented; it is necessary to store merely the last two values of the switching functions and perform a simple test with the current value to check if the null space of the derivative of a switching hyperplane has been crossed. A further advantage is that a smoother system response results owing to the reduction of chatter motion that is associated with SmVSC.

A simple second-order example illustrates the scalar CSAC and this is followed by the simulation of a fifth-order crane model to highlight possibilities in the higher-order scalar system. Two multivariable CSAC examples demonstrate both the defined cases.

Example 6.1: Consider the double integrator plant

$$\dot{\boldsymbol{x}} = \begin{bmatrix} 0 & 1 \\ 0 & 0 \end{bmatrix} \boldsymbol{x} + \begin{bmatrix} 0 \\ 1 \end{bmatrix} u \tag{15.79}$$

The initial design is such that the eigenvalue during ideal sliding motion is given as $\sigma = -1$; this implies that

$$G = [1, \quad 1] \tag{15.80}$$

which expressed adaptively is simply

$$G^{(0)} = G = [1g, \quad 1] \tag{15.81}$$

The initial condition of the system is

$$x(0) = [0{\cdot}5, \quad -0{\cdot}25]^T \tag{15.82}$$

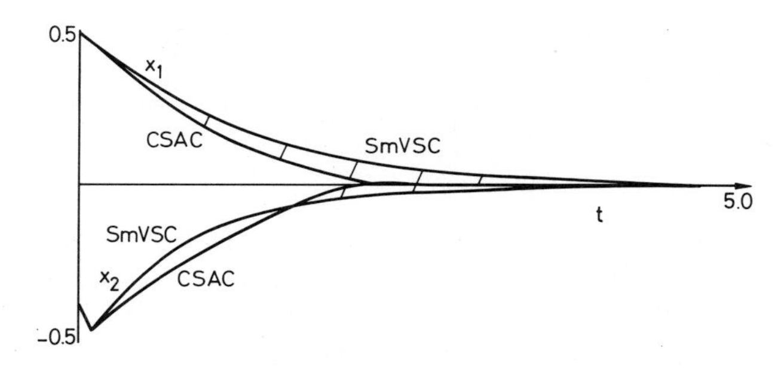

Fig. 15.6A *State versus time – SmVSC and CSAC for second-order system*

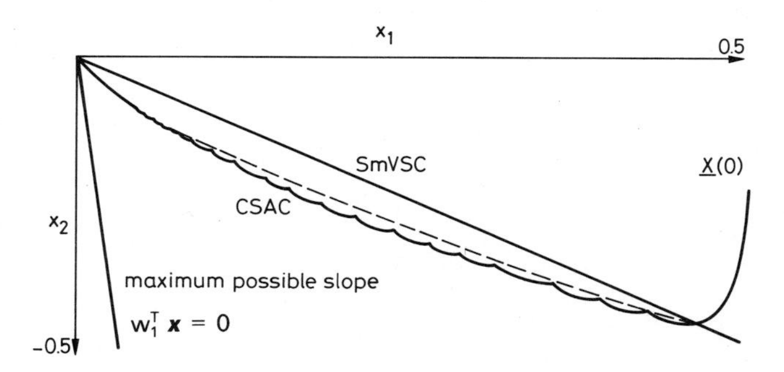

Fig. 15.6B *State space response of SAC and CSAC in the second-order system*
---- ideal SAC

The system is simulated with $k = 50$ in the control function for sufficient accuracy. The improvement over the SmVSC response (with $G^{(0)}$) is shown in Fig. 15.6A, and the phase plane responses of CSAC and ideal SAC are shown in Fig. 15.6B, where the maximum possible slope of the switching line is also indicated. The progressive values of $g^{(m)}$ and the time intervals of adaptation for this system are tabulated in Table 15.1.

Table 15.1 *Progressive values of $g^{(m)}$*

m	τ		$g^{(m)}$	
0	–		1·00	
1	0·100	(0·081)	1·02	(1·020)
2	0·090	(0·080)	1·05	(1·041)
3	0·088	(0·080)	1·08	(1·062)
4	0·086	(0·080)	1·01	(1·085)
5	0·083	(0·079)	1·13	(1·108)
6	0·079	(0·079)	1·16	(1·133)
7	0·076	(0·079)	1·19	(1·160)
8	0·073	(0·078)	1·22	(1·186)
9	0·070	(0·077)	1·25	(1·214)

(·) theoretical values from eqns. (15.25) and (15.42)

Example 6.2 (the crane problem): Effective control of an overhead crane system can provide economic savings in load transfer time and SAC has been applied to this problem (Zinober, 1983 and 1986). The SAC algorithm is able to cope with the wide ranging parameter variations, such as wind gusts or larger angular deviations than are allowable in the linearised model, which cause optimal control techniques to break down, because of nonlinear effects.

The idealised linear system equations can be written, after linearisation and normalisation, as

$$\dot{x} = Ax + Bu \tag{15.82}$$

where

$$A = \begin{bmatrix} 0 & 1 & 0 & 0 \\ 0 & 0 & 1 & 0 \\ 0 & 0 & 0 & 1 \\ 0 & 0 & \dfrac{-g(m+M)}{Ml} & 0 \end{bmatrix} B = \begin{bmatrix} 0 \\ 0 \\ 0 \\ \dfrac{M}{1} \end{bmatrix} \tag{15.83}$$

The crane hoist has length l, the trolley mass is M and the load has unknown mass m. It is desired to move the trolley from some initial condition $x_1 = -X$ to the state origin representing a stationary system. Results via control through other schemes are given in Zinober (1983) and certain pathological cases are highlighted. This system is used here to demonstrate the application of CSAC to higher-order scalar systems. The results are displayed in Fig. 15.7 for a nominal set of parameters.

The advantage gained over SmVSC and ideal VSC are an improvement in response time, and smoother state portraits than in SAC. Improvement is subject to the duration of time for which fine tuning takes place; in this case the system is subject to ordinary SmVSC for some switching hyperplane for a large section of the control period and so does not take full advantage of the possible adaptation of the switching hyperplane. A suggested remedy for this is to rotate the switching function slightly further than required and so attempt to speed up the adaptation process.

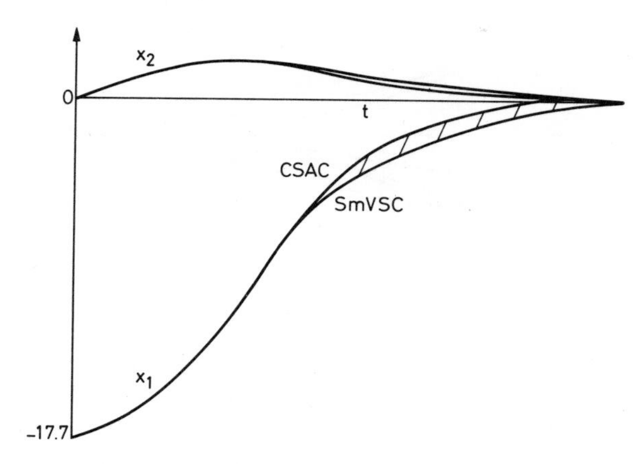

Fig. 15.7 *CSAC of fourth-order nonlinear crane*

Example 6.3: The simplest multivariable CSAC system can be implemented on a system of order 3 with two control functions. The system $\sigma(A) = \{-1, -2, -3\}$ is chosen, and for simplicity the submatrix of the input matrix B is chosen as the identity matrix, i.e.

$$\dot{\boldsymbol{x}} = \begin{bmatrix} 0 & 1 & 0 \\ 0 & 0 & 1 \\ -6 & -11 & -6 \end{bmatrix} \boldsymbol{x} + \begin{bmatrix} 0 & 0 \\ 1 & 0 \\ 0 & 1 \end{bmatrix} \boldsymbol{u} \tag{15.84}$$

The initial desired eigenvalue during ideal sliding motion is $\sigma = -1$, and so via (15.60) and (15.62) a hyperplane matrix of the form

$$G = \begin{bmatrix} 1 & 1 & 0 \\ 2 & 0 & 1 \end{bmatrix} \tag{15.85}$$

is generated. The first element of the second switching plane has no significance and is chosen at random as the corresponding element of the input matrix is zero. Expressed adaptively, according to scheme (15.63), this is then

$$G^{(0)} = \begin{bmatrix} 1g_{a_1} & 1 & 0 \\ 2 & 0 & 1 \end{bmatrix} \quad \text{i.e. } c_{11} = 1,\ g_{a_1}(0) = 1 \tag{15.86}$$

On attainment of $\dot{s}_1 = 0$ the adaptive parameter is recalculated. The resulting improvement in the SmVSC response, for initial condition $\boldsymbol{x}(0) = (1, 0, 0)^T$, is shown in Fig. 15.8.

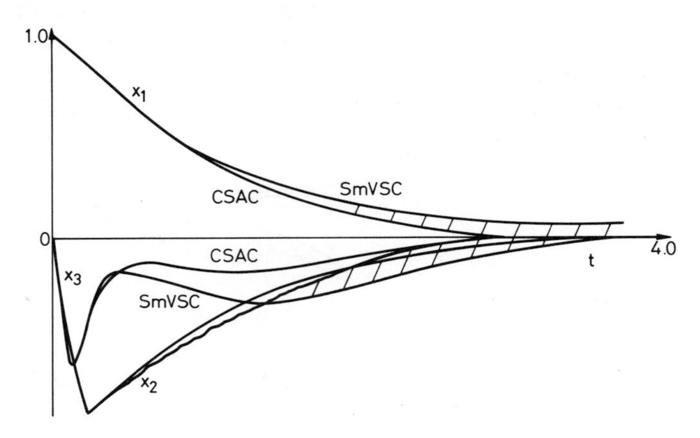

Fig. 15.8 *State versus time – SmVSC and CSAC of third-order multivariable control system*

Example 6.4: Finally the case $n - m = 2$ is illustrated with the system

$$\dot{\boldsymbol{x}} = \begin{bmatrix} 0 & 1 & 0 & 0 & 0 \\ 0 & 0 & 1 & 0 & 0 \\ 0 & 0 & 0 & 1 & 0 \\ 0 & 0 & 0 & 0 & 1 \\ -1 & -5 & -10 & -10 & -5 \end{bmatrix} \boldsymbol{x} + \begin{bmatrix} 0 & 0 & 0 \\ 0 & 0 & 0 \\ 1 & 0 & 1 \\ 0 & 1 & 0 \\ 0 & 0 & 1 \end{bmatrix} \boldsymbol{u} \tag{15.87}$$

and initial condition $\boldsymbol{x}(0) = (-0{\cdot}25, 0, 0, 0, 0)^T$. Initial design specifies eigenvalues during ideal sliding motion of $\sigma_1 = -1$, $\sigma_2 = -1$; therefore from (15.60) and (15.62) the hyperplane matrix must have the form

$$G = \begin{bmatrix} 0{\cdot}5 & 1{\cdot}5 & 1 & 0 & -1 \\ 3 & 1 & 0 & 1 & 0 \\ 0{\cdot}5 & 0{\cdot}5 & 0 & 0 & 1 \end{bmatrix} \tag{15.88}$$

The first two elements of the second row of the switching hyperplane matrix are chosen at random as the respective elements in B are zero. The matrix is expressed adaptively according to the general formulation (15.53) and (15.63) as

$$G^{(0)} = G = \begin{bmatrix} 0{\cdot}5g_{a_1}^2 & 1{\cdot}5g_{a_1} & 1 & 0 & -1 \\ 3 & 1 & 0 & 1 & 0 \\ 0{\cdot}5g_{a_3}^2 & 0{\cdot}5g_{a_3} & 0 & 0 & 1 \end{bmatrix} \tag{15.89}$$

$$\text{i.e. } c_{11} = 0{\cdot}5,\ c_{12} = 1{\cdot}5$$

$$c_{31} = 0{\cdot}5,\ c_{32} = 0{\cdot}5$$

$$g_{a_1}(0) = 1,\ g_{a_3}(0) = 1$$

No adaptive parameter is present in the second hyperplane. The system response is shown in Figs. 15.9*a* and *b* for CSAC and SmVSC. In the CSAC system each

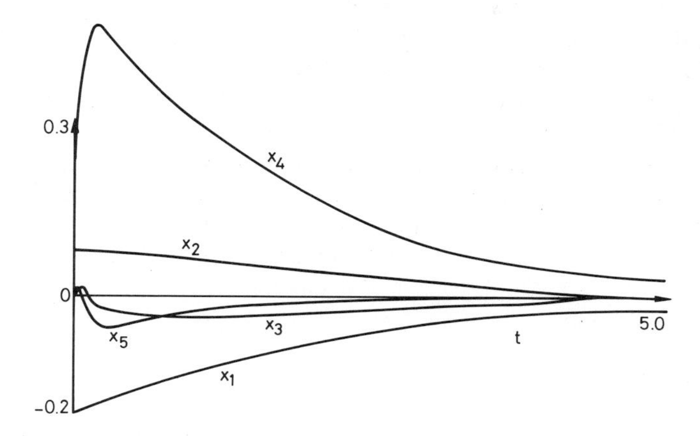

Fig. 15.9A *SmVSC of fifth order multivariable control system*

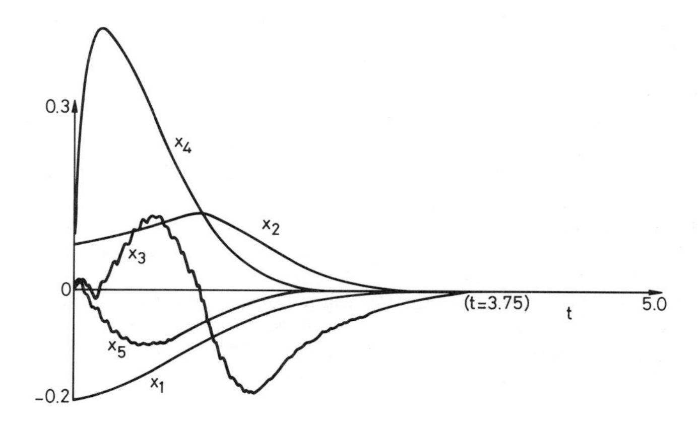

Fig. 15.9B *CSAC of fifth order multivariable control system*

hyperplane possesses separate adaptive parameters. The adaptive parameters increase individually as each switching function reaches the null space of the first derivative associated with it. Note the faster response using CSAC.

15.7 Conclusion

With the development of CSAC the control system designer can, for all orders of system, 'fine tune' an initial VSC system design and therefore improve the speed of response of the system whilst in the sliding mode. The CSAC possesses a degree of robustness since it is based upon robust VSC theory. Chatter motion associated with ideal sliding motion in an ideal VSC system is eliminated as the CSAC system is developed from SmVSC, and CSAC is easily implemented.

By introducing the CSAC system, the VSC system is now split into three distinct phases: the attainment (hitting) of the sliding mode, the sliding mode itself and then an improvement (adaptation) of the response in that sliding mode, should that be required.

There are many degrees of freedom available in prescribing the rate of adaptation in the multivariable CSAC system and this degree greatly increases as the order of the reduced state space of the sliding mode increases. Further investigation into CSAC and its applications should provide additional interesting results.

15.8 References

AMBROSINO, G., CELENTANO, G., and GAROFALO, F. (1984): 'Variable structure model reference adaptive control systems', *Int. J. Control,* **39,** pp. 1339–1349

BURTON, J. A. (1987): 'Adaptive control using variable structure control theory', Ph.D. Thesis, Sheffield University

BURTON, J. A., and ZINOBER, A. S. I. (1986): 'Continuous approximation of variable structure control', *Int. J. Systems Sci.,* **17,** pp. 875–885

DRAŽENOVIĆ, B. (1969): 'The invariance condition in variable structure systems', *Automatica,* **5,** pp. 287–295

EL-GHEZAWI, O. M. E., ZINOBER, A. S. I., and BILLINGS, S. A. (1983*a*): 'Analysis and design of variable structure systems using a geometrical approach', *Int. J. Control,* **38,** pp. 657–671

EL-GHEZAWI, O. M. E., ZINOBER, A. S. I., and BILLINGS, S. A. (1983*b*): 'Variable structure systems and system zeros', *Proc. IEEE,* **130D,** pp. 1–5

ITKIS, U. (1976): 'Variable structure systems with sliding modes', *IEEE Trans.,* **AC-22,** pp. 212–222

UTKIN, V. I. (1978): 'Sliding modes and their application in variable structure systems' (MIR Publishers, Moscow)

WILKINSON, J. H. (1965): 'The algebraic eigenvalue problem' (Clarendon Press, Oxford)

YOUNG, K.-K. D., KOKOTOVIC, P. V., and UTKIN, V. I. (1977): 'A singular perturbation analysis of high gain feedback systems', *IEEE Trans.,* **AC-22,** pp. 931–937

ZINOBER, A. S. I. (1975): 'Adaptive relay control of second-order systems', *Int. J. Control,* **21,** pp. 81–98

ZINOBER, A. S. I. (1977): 'Analysis of a third-order relay control system using nonlinear switching surface theory', *Proc. Royal Soc. Edinburgh,* **76A,** pp. 239–254

ZINOBER, A. S. I. (1981): 'Controller design using the theory of variable structure systems' *in* BILLINGS, S. A. and HARRIS, C. (Eds.): 'Self tuning and adaptive control' (Peter Penegrinus) pp. 206–224

ZINOBER, A. S. I. (1983): 'Robust control of the nonlinear overhead crane problem. Proc. IASTED Symp. on Automatic Control and Identification, Copenhagen, Denmark

ZINOBER, A. S. I. (1986): 'Self-adaptive control of the time-varying nonlinear crane problem with disturbances and state constraints', *Control & Computers,* **14,** pp. 103–109

Chapter 16

State observation of nonlinear control systems via the method of Lyapunov

S. H. Żak

Purdue University

B. L. Walcott

University of Kentucky

16.1 Introduction

Recently, a considerable number of papers have been devoted to the topic of control of nonlinear/uncertain systems (see Baumann & Rugh, 1986; IEEE Workshop, 1986; Chouinard *et al.*, 1985; Corless & Leitmann, 1981; Draženović, 1969; El Ghezawi *et al.*, 1983; Gutman and Palmor, 1982; Hunt *et al.*, 1983; Utkin, 1977, 1978; Utkin & Vostrikov, 1978; Young *et al.*, 1977; De Carlo *et al.*, 1988; Żak & De Carlo, 1987 among many other papers and books on the subject). It is convenient during the design process to assume that the entire state vector of the plant to be controlled is available through measurement. If the state vector cannot be measured, then either a new approach to the design must be devised or a suitable estimate of the state vector must be determined that can be utilized in the control law. If an approximate state vector will be utilized for the inaccessible state, then a control design problem can be split into three phases. The first phase is the design of the controller assuming availability of all the elements of the state vector. The second phase is the design of a system that generates an estimate of the state vector when provided with the measurements of the plant input and output. The final step consists of combining the control strategy and the estimator. This estimator for time-invariant deterministic continuous-time linear systems is labelled as the Luenberger observer, in honour of D. G. Luenberger who proposed and developed the estimator in 1964 (for a lucid exposition of the early stages of the observer theory consult Luenberger, 1971). Since then observer theory has been extended and generalised to include a more general class of systems. In particular, consider the method for designing asymptotic observers for a class of nonlinear systems as presented by Bestle and Zeitz (1983), Zeitz (1987) and Krener and Respondek (1985). In this approach one first transforms the nonlinear plant into observer canonical form from where the design of an observer is facilitated. The error between the state of the plant and the state of the observer in these new co-ordinates evolves linearly and can be made to decay arbitrarily exponentially fast. However, finding an appropriate nonlinear, one-to-one transformation is highly nontrivial. Moreover, knowledge of the nonlinearities of the plant must be exact since it is needed in the computation of the observer dynamics.

In another approach, Baumann and Rugh (1986) utilise an extended linearization technique to produce an observer which, when linearised about any of a family of equilibrium points, has locally invariant eigenvalues. Here again, exact knowledge of the plant's nonlinearities and, furthermore, the first derivatives of these nonlinearities must be known in order to calculate the gain function of the observer.

Another approach to observing the states of a nonlinear system was offered by Thau (1973), whose results were later generalised by Kou *et al.* (1975). This method does not include a systematic technique for the construction of the observer, but rather gives a sufficient condition for asymptotic stability of the error differential equation.

A shortcoming common to the above approaches is that the nonlinearities of the plant must be incorporated, either directly or indirectly, into the dynamics of the observer.

In the design recently proposed by Walcott and Żak (1987), only the bounds of the nonlinearities of the plant are used in the observer dynamics. Moreover, these dynamics may easily be implemented in continuous time on an analogue computer with comparators, or in discrete time via a microprocessor. However, the drawback of this approach is that nonlinearities/uncertainties must satisfy the so-called matching condition.

A comparative study of the above four techniques for observing the states of nonlinear systems is the subject of the paper by Walcott *et al.* (1987).

In this chapter we concentrate our attention on the problem of state observation of nonlinear dynamical systems whose nonlinearities/uncertainties are bounded. The main tool used in the design of observers for such systems is the Lyapunov method. We will begin with Thau's observer followed by a class of min–max type of observers which will be analyzed along with their sliding mode properties. Finally, we will discuss a way of improving the performance of variable structure controllers in the presence of neglected sensor and actuator dynamics via a state observer.

16.2 System description and notation

Consider a dynamical system modelled by the following equations:

$$\begin{aligned} \dot{x} &= Ax + f(x) + Bu \\ y &= Cx \end{aligned} \tag{16.1}$$

where $x \in \mathscr{R}^n$, $u \in \mathscr{R}^m$, $y \in \mathscr{R}^p$, $p \geqq m$. The matrices A, B and C are of appropriate dimensions and B and C are of full rank. The function $f: \mathscr{R}^n \rightarrow \mathscr{R}^n$ can be construed as the uncertainties or nonlinearities in the plant. We require that f is continuous and satisfies the Lipschitz condition.

The following notation is utilised throughout this chapter. Given a vector $x \in \mathscr{R}^n$, then $\|x\| \triangleq (x^T x)^{1/2}$. This induces the matrix norm $\|M\| \triangleq [\lambda_{max}(M^T M)]^{1/2}$, where $\lambda_{max}(\cdot)$ denotes the operation of taking the largest eigenvalue. If P is a symmetric (square) matrix then $P > 0$ $(P < 0)$ is taken to mean that P is positive-definite (negative-definite). We will often utilise the Rayleigh principle in the following form: If $P = P^T > 0$ then

$$\lambda_{min}(P) \| x \|^2 \leqq x^T Px \leqq \lambda_{max}(P) \| x \|^2$$

where $\lambda_{min}(\cdot)$ denotes the operation of taking the smallest eigenvalue.

The problem is to design an observer with inputs y and u whose output $\bar{x}$ will converge to x, i.e.

$$\lim_{t \to \infty} (\bar{x}(t) - x(t)) = 0.$$

The initial state $x_0 = x(t_0)$ of (16.1) is unknown while the initial state $\bar{x}_0 = \bar{x}(t_0)$ of the observer can be assigned arbitrarily. Hence the error between x_0 and $\bar{x}_0$ is still unknown even if we know $\bar{x}_0$. We assume the following:

Assumption A1: The pair (A, C) is completely observable.

Therefore we can find $K \in \mathscr{R}^{n \times p}$ such that the eigenvalues of $A_0 = A - KC$ are in the open left half-plane (LHP).

In the next section we will discuss the exponential observer for (16.1) which was proposed by Thau (1973). Extension of Thau's results can be found in Kou *et al.* (1975).

16.3 The Thau observer

Consider a nonlinear system modelled by (16.1) and a dynamic system described by the following equation:

$$\dot{\bar{x}} = A_0 \bar{x} + f(\bar{x}) + Ky + Bu. \tag{16.2}$$

Let e denote

$$e = \bar{x} - x$$

Thus e satisfies the differential equation

$$\dot{e} = A_0 e + f(\bar{x}) - f(x) = A_0 e + f(x + e) - f(x) \tag{16.3}$$

Since the spectrum of A_0 is contained in the open LHP, for any given positive-definite $Q \in \mathscr{R}^{n \times n}$ there exists a unique positive-definite $P \in \mathscr{R}^{n \times n}$ such that

$$A_0^T P + PA_0 = -2Q$$

Next consider the following positive-definite Lyapunov-function candidate

$$V(e) = e^T P e$$

The derivative of $V(e)$ evaluated along the solution of the error differential equation (16.3) is given by

$$\dot{V}(e) = \dot{e}^T Pe + e^T P\dot{e} = -2e^T Qe + 2e^T P[f(x + e) - f(x)] \tag{16.4}$$

The function $f(\cdot)$ is Lipschitz., i.e., there exists a positive constant L such that

$$\|f(x_1) - f(x_2)\| \leqq L \|x_1 - x_2\|$$

for all x_1, x_2. Therefore the following inequalities are valid:

$$\dot{V}(e) \leqq -2e^T Qe + 2L \|Pe\| \|e\| \leqq (-2a + 2L \|P\|)\|e\|^2$$

where a is the minimum eigenvalue of Q and $\|P\|$ is the maximum eigenvalue of P. Hence if

$$\frac{\lambda_{min}(Q)}{\lambda_{max}(P)} > L \tag{16.5}$$

then $e = 0$ is an asymptotically stable equilibrium point of (16.3). From Patel and Toda (1980), the maximum value of the ratio $\lambda_{min}(Q)/\lambda_{max}(P)$ occurs when the matrix Q is the identity matrix. Hence a sufficient condition for $e = 0$ to be an asymptotically stable equilibrium point of (16.3) is

$$\frac{1}{\lambda_{max}(P)} > L \tag{16.6}$$

Notice that an exact knowledge of the nonlinearities of the plant is needed in the computation of the dynamics of the observer (16.2). In what follows we present an observer design which does not necessitate exact knowledge of the system nonlinearities. This aim is accomplished by utilizing techniques prevalent in variable structure systems (VSS) theory and min–max control.

16.4 Observation of dynamical systems in the presence of bounded uncertainties

Consider a dynamical system as modelled by eqn. (16.1) and the following additional assumptions pertaining to the system under consideration:

Assumption A2: There exists a function $\zeta(\cdot)$: $\mathscr{R}^n \rightarrow \mathscr{R}^m$ such that $B\zeta = f$. That is, the uncertainties f satisfy the matching condition.

Assumption A3: There exist continuous positive real-valued functions $\varrho(\cdot)$ and $\eta(\cdot)$ such that

$$\| \zeta(x) \| \leqq \varrho(x)$$

and

$$\| \zeta(x) \| \leqq \eta(y)$$

Assumption A4: There exists a real symmetric positive definite matrix $Q \in \mathscr{R}^{n \times n}$ and a matrix $F \in \mathscr{R}^{m \times p}$ such that

$$FC = B^T P$$

where the matrix P is the unique, real symmetric positive definite solution of the Lyapunov matrix equation

$$(A - KC)^T P + P(A - KC) = -Q$$

Taking into account assumption A2 we can represent the model (16.1) in the form

$$\begin{aligned} \dot{x} &= Ax + B[u + \zeta] \\ y &= Cx \end{aligned} \qquad (16.7)$$

Consider now the following dynamical system:

$$\dot{\bar{x}} = [A - KC]\,\bar{x} - M(e,\eta) + Ky + Bu \qquad (16.8)$$

where

$$M(e,\eta) = \begin{cases} \dfrac{BFCe}{\|FCe\|}\,\eta & \text{for all } e \notin S \\ \{Bv \mid v \in \mathscr{R}^m,\ \|v\| \leqq \eta\} & \text{for all } e \in S, \end{cases}$$

and $S = \{e \mid \| FCe\| = 0\}$.

Theorem 1 (Walcott & Żak, 1987): Given the system (7) and the observer governed by eqn. (16.8), if assumptions A1—A4 are valid, then

$$\lim_{t \to \infty} (\bar{x}(t) - x(t)) = \lim_{t \to \infty} e(t) = 0$$

Proof: The error difference between the output of the observer and the true state obeys the following differential equation:

$$\dot{e} = (A - KC)e - \frac{BFCe}{\|FCe\|}\eta - B\zeta \tag{16.9}$$

for all $e \notin S$. Consider the following Lyapunov function candidate for the error equation:

$$V(e) = e^T Pe$$

where P is defined in assumption A4. The derivative of $V(e)$ with respect to time evaluated on the trajectories of the error equation is

$$\dot{V}(e) = e^T [(A - KC)^T P + P(A - KC)]e - 2\frac{e^T PB\, FCe}{\|FCe\|}\eta - 2\, e^T PB\zeta \tag{16.10}$$

Taking into account assumption A4, the above equation simplifies to the following one:

$$\dot{V}(e) = -e^T Qe - 2\|FCe\|\,\eta - 2e^T C^T F^T \zeta$$

Hence

$$\dot{V}(e) \leqq -e^T Qe - 2\|FCe\|\,\eta + 2\|FCe\|\,\eta < 0$$

For $e \in S$ and $e \neq 0$

$$\begin{aligned}\dot{V}(e) &= -e^T Qe - 2e^T PBv - 2e^T PB\zeta \\ &= -e^T Qe - 2e\, C^T F^T v - 2e^T C^T F^T \zeta \\ &= -e^T Qe < 0\end{aligned}$$

Thus

$$\lim_{t \to \infty} e(t) = 0$$

It is significant to observe that the origin $e = 0$ is not merely a globally, uniformly, asymptotically stable equilibrium point of (16.9), but the norm $\|e(t)\|$ decays with a known exponential rate to zero. To show this, consider the following inequality:

$$\frac{-\dot{V}}{V} \geqq \frac{e^T Qe}{e^T Pe}$$

or

$$\frac{-\dot{V}}{V} \geqq \mu$$

where μ is defined to be

$$\mu \triangleq \min_{e} \{e^T Qe\} \text{ such that } e^T Pe = 1$$

Utilizing the method of Lagrange multipliers to solve the above minimisation problem and letting v represent the Lagrange multiplier yields the equivalent minimisation problem

$$\min_{e,v} \{e^T Qe - v(e^T Pe - 1)\}$$

The value of x which minimises this expression is such that

$$(Q - vP)\, e_{min} = 0 \tag{16.11}$$

and therefore

$$\mu = v\, e_{min}^T Pe_{min} = v$$

The above equation is minimum if v is minimum, but from eqn. (16.11), v is an eigenvalue of $P^{-1}Q$. Consequently,

$$V(e(t)) \leqq V(e(t_0)) \exp[-\mu(t - t_0)]$$

where μ is the minimum eigenvalue of $P^{-1}Q$. Note that,

$$\|e\|^2 \leqq \frac{e^T Pe}{\lambda_{min}(P)} \leqq \frac{e_0^T Pe}{\lambda_{min}(P)} \exp[-\mu(t - t_0)]$$

which implies that $\|e(t)\|^2$, decays as fast as $\exp(-\mu t)$ or that $\|e(t)\|$ approaches the origin at least as fast as $\exp(-\mu/2\, t)$ and $e(t)$ is bounded.

Note that if $m = p$, i.e. the number of inputs and outputs in the plant (16.7) is the same, then we can take $F = I_m$ and the condition $B^T P = FC$ becomes $B^T P = C$. For a system theoretic interpretation of the condition $B^T P = FC$ (or $B^T P = C$) and its application (see Galimidi & Barmish, 1986; Steinberg & Corless, 1985).

Subsequently, we will refer to the observer (16.8) as the min–max observer. The reason for this name will become obvious in Section 16.6.

16.5 Sliding mode properties of min–max observers

Let us look at eqn. (16.9) which governs the error difference between the observer and the true state. We note that the error exhibits a switching nature with respect to a naturally defined surface dependent on the solution of the Lyapunov equation

$$(A - KC)^T P + P(A - KC) = -Q$$

Consider now the following surface:

$$S = \{e \mid \sigma \triangleq FCe = 0\}$$

Definition 1: A surface $S \subset \mathscr{R}^n$ is asymptotically attractive if (i) any trajectory starting on the surface remains there, and (ii) any trajectory starting outside the surface tends to it at least asymptotically.

Our goal is to find conditions under which the trajectory of (16.9) is forced onto S and then stays there for all subsequent time. A trajectory confined to the switching surface is said to be in a sliding mode. If one defines the generalised Lyapunov function $W = \frac{1}{2}\,\sigma^T\,\sigma = \frac{1}{2}\,\|\sigma\|^2$, then a sufficient condition for attractivity to S is

$$\dot{W} = \frac{d}{dt}\left(\frac{1}{2}\,\|\sigma\|^2\right) = \sigma^T\,\dot{\sigma} < 0$$

in a neighborhood of S (Utkin, 1977, 1978).

The attractivity regions of S can be characterized in terms of the observer gain η. Specifically, consider the observer described by (16.8) where

$$M = \begin{cases} \dfrac{BFCe}{\|FCe\|}\,\bar{\eta} & \text{for all } e \in S \\ \{Bv\;|v \in \mathscr{R}^m,\ \|v\| \leqq \bar{\eta} & \text{for all } e \in S \end{cases} \tag{16.12}$$

and $\bar{\eta}$ is to be determined. The equation describing the error difference between the output of the observer and the true state now becomes

$$\bar{e} = A_0 e - \frac{BFCe}{\|FC\|}\,\bar{\eta} - B\zeta \tag{16.13}$$

for all $e \notin S$, where $A_0 \triangleq A - KC$. Let us evaluate now $\dot{W}$ on the solutions of (16.13). We have

$$\begin{aligned} \dot{W} &= \sigma^T\,\dot{\sigma} = \sigma^T\,(FC\dot{e}) \\ &= \sigma^T\left(FCA_0e - FC\,\frac{BFCe}{\|FCe\|}\,\bar{\eta} - FCB\zeta\right) \end{aligned}$$

Taking into account assumption A4, we can represent $\dot{W}$ as follows:

$$\begin{aligned} \dot{W} &= \sigma^T\left(B^TPA_0e - B^TPB\,\frac{\sigma}{\|\sigma\|}\,\bar{\eta} - B^{\mathrm{T}}PB\zeta\right) \\ &= \sigma^TB^TPA_0e - \sigma^T\,(B^TPB)\,\frac{\sigma}{\|\sigma\|}\,\bar{\eta} - \sigma^T(B^TPB)\zeta \\ &\leqq \frac{1}{2}\,e^T\,\{(PBB^T\,P)\,A_0 + A_0^T\,(PBB^T\,P)\}e \\ &\quad - \lambda_{min}\,(B^TPB)\|\sigma\|^2\,\frac{\bar{\eta}}{\|\sigma\|} \\ &\quad + \lambda_{max}\,(B^T\,PB)\,\|\sigma\|\,\eta < 0 \end{aligned} \tag{16.14}$$

Define $\hat{Q}$ as

$$-\hat{Q} = (PBB^T P)\, A_0 + A_o^T\, (PBB^T\, P) \tag{16.15}$$

Hence from (16.14) S is globally asymptotically attractive if

(i) $\hat{Q}$ is positive semidefinite and

$$\bar{\eta} > \frac{\lambda_{max}\,(B^T\, PB)}{\lambda_{min}\,(B^T\, PB)}\,\eta$$

or

(ii) $\hat{Q}$ is not positive semidefinite, but

$$\bar{\eta} > \frac{\lambda_{max}\,(B^T PB)}{\lambda_{min}\,(B^T PB)}\,\eta + \frac{\|\, B^T\, PA_0\, e\|}{\lambda_{min}\,(B^T PB)}$$

From the discussion in Section 16.4 it follows that $e(t)$ is bounded. Hence $\bar{\eta}$ can be made independent of e if we know $\|\, e(t_0)\| = \|\, \bar{x}_0 - x_0\|$. Note that the type of behaviour of the observer (16.8) depends on the positive semi-definiteness of $\hat{Q}$ and the gain η. Thus if $\hat{Q}$ is positive semidefinite, the observer requires less energy to force the error on the surface S.

Conditions under which the matrix $\hat{Q}$ is positive semidefinite will be discussed in Section 16.7.

16.6 Duality between min–max controllers and observers

Two recently developed methods of robust control and stabilisation of dynamical systems in the presence of modelling errors and uncertain parameters are variable structure control (VSC) and deterministic control of uncertain systems. VSC control uses a switched control strategy to drive the plant trajectory onto a prespecified so-called switching surface in the state space and maintains the trajectory on this surface for all subsequent time. A trajectory, confined to such a switching surface, is said to be in a sliding mode, in which case the system performance is insensitive to bounded disturbances confined to a subspace of the state space. Determination of the control parameters which in fact force this behaviour utilise the generalised Lyapunov second method. In the deterministic control of uncertain systems the plant uncertainties are represented by bounds on model parameters. The min–max controller structures (Chouinard *et al.*, 1985; Gutman and Palmor, 1982) which result in robust stabilisation of these 'uncertain' plants utilise the second method of Lyapunov for their specification. In addition, these control strategies exhibit a switching nature with respect to a naturally defined surface dependent on the solution to a Lyapunov equation. Hence one expects a parallelism between the two control methodologies. This parallelism was also seen by Gutman and Palmor (1982) and Chouinard *et al.*

(1985) where the attractivity to the naturally defined switching surface was considered.

Consider a state model as given by (16.7). The nominal system

$$\dot{x}(t) = Ax(t)$$

is assumed to be asymptotically stable, in which case the matrix Lyapunov equation

$$A^T P_c + P_c A = -Q_c$$

has a solution P_c for any real symmetric positive definite matrix $Q_c = Q_c^T > 0$. Thus one can show that the closed-loop system is uniformly asymptotically stable in the large for any admissible uncertainty $\zeta(\cdot)$ under the min-max control strategy

$$u = u(x) = \begin{cases} -\dfrac{B^T P_c x}{\|B^T P_c x\|}\varrho(x) & \text{if } \| B^T P_c x\| \neq 0 \\ \{\hat{u} \in R^m \mid \| \hat{u}\| \leqq \varrho(x)\} & \text{if } \| B^T P_c x\| = 0 \end{cases} \quad (16.16)$$

It is important to note that the surface

$$S = \{x \mid \sigma = B^T P_c x = 0\}$$

acts as a switching surface. It is our purpose to consider the attractivity of a state to this surface and develop sufficient conditions for global attractivity.

If one defines the generalised Lyapunov function $W_c = \frac{1}{2}\sigma^T \sigma$, then a sufficient condition for attractivity to S is

$$\dot{W}_c = \frac{d}{dt}\left(\frac{1}{2}\|\sigma\|^2\right) < 0 \quad (16.17)$$

in a neighborhood of S (Utkin, 1977). Hence, analogous to the control structure of (16.16) with respect to the surface S, let us consider the structure

$$u = u(x) = \begin{cases} -\dfrac{\sigma}{\|\sigma\|}\hat{\varrho} & \text{for } \|\sigma\| \neq 0 \\ \{\hat{u} \in \mathscr{R}^m \mid \|\hat{u}\| \leqq \hat{\varrho}\} & \text{for } \|\sigma\| = 0 \end{cases} \quad (16.18)$$

where $\hat{\varrho}$ is to be determined. Thus to determine the attractivity to the surface with control (16.18) it is sufficient to show that

$$\begin{aligned} \dot{W}_c = \sigma^T \dot{\sigma} \leq \tfrac{1}{2} x^T [(PBB^T P)A + A^T(PBB^T P)]x \\ - \lambda_{min} (B^T PB) \|\sigma\| \hat{\varrho} + \lambda_{max} (B^T PB) \|\sigma\| \varrho < 0. \end{aligned} \quad (16.19)$$

Define $\hat{Q}$ as

$$\hat{Q} \triangleq -((PBB^T P)A + A^T(PBB^T P)) \quad (16.20)$$

Hence from (16.19), S is globally asymptotically attractive if:

(i) $\hat{Q}$ is positive semidefinite and

$$\hat{\varrho} > \frac{\lambda_{max}(B^T PB)}{\lambda_{min}(B^T PB)} \varrho \qquad (16.21)$$

or

(ii) $\hat{Q}$ is not positive semidefinite, but

$$\hat{\varrho} > \frac{\lambda_{max}(B^T PB)}{\lambda_{min}(B^T PB)} \varrho + \frac{\|B^T P\ Ax\|}{\lambda_{min}(B^T PB)} \qquad (16.22)$$

Such a VSC type of behaviour for the min–max control structure of (16.18) depends on the positive semidefiniteness of $\hat{Q}$. That is, if $\hat{Q}$ is positive semidefinite, then the controller requires less energy (as per eqn. (16.21) versus (16.22)) to achieve stability to the surface.

Note that the structures of the matrices (16.15) and (16.20) are the same. Therefore, the control problem and observer problem demonstrate their duality once again.

16.7 Sufficiency conditions for the positive semi-definiteness of the matrix $\hat{Q}$

In this section we will derive sufficient conditions for the matrix $\hat{Q}$ as given by (16.20) to be positive semidefinite. If one then makes the associations $A := A_0$, $B^T P: = FC$ and $x: = e$, this will lead to sufficient conditions for the positive semidefiniteness of the matrix $\hat{Q}$ given by (16.15).

To determine the desired sufficiency conditions first observe that $\hat{Q} \geqq 0$ if and only if $\sigma^T B^T\ PAx \leqq 0$, where as before $\sigma = B^T\ Px$.

Consider now the nominal system

$$\dot{x} = Ax, \qquad x(0) = x_0$$

The switching surface $\sigma = B^T Px$ is asymptotically attractive for the nominal system if

$$\sigma^T \dot{\sigma} = \sigma^T\ B^T P\dot{x} = \sigma^T B^T PAx < 0 \qquad \forall\ x \notin N(B^T P)$$

where $N(\cdot)$ denotes 'the null space of'. Thus the surface S is asymptotically attractive for the nominal system if $\hat{Q} \geqq 0$.

Suppose now that we were able to find a solution P to the matrix Lyapunov equation such that the rows of the matrix $B^T P$ are the left eigenvectors of the matrix A corresponding to the real eigenvalues $\lambda_i\ (i = 1, \ldots, m)$. Then, we can present the following propositions, and theorem:

Proposition 1 (Żak & De Carlo, 1987): If the rows of the matrix $B^T P$ are the left eigenvectors of the matrix A corresponding to real eigenvalues $\lambda_i (i = 1, \ldots, m)$, then $\hat{Q} \geqq 0$.

Proof: Let

$$B^T P \triangleq \begin{bmatrix} c_1^T \\ \cdot \\ \cdot \\ \cdot \\ c_m^T \end{bmatrix}, \quad c_i^T \in \mathscr{R}^{1\times n}.$$

By assumption

$$c_i^T A = \lambda_i c_i^T, \quad i = 1, \ldots, m$$

We have

$$-\tfrac{1}{2} x^T \hat{Q} x = \sigma^T B^T P A x = \sigma^T \begin{bmatrix} \lambda_1 & & 0 \\ & \ddots & \\ 0 & & \lambda_m \end{bmatrix} B^T P x$$

$$= \sigma^T \begin{bmatrix} \lambda_1 & & 0 \\ & \ddots & \\ 0 & & \lambda_m \end{bmatrix} \sigma = [\lambda_1 \sigma_1, \ldots, \lambda_m \sigma_m] \begin{bmatrix} \sigma_1 \\ \cdot \\ \cdot \\ \cdot \\ \sigma_m \end{bmatrix}$$

$$= \sum_{i=1}^{m} \lambda_i \sigma_i^2 \leqq \lambda_{imax} \|\sigma\|^2 \leqq 0$$

where

$$\lambda_{imax} = \max\{\lambda_i, i = 1, \ldots m\}$$

and $\lambda_{imax} = \lambda_{imax}(A) < 0$ since the matrix A is asymptotically stable.

Proposition 2 (Żak & De Carlo, 1987): Let the row space of $B^T P$ be spanned by a set of m left eigenvectors of A labelled $\{c_1, \ldots c_m\}$. Then $N[B^T P A] = N[B^T P]$, i.e. $N[B^T P]$ is A-invariant.

Proof: By assumption there exists a nonsingular matrix M such that

$$(MB^T P)A = \begin{bmatrix} c_1^T \\ \cdot \\ \cdot \\ \cdot \\ c_m^T \end{bmatrix} A = \begin{bmatrix} \lambda_1 & & 0 \\ & \ddots & \\ 0 & & \lambda_m \end{bmatrix} \begin{bmatrix} c_1^T \\ \cdot \\ \cdot \\ \cdot \\ c_m^T \end{bmatrix} = \begin{bmatrix} \lambda_1 & & 0 \\ & \ddots & \\ 0 & & \lambda_m \end{bmatrix} MB^T P$$

We shall now demonstrate that if $N(B^T P)$ is A-invariant then

$$\sigma^T B^T P A x \ < \ 0 \qquad \forall\, x \notin N(B^T P)$$

that is $\hat{Q} \geqq 0$.

Theorem 2 (Żak and De Carlo, 1987): If $N(B^T P)$ is A-invariant then $S = \{x|\sigma(x) = B^T P x = 0\}$ is globally attractive for the nominal system.

Proof: Since $N[B^T P] \oplus \mathrm{Im}(B) = R^n$, $x = b + v$, where $b \in \mathrm{Im}(B)$ and $v \in N[B^T P]$. As such there exists α such that $b = B\alpha$. Since $(B^T PB)$ is nonsingular there exists β such that $b = B(B^T PB)^{-1}\beta$. Multiplying on the left by $B^T P$, a solution β to this equation is given by $\beta = B^T Pb$. However, since $v \in N[B^T P]$ we can choose $\beta = B^T P(b + v) = B^T Px$, i.e. $b = B(B^T PB)^{-1} B^T Px$. Therefore $x = b + v = B(B^T PB)^{-1} B^T Px + v$. This will be used below.

Using the generalised Lyapunov function $W = \frac{1}{2}\sigma^T (B^T PB)^{-1}\sigma$, observe that

$$\begin{aligned}
\dot{W} &= \sigma^T (B^T PB)^{-1} B^T PAx \\
&= \sigma^T (B^T PB)^{-1} B^T PA[B(B^T PB)^{-1}\sigma + v] \\
&= [\sigma^T (B^T PB)^{-1} B^T](PA)[B(B^T PB)^{-1}\sigma] \text{ (by } A\text{-invariance)} \\
&= \tfrac{1}{2}[\sigma^T (B^T PB)^{-1} B^T](A^T P + PA)[B(B^T PB)^{-1}\sigma] \\
&= -\tfrac{1}{2}[\sigma^T (B^T PB)^{-1} B^T] Q [B(B^T PB)^{-1}\sigma] \ < \ 0 \ \forall\, x \notin N(B^T P)
\end{aligned}$$

where Q is as defined as the solution of the equation

$$A^T P + PA \ = \ -Q$$

Corollary 1: If $N(B^T P)$ is A-invariant and $B^T PB = I_m$, then $\sigma^T B^T PAx < 0$ $\forall\, x \notin N(B^T P)$.

Observe that theorem 2 and corollary 1 determine sufficient conditions for attractivity to the surface in terms of the A-invariance of the null space of $B^T P$.

16.8 VSC performance in the presence of small unmodelled inertias

In this section we examine the effect of fast dynamics that are not included in the system model on VSC performance. We first consider the case when there is full information about the system state vector and small inertias due to unmodelled actuator or sensor dynamics are present in the system.

Consider the system modelled by (16.7). Let the system state be measured by a *sensor circuit* modelled by the equation

$$\mu\dot{z} \ = \ Dz + Hx \qquad (16.23)$$

where $z \in \mathscr{R}^q$, $D \in \mathscr{R}^{q\times q}$, $H \in \mathscr{R}^{q\times n}$, $q \leqq n$, and the scalar μ describes the speed of the sensor. We assume that the matrix D is asymptotically stable.

The control u is a variable structure control with elements $u_i(x, z)$ where

$$u_i(x, z) = \begin{cases} u_i^+(x, z) & \text{if } \sigma_i(x, z) > 0 \\ u_i^-(x, z) & \text{if } \sigma_i(x, z) < 0, \end{cases} \tag{16.24}$$

$\sigma(x, z) = [\sigma_1(x, z), \ldots, \sigma_m(x, z)]^T = K_\sigma x + R_\sigma z$, $K_\sigma \in \mathscr{R}^{m\times n}$ and $R_\sigma \in \mathscr{R}^{m\times q}$. The manifold $\sigma(x, z) = 0$ is a linear switching surface composed of the accessible components of x and z.

If the system (16.7) enters the sliding mode at $t = t_0$ then $\sigma(x, z) = 0$ for all $t \geqq t_0$. This implies that $d\sigma/dt = 0$ for all $t \geqq t_0$. Thus, from (16.7) and (16.23)

$$\begin{aligned} d\sigma/dt &= K_\sigma \dot{x} + R_\sigma z = K_\sigma Ax + K_\sigma Bu + K_\sigma B\zeta \\ &\quad + 1/\mu R_\sigma Dz + 1/\mu R_\sigma Hx = 0 \end{aligned} \tag{16.25}$$

For a sliding mode to exist (16.25) must be solvable for the equivalent control $u = u_{eq}$. This is only possible if $[K_\sigma B]^{-1}$ exists, which is not generally satisfied even for systems without uncertainties (Bondarev *et al.*, 1985).

Now consider the case in which unmodelled *actuator dynamics* make an appearance. Specifically, consider the system modelled by

$$\dot{x} = Ax + B(\tilde{u} + \zeta) \tag{16.26}$$

where $\tilde{u} \in \mathscr{R}^m$. The vector $\tilde{u}$ represents the control after being partially distorted by the actuator dynamics which are modelled by

$$\tau\dot{z} = Dz + B_u u \tag{16.27}$$

where $z \in \mathscr{R}^q$, $D \in \mathscr{R}^{q\times q}$, $B_u \in \mathscr{R}^{q\times m}$, and the scalar τ describes the speed of the actuator and u is the undistorted control. Therefore, we may write

$$\tilde{u} = K_z z + R_u u \tag{16.28}$$

where $K_z \in \mathscr{R}^{m\times q}$ and $R_u \in \mathscr{R}^{m\times m}$.

We assume we have access to the full state vector x, so let us choose the switching surface $\sigma(x) = Sx$ where $S \in \mathscr{R}^{m\times n}$. When the system (16.26) is in the sliding mode then

$$d\sigma/dt = SAx + SBK_z + SBR_u u + SB\zeta = 0 \tag{16.29}$$

For a sliding mode to exist (16.29) must be solvable for $u = u_{\mathrm{eq}}$. This is only possible if $[SBR_u]^{-1}$ exists which is generally not the case (Bondarev *et al.*, 1985).

The conclusion of this section is that the neglected actuator or sensor dynamics may prevent the occurrence of the sliding mode.

In their paper, Bondarev *et al.*, (1985) showed that a Luenberger state observer may improve the performance of VSC laws for systems without uncertainties when small unmodelled inertias are present in the system.

16.9 Sliding modes in uncertain systems with asymptotic observers

Consider the system formed by the plant (16.7) where the control u utilises the observer state $\bar{x}$ described by (16.8) rather than the actual state x. We use a switching surface that is a function of the observer state, namely

$$\sigma(\bar{x}) = S\bar{x}$$

In the sliding mode $\dot{\sigma}(\bar{x}) = 0$. Substituting (16.8) into this condition yields

$$\frac{d\sigma}{dt} = SA\bar{x} - SKCe - S\frac{BFCe}{\|FCe\|}\eta + SBu_{eq} = 0 \tag{16.30}$$

for all $e \notin S$. For a sliding mode to exist (16.30) must be solvable for the equivalent control $u = u_{eq}$, which implies that $[SB]^{-1}$ must exist. This is easily satisfied. The state trajectory in the sliding mode will not be precisely along $Sx = 0$, as it would if the observer was not used, but along $\sigma(\bar{x}) = S(x + e) = 0$ since $\bar{x} = x + e$. This is a minor difference since the error e can be made to damp to zero as quickly as desired by choosing the design parameters of the observer in an appropriate way. This aspect of the design was discussed in Section 16.5.

16.10 Improved performance through use of a state observer

16.10.1 Dynamical sensor and asymptotic observer

Let us return to the case of the system (16.7) whose state is measured by the sensor (16.23). However, instead of synthesising a control from the plant state x and the sensor state z as in (16.24), we now utilise the state observer (16.8). Specifically, the inputs to the observer are

$$y = R_z z + K_x x \tag{16.31}$$

where $y \in \mathscr{R}^p$, $R_z \in \mathscr{R}^{p \times q}$, and $K_x \in \mathscr{R}^{p \times n}$. In the ideal case ($\mu = 0$ in (16.23)), the sensor circuit would have no effect on the system (16.7). In this case, from (16.23)

$$z_0 = -D^{-1}Hx \tag{16.32}$$

Substituting (16.32) into (16.31) yields

$$y_0 = [-R_z D^{-1}H + K_x]x \tag{16.33}$$

Furthermore, since the output of the system (16.7) with an ideal sensor is the same as the output of (16.7) with no sensor, we compare (16.7) and (16.33) and find

$$-R_z D^{-1}H + K_x = C \tag{16.34}$$

Let us now define a new variable λ which describes the difference between the actual and ideal sensor

$$\lambda = z - z_0 = z + D^{-1}Hx \tag{16.35}$$

Differentiating this equation and using (16.7) and (16.23) yields

$$\mu\dot{\lambda} = Dz + Hx + \mu D^{-1}H[Ax + B(u + \zeta)]$$

So, from (16.35)

$$\mu\dot{\lambda} = D\lambda + \mu D^{-1}H[Ax + B(u + \zeta)] \tag{16.36}$$

Substituting (16.35) into (16.31) yields

$$y = R_z\lambda + [-R_zD^{-1}H + K_x]x$$

so that from (16.34) we find an expression for the output of the form

$$y = R_z\lambda + Cx \tag{16.37}$$

Collecting the above equations we can describe the sysem (16.7) with the sensor (16.23) and observer (16.8) in terms of x, e and λ as

$$\begin{aligned}\dot{x} &= Ax + B(u + \zeta)\\ \dot{e} &= [A - KC]e + KR_z\lambda - M(e, \eta) - B\zeta\\ \mu\dot{\lambda} &= D\lambda + \mu D^{-1}\mathrm{H}[Ax + B(u + \zeta)]\end{aligned} \tag{16.38}$$

We use the switching surface $\sigma(\bar{x}) = S\bar{x} = S(e + x)$. In the sliding mode

$$d\sigma/dt = S\dot{e} + S\dot{x} = 0$$

is satisfied. Using (16.38) this condition can be written as

$$d\sigma/dt = S[A - KC]e + SKR_z\lambda - SM(e, \varrho) + SAx + SBu = 0$$

For a sliding mode to exist we must be able to solve for the equivalent control, $u = u_{eq}$. This is only possible if $[SB]^{-1}$ exists, a condition which is independent of the sensor and observer. The sliding mode will not be along $Sx = 0$, but along $S(x + e) = 0$.

16.10.2 Dynamical actuator and asymptotic observer

Now, let us again consider the case in which dynamics are neglected in system modelling. This time, though, let us use a state observer to estimate the plant state. The error between the observer and plant state is given by $e = \bar{x} - x$. We use (16.26), (16.28) and (16.8) to find

$$\dot{e} = \dot{\bar{x}} - \dot{x} = [A - KC]e + B[I - R_u]u - BK_zz - M(e, \eta) - B\zeta \tag{16.39}$$

Combining eqn. (16.39) with (16.27), the system with unmodelled actuator dynamics and a state observer is described by

$$\left.\begin{aligned} \dot{x} &= Ax + BK_z z + BR_u u + B\zeta \\ \dot{e} &= [A - KC]e + B[I - R_u]u - BK_z z - M(e, \varrho) - B\zeta \\ \tau\dot{z} &= Dz + B_u u \end{aligned}\right\} \quad (16.40)$$

We use the switching surface $\sigma(\bar{x}) = S\bar{x}$ so that from (16.40) in the sliding mode

$$\begin{aligned} \frac{d}{dt}\sigma(\bar{x}) &= S\dot{x} + S\dot{e} \\ &= SAx + SBu_{eq} + S[A - KC]e - SM(e, \varrho) = 0 \end{aligned}$$

This implies that $[SB]^{-1}$ must exist for the sliding mode to exist. This condition is independent of the unmodelled motions and the observer. As before, the sliding mode will be along $\sigma(\bar{x}) = S(x + e) = 0$.

This shows that when small unmodelled inertias degrade performance in systems with VSC a state observer may be used to restore the sliding mode.

We point out that when there are no uncertainties in the plant then instead of using the state observer (16.8) we can use a Luenberger observer to preserve the sliding mode when small unmodelled inertias are present in the system. In this case we just set $\zeta = 0$ and $M(e, \eta) = 0$ and we obtain the results similar to those arrived at by Bondarev *et al.* (1985).

16.11 References

BAUMANN, W. T., and RUGH, W. J., (1986): 'Feedback control of nonlinear systems by extended linearization', *IEEE Trans.* **AC-31**, pp. 40–46.

BESTLE, D., and ZEITZ, M., (1983): 'Canonical form observer design for non-linear time-variable systems,' *Int. J. Control*, **38**, pp. 419–431.

BONDAREV, A. G., BONDAREV, S. A., KOSTYLEVA, N. E., and UTKIN, V. I., (1985): 'Sliding modes in systems with asymptotic state observers,' *Automation & Remote Control*, V. **46**, Pt. 1, pp. 679–684.

'Challenges to control: A collective view' (1987): Report of Workshop held at the University of Santa Clara, 18–19 Sept. 1986 *IEEE Trans.* **AC-32**, Vol. 4, pp. 275–285.

CHOUINARD, L. G., DAUER, J. P. and LEITMANN, G. (1985): 'Properties of matrices used in uncertain linear control systems,' *SIAM J. Control and Optimiz.*, V. **23**, pp. 381–389.

CORLESS, M. J., and LEITMANN, G., (1981): 'Continuous state feedback guaranteeing uniform ultimate boundedness for uncertain dynamical systems,' *IEEE Trans.* **AC-26**, pp. 1139–1144.

DE CARLO, R. A., ŻAK, S. H. and MATTHEWS, G. P., (1989): 'Variable structure control of nonlinear multivariable systems: A tutorial', *Proc. IEEE V.* **76**, pp. 212–232.

DRAŽENOVIĆ, B. (1969): 'The invariance conditions in variable structure systems,' *Automatica*, V. **5**, pp. 287–295.

EL-GHEZAWI, O. M. E., ZINOBER, A. S. I., and BILLINGS, S. A. (1983), 'Analysis and design of variable structure systems using a geometric approach', *Int. J. Control*, V. **38**, pp. 657–671.

GALIMIDI, A. R. and BARMISH, B. R. (1986): 'The constrained Lyapunov problem and its application to robust output feedback stabilization,' *IEEE Trans.* **AC-31**, pp. 410–419.

GUTMAN, S., and PALMOR, Z., (1982): 'Properties of min-max controllers in uncertain dynamical systems,' *SIAM J. Control & Optimiz.*, V. **20**, pp. 850–861.

HUNT, L. R., SU, R., and MEYER, G., 'Global transformations of nonlinear systems,' *IEEE Trans.* V. **AC-28**, pp. 24–31.

KRENER, A. J., and RESPONDEK, W., (1985): 'Nonlinear observers with linearizable error dynamics,' *SIAM J. Control & Optimiz.*, **23**, pp. 197–216.

KOU, S. R., ELLIOTT, D. L., and TARN, T. J., (1975): 'Exponential observers for nonlinear dynamic systems,' *Information & Control*, **29**, pp. 204–216.

LUENBERGER, D. G., (1971): An introduction to observers,' *IEEE Trans.* **AC-16**, pp. 596–602.

LEITMANN, G., RYAN, E. P. and STEINBERG, A. (1986): 'Feedback control of uncertain systems: Robustness with respect to neglected actuator and sensor dynamics,' *Int. J. Control*, **43**, pp. 1243–1256.

PATEL, R. V. and TODA M., (1980): 'Quantitative measures of robustness for multivariable systems' Proc. Joint Automatic Control Conf. (JACC), San Francisco, CA, pp. TP8-A.

STEINBERG, A., and CORLESS, M., (1985): 'Output feedback stabilization of uncertain dynamical systems,' *IEEE Trans.* **AC-30**, pp. 1025–1027.

SLOTINE, J.-J. E., HEDRICK, J. K., and MISAWA, E. A., (1986): 'On sliding observers for nonlinear systems,' Proc. American Control Conf., Seattle, WA, pp. 1794–1800.

THAU, F. E., (1973): 'Observing the state of non-linear dynamic systems,' *Int. J. Control*, **17**, pp. 471–479.

UTKIN, V. I., (1978): '*Sliding modes and their application in variable structure systems*', (Mir Publishers, Moscow).

UTKIN, V. I., (1977): 'Variable structure systems with sliding modes,' *IEEE Trans.* **AC-22**, pp. 212–222.

UTKIN, V. (1978): 'Application of equivalent control method to the systems with large feedback gain,' *IEEE Trans.* **AC-23** pp. 484–486.

UTKIN, V. I., and VOSTRIKOV, A. S., (1978): 'Control systems with decoupling motions', Proc. 7th Triennial World Congress of the IFAC, Vol. 2, Helsinki, Finland, pp. 967–973.

WALCOTT, B. L., CORLESS, M. J., and ŻAK, S. H., (1987): 'Comparative study of non-linear state-observation techniques,' *Int. J. Control*, **45**, pp. 2109–2132.

WALCOTT, B. L., and ŻAK, S. H., (1987): 'State observation of nonlinear uncertain dynamical systems', *IEEE Trans.* **AC-32**, pp. 166–170.

YOUNG, K.-K. D., KOKOTOVIĆ, P. V., and UTKIN, V. I., (1977): 'A singular perturbation analysis of high-gain feedback systems,' *IEEE Trans*, **AC-22**, pp. 931–938.

ZEITZ, M., (1987): 'The extended Luenberger observer for nonlinear systems,' *Systems & Control Lett.*, **9**, pp. 149–156.

ŻAK, S. H., BREHOVE, J. D., and CORLESS, M. J., (1987): 'Control of uncertain systems with unmodelled actuator and sensor dynamics and incomplete state information', Proc. IEEE Int. Conf. Systems, man and cybernetics, Alexandria, VA, pp. 222–226.

ŻAK, S. H. and DECARLO, R. A., (1987): 'Sliding mode properties of min-max controllers and observers for uncertain systems,' Proc. 25th Annual Allerton Conf. Communication, control and computing, Monticello, IL, pp. 813–820.

Chapter 17

Control of infinite-dimensional plants

V. I. Utkin

Institute of Control Sciences, Moscow

17.1 Introduction

In sliding mode control theory the major attention has been paid to finite-dimensional systems described by ordinary differential equations. However, mathematical models of a wide range of processes in modern technology are partial differential equations, integro-differential equations and equations with delays. Attempts at theoretical generalisations and applications of sliding modes to control of infinite-dimensional plants show that control scientists and engineers are faced the challenge of increased complexity. Practically all the concepts of discontinuous control theory should be completely revised. Even the basic concepts – discontinuity surface, sliding mode, component-wise design – should be clarified or re-introduced. Nevertheless the first results obtained in the field seem encouraging.

17.2 Examples

Let us show the potential of sliding mode control for providing the desired spatial distribution of temperature Θ under uncertainty conditions for the plant governed by the one-dimensional heat equation

$$\frac{\partial \Theta(x,t)}{\partial t} = \frac{\partial^2 \Theta(x,t)}{\partial x^2} - \Theta(x,t) + u(x,t) + f(x,t) \tag{17.1}$$

$$0 < x < 1,\ t > 0,\ \Theta(x,0) = \Theta_0(x),$$

$$\Theta'(0,t) = \Theta'(1,t) = \text{for } t \geqslant 0 \text{ (the process with heat insulated ends)} \tag{17.2}$$

with distributed control $u(x,t)$ and unmeasured and bounded disturbance $|f(x,t)| \leqslant C$, the constant.

If proper constraints are imposed on the smoothness of $\Theta^*(x,t)$, the desired temperature field satisfying boundary conditions (17.2), without loss of generality only the case $\Theta^*(x,t) = 0$ need be studied and the goal of control consists in zeroing temperature $\Theta(x,t)$.

The design procedure is based on the Lyapunov functional, which is the deviation of heat field energy from the desired value

$$V(t) = \frac{1}{2}\int_0^1 \Theta^2(x,t)dx \tag{17.3}$$

Find the time derivative of $V(t)$ on the solutions of (17.1) with boundary conditions (17.2):

$$\begin{aligned}\dot{V}(t) &= \int_0^1 \Theta\Theta''\,dx - \int_0^1 \Theta^2 dx + \int_0^1 \Theta(u+f)dx \\ &= -\int_0^1 (\Theta')^2\,dx - \int_0^1 \Theta^2 dx + \int_0^1 \Theta(u+f)dx\end{aligned}$$

Discontinuous control

$$u = -M \ \mathrm{sign}\ \Theta(x,t),\ M > C \tag{17.4}$$

results in negativeness of $\dot{V}(t)$ for $V(t) \neq 0$; therefore $V(t) \to 0$ with $t \to \infty$.

Convergence of $V(t)$ to zero means convergence of the distributed heat field to zero in the metric $\mathscr{L}_2$. Let us assume that pointwise convergence takes place as well. If at some t

$$\min_{x\in[0,1]} |\Theta(x,t)| = \eta_1(t) \neq 0,$$

then $\dot{V}(t) \leqslant -\eta_1^2(t) - (M - C)\eta_1(t)$ and the function $V > 0$ diminishes with finite speed. If for some x_0 and t

$$\Theta(x_0,t) = 0,\ \max_{x\in[0,1]} |\Theta(x,t)| = \eta_2(t) \neq 0 \tag{17.5}$$

and this maximum is attained at some point x_1, then

$$\left|\int_{x_0}^{x_1} \frac{\partial\Theta}{\partial x}dx\right| = \eta_2(t)$$

In accordance with the Schwartz inequality

$$\eta_2^2(t) = \left(\int_{x_0}^{x_1} \frac{\partial\Theta}{\partial x}dx\right)^2 \leqslant \left(\int_{x_0}^{x_1}\left(\frac{\partial\Theta}{\partial x}\right)^2 dx\right)\left(\int_{x_0}^{x_1} dx\right)$$

and

$$\int_0^1 \left(\frac{\partial\Theta}{\partial x}\right)^2 dx \geqslant \eta_2^2(t)$$

Since the time derivative of $V(t)$ depends on $-\int_0^1 (\partial\Theta/\partial x)^2\,dx$, the functional $V(t)$ diminishes with finite speed for the case (17.5). Thus the process will be completed (maybe asymptotically) only at $\eta_1(t)$ and $\eta_2(t)$ equal to zero, which means $\Theta(x, t) = 0$ or pointwise convergence.

Since $\Theta(x, t) = 0$ corresponds to the set of discontinuity points of the control (17.4), the desired state is achieved owing to the origination of the sliding mode. To make the system behaviour invariant with respect to the disturbance, only the range of its variation should be known.

Implementation of distributed control may prove to be difficult in practice. One of the promising fields in the theory of distributed systems is mobile control with the intensity and spatial trajectory as control variables (Butkovsky, 1979).

An idealised model of mobile control in (17.1) is of the form

$$u(x, t) = p(t)\,\delta[x - s(t)], \tag{17.6}$$

where intensity $p(t)$ and position $s(t)$ are controls and $\delta(\cdot)$ is the δ-function. Control variables $p(t)$ and $s(t)$ should be designed to provide zero distribution of the temperature field $\Theta(x, t)$ in (17.1) and (17.2). Suppose that $s(t)$ and $p(t)$ may be chosen arbitrarily from the ranges $s(t) \in [0, 1]$, $|p(t)| \leqslant M$. The principle of our approach consists in location of the mobile control at the point of maximal deviation of temperature from the desired value, the intensity being a discontinuous function of the deviation

$$u(x, t) = -M(\text{sign}\,\Theta(x_m(t), t)\,\delta(x - x_m(t)), \tag{17.7}$$

$$|\Theta(x_m(t), t)| = \max_{x \in [0, 1]} |\Theta(x, t)|$$

An upper estimate of the time derivative of the Lyapunov functional (17.3) on the solutions of (17.1), (17.2) with control (17.7) and $M > C$ is

$$\dot{V} = -\int_0^1 (\Theta')^2 dx - \int_0^1 \Theta^2 dx -$$

$$\int_0^1 \Theta[M(\text{sign}\,\Theta(x_m(t), t))\delta(x - x_m(t) + f(x, t)]dx$$

$$\leqslant -\int_0^1 (\Theta')^2 dx - \int_0^1 \Theta^2 dx - M|\Theta(x_m(t))|\left(1 - \frac{C}{M}\right)$$

is negative. Similarly to the previous case, it means both convergence of $\Theta(x, t)$ to zero in the metric $\mathscr{L}_2$ and pointwise convergence. Any deviation from the desired state $\Theta(x, t) = 0$ may lead to jumpwise change of the intensity and/or position of the mobile control. In other words, the desired state consists of discontinuity points of control, and it is maintained in the sliding mode.

In the above examples the problem of the mathematical description of sliding modes does not arise: origination of sliding mode means $\Theta(x, t) = 0$, which is the goal of the control.

We shall next consider discontinuous control of distributed plants with dynamic behaviour in sliding modes. The differential equations of the motion will be taken in accordance with the methods developed for finite-dimensional systems. In the final Section the problem of substantiating their validity will be studied.

17.3 Control of oscillation process

Let the behaviour of mechanical distributed systems with fixed ends be governed by the equation

$$\begin{aligned}\ddot{\Theta}(x,t) &= \Theta''(x,t) - \gamma\Theta(x,t) + u(x,t) + f(x,t),\\ 0 &< x < 1,\ t > 0,\ \gamma > 0, \\ \Theta(0,t) &= \Theta(1,t) = 0,\ \Theta(x,0) = \Theta_0(x),\\ \dot{\Theta}(x,0) &= \Theta_1(x)\end{aligned} \tag{17.8}$$

The goal of control is to steer the deviation $\Theta(x, t)$ to zero. Distributed control $u(x, t)$ should be found as a function of the state under the assumption that only the range of disturbance variation i.e. $|f(x, t| \leqslant C)$ is known.

The design procedure is based on the Lyapunov functional

$$V(t) = \frac{1}{2}\int_0^1 (s^2 + (\Theta')^2)dx \tag{17.9}$$

with $s = h\Theta + \dot{\Theta}$, $h > 0$. Compute its time derivative on the solutions of (17.8):

$$\begin{aligned}\dot{V}(t) &= \int_0^1 (s\dot{s} + \Theta'\dot{\Theta}')dx = \int_0^1 s(h\dot{\Theta} + \ddot{\Theta})dx + \int_0^1 \Theta'\dot{\Theta}'\,dx\\ &= \int_0^1 sh\dot{\Theta}\,dx + \int_0^1 s(\Theta'' - \gamma\Theta + u + f)dx + \int_0^1 \Theta'\dot{\Theta}'dx\\ &= \int_0^1 s(h\dot{\Theta} - \gamma\Theta + u + f)dx + \int_0^1 (h\Theta + \dot{\Theta})\Theta''dx + \int_0^1 \Theta'\dot{\Theta}dx\\ &= \int_0^1 s(h\dot{\Theta} - \gamma\Theta + u + f)dx + (h\Theta + \dot{\Theta})\Theta'|_0^1 - \int_0^1 h(\Theta')^2dx\\ &\quad - \int_0^1 \Theta'\dot{\Theta}'dx + \int_0^1 \Theta'\dot{\Theta}'dx = \int_0^1 s(h\dot{\Theta} - \gamma\Theta + u + f)dx\\ &\quad - \int_0^1 h(\Theta')^2dx\end{aligned}$$

For distributed discontinuous control

$$u = -(M + R|\dot{\Theta}| + N|\Theta|)\operatorname{sign} s \text{ with } M > C,\ R > h,\ N > \gamma \tag{17.10}$$

the time derivative of the Lyapunov functional is negative: $\dot{V}(t) < 0$ if $V(t) \neq 0$

and consequently lim $V(t) = 0$ with $t \to \infty$. Similarly to thermal processes, pointwise convergence results from the convergence $V(t)$ to zero. If at some point $x_0 \in (0,1)$ the value $|\Theta(x_0,t)| = \eta(t)$ then $|\int_0^{x_0} \Theta'(x,t)dx| = \eta(t)$, by the Schwartz inequality $\eta^2 = (\int_0^{x_0} \Theta' dx)^2 \leqslant (\int_0^{x_0} (\Theta')^2 dx)(\int_0^{x_0} dx)$ or $\int_0^1 (\Theta')^2 dx \geqslant \eta^2$. Then $\dot{V} \leqslant -h\eta^2$; i.e. the process is completed only if $\Theta(x,t) = 0$ at all $x \in (0,1)$. The control (17.10) is a discontinuous function of the state with a set of discontinuity points $s(\Theta, \dot{\Theta}) = 0$. Since the time derivative of $s, \dot{s} = \Theta'' - \gamma\Theta + u + f + h\dot{\Theta}$, is discontinuous, the sliding mode may arise with state trajectories within this set, if s and $\dot{s}$ have different signs in its vicinity*.

Formally the motion equation immediately results from $s = 0$:

$$\dot{\Theta} + h\Theta = 0, \; \Theta(x,0) = \Theta_0(x) \tag{17.11}$$

Apparently in the sliding mode for any x deviation, $\Theta(x,t)$ decays exponentially with time constant $1/h$.

Thus the discontinuous control (17.10) steers the system state to zero for any disturbances within some known range.

17.4 Control of interconnected heat processes

Let the equation of motion of the system controlling the thermal fields of a set of plants be of the form

$$\frac{\partial \Theta}{\partial t} = \frac{\partial^2 \Theta}{\partial x^2} + D\Theta + Fu, \; 0 < x < 1, \; t > 0, \tag{17.12}$$

$$\Theta'(0,t) = \Theta'(1,t) = 0, \; t \geqslant 0,$$

$$\Theta(x,0) = \Theta_0(x), \; 0 \leqslant x \leqslant 1,$$

where $\Theta(x,t) \in R^n$, $u(x,t) \in R^m$ with fixed t and $x \in R^1$, D and F are constant matrices, rank $F = m$.

From the physical viewpoint the problem consists in heating n plants of a single type using m distributed sources; a matrix D characterises heat exchange with the environment and between the plants (in the special case of no interaction between the plants D is a diagonal matrix).

Let the control be discontinuous on the manifold

$$s(\Theta) = c\Theta(x,t) = c_1\Theta_1(x,t) + c_2\Theta_2(x,t) = 0,$$

$$\Theta_1(x,t) \in R^{n-m}, \; \Theta_2(x,t) \in R^m, \text{ const, det } c_2 \neq 0 \tag{17.13}$$

* The problem of sliding mode existence is still open for distributed systems. The difficulty is checking conditions sign $s = -$ sign $\dot{s}$, since discontinuity of control may lead to discontinuity not only of $\dot{s}$ but also Θ''.

and consider eq. (17.12) with respect to Θ_1 and s

$$\dot{\Theta}_1 = \Theta_1'' + D_{11}\Theta_1 + D_{12}s + F_1 u,$$

$$\dot{s} = s'' + D_{21}\Theta_1 + D_{22}s + cFu, \qquad (17.14)$$

where $D_{i,j}(i,j = 1,2)$, F_1 are constant matrices.

To design the control steering the state to the discontinuity manifold $s = 0$, and as a result generate sliding motion, compute the time derivative of the functional $V_s = 1/2 \int_0^1 s_1^T s_1 dx$ with $s_1 = (cF)^{-1}s$ on the solutions of (17.14) (matrix cF is supposed to be invertible)

$$\dot{V}_s(t) = \int_0^1 s_1^T \dot{s}_1 dx = \int_0^1 s_1^T [s_1'' + (cF)^{-1} cD\Theta + u]dx =$$

$$- \int_0^1 (s')^T s' \, dx + \int_0^1 s_1^T [(cF)^{-1} cD\Theta + u]dx$$

We obtain $\dot{V}_s(t) < 0$ for a sufficiently large value of the scalar M_0 in discontinuous control

$$u = - M(\Theta) \text{ sign } s_1, \; M(\Theta) = M_0 \|\Theta\|$$

where $\|\Theta\|$ is the Euclidian norm. This equality is implied for each component. This fact means the convergence of the state vector $\Theta(x, t)$ to the manifold (17.13) in L_2 (and pointwise convergence which follows from Section 17.2). It should be noted that increase of M_0 leads to a higher rate of convergence, and convergence may be provided for a bounded domain of initial conditions using bang-bang control ($M(\Theta)$ const).

The problem of deriving sliding mode equations for distributed systems needs special treatment. Let us use the so-called equivalent control method developed for finite-dimensional systems (Utkin, 1971). The principle of the method may be easily explained using geometrical considerations. Since sliding mode trajectories lie in the manifold $s = 0$, discontinuous control is replaced by a continuous one referred to as the equivalent control, that the state velocity vector lies in the tangential manifold. It means that the equivalent u_{eq} should satisfy the equation

$$\dot{s} = s'' + D_{21}\Theta_1 + D_{22}s + cFu_{eq} = 0$$

under the condition $s \equiv 0$. Substitution of

$$u_{eq} = - (cF)^{-1} D_{21}\Theta_1$$

into (17.14) results in the sliding mode equation

$$\dot{\Theta}_1 = \Theta_1'' + R\Theta_1, \; R = D_{11} - F_1(cF)^{-1}D_{21} \qquad (17.15)$$

The validity of the eq. (17.15) will be discussed in Section 17.5.

The discontinuity manifold $s = 0$ is chosen to provide the prescribed properties of motion in the sliding mode. For the finite-dimensional system $\dot{\Theta} = D\Theta + Fu$ the equation of sliding along the manifold $s = c\Theta(t) = 0$ has the form $\dot{\Theta}_1 = R\Theta_1$ by virtue of the equivalent control method. It is known that matrix c for the controllable system may be chosen so that $\det(cF) \neq 0$ and the eigenvalues of matrix R in the sliding equation take up the desired values (Utkin, 1981). Using this fact we may show that the required rates of convergence of the state to equilbrium $\Theta = 0$ may be achieved in the distributed system (17.15) if the pair (D, F) is controllable. Assuming $\mathrm{Re}\lambda\{R\} < 0$ we may write the Lyapunov functional as

$$V(t) = \frac{1}{2}\int_0^1 \Theta_1^T(x,t)W\Theta_1(x,t)dx \tag{17.16}$$

where W is the positive definite solution of the Lyapunov equation $R^T W + WR = -I$. Find the time derivative of this function along the sliding mode equation trajectories

$$\dot{V}(t) = -\int_0^1 (\Theta_1')^T W\Theta_1' dx - \int_0^1 \Theta_1^T\Theta_1 dx \tag{17.17}$$

Bearing in mind that $\Theta_1^T W\Theta_1 \leqslant \lambda_{max}\, \Theta_1^T\Theta_1$($\lambda_{max}$ is the maximal eigenvalue of matrix W) we obtain from (17.17)

$$\dot{V}(t) \leqslant -\frac{1}{\lambda_{max}} V(t) \tag{17.18}$$

Matrix $W = \int_0^\infty \exp\{R^T t\} \exp\{Rt\}dt$ (Kwakernaak and Sivan, 1972); therefore by choosing eigenvalues of matrix R with negative real parts sufficiently large in magnitude, the norm of matrix W, and consequently λ_{max}, may be made as small as desired. As stated above, the appropriate choice of matrix c permits arbitrary assignment of eigenvalues of R and, by virtue of (17.18), convergence of the Lyapunov functional (17.16) or vector Θ_1 in metric $\mathscr{L}_2$(and pointwise convergence as well) to zero at a desired rate.

Thus as for finite-dimensional systems (Utkin, 1978, 1981) the sliding mode control approach leads to decoupling system design into two independent steps: design of the desired dynamics in sliding mode by a proper choice of the discontinuity manifold (or matrix c) and design of discontinuous control to provide sliding mode existence.

The Lyapunov functional-based design method discussed in Section 17.2 may be applied to solve a zero distribution problem for a set of interconnected heat processes. In the equation of motion of n plants with m controls,

$$\frac{\partial\Theta}{\partial t} = \frac{\partial^2\Theta}{\partial x^2} + D\Theta + Fu + Gf,\ 0 < x < 1,\ t > 0,$$

$$\Theta(0,t) = \Theta(1,t) = 0,\ \Theta(x,0) = \Theta_0(x),\ t \geqslant 0,\ 0 \leqslant x \leqslant 1,$$

$$f(x,t) \in R^l \tag{17.19}$$

Let the components of the l-dimensional disturbance be bounded. The disturbances are supposed 'to act' in the control space, or G is representable as

$G = FA$ where$A$ is an $m$xl matrix. A necessary condition of invariance with respect to disturbances for finite-dimensional systems (Draženović, 1969).

Assume that matrix D, which characterises heat exchange between the system elements and the environment, is Hurwitz; therefore the Lyapunov equation $D^T W + WD = -S(S > 0)$ has a positive definite solution $W > 0$.

Choosing control in the form of a discontinuous system state function $u = -M \text{ sign } (F^T W\Theta)$, $M > 0$, compute the time derivative of the positive definite functional $V(t) = 1/2 \int_0^1 \Theta^T W\Theta dx$ on the trajectories of (17.19)

$$\dot{V}(t) = -\int_0^1 (\Theta')^T W\Theta' dx - \int_0^1 \Theta^T S\Theta dx - M\int_0^1 |\Theta^T WF| \times \left(1 - \frac{\Theta^T WFAf}{M|\Theta^T WF|}\right) dx^{\dagger}$$

Since the disturbance$f(x, t)$ is assumed to be bounded (and thus the components of vector $Af(x, t)$ are bounded), there will be a number M such that $\dot{V} < 0$. This means that the distribution of $\Theta(x, t)$ converges to zero both in $\mathcal{L}_2$ and pointwise (the justification is similar to the consideration of Section 17.2). If the pair (D, F) is controllable (in this case matrix D does not need to be Hurwitz), one can obtain in (17.19) the invariance to disturbances and the desired transients by taking control in the form of $u = K\Theta - M \text{ sign } F^T W\Theta$ and making a suitable choice of the eigenvalues of matrix $D + FK$. Under the assumption that the plant is stable and the corresponding Lyapunov functional is known, discontinuous control is chosen to suppress the effect of disturbances and to preserve the sign of the Lyapunov functional time derivative. After the completion of the transient the system is certainly in sliding mode because the argument of discontinuous control function is zero and corresponds to the discontinuity points. During the transient, sliding may occur in the discontinuity manifold $s = cx = F^T W\Theta = 0$. Since $\det(F^T WF) \neq 0$ the equivalent control exists and the motion equation may be found uniquely. Under the invariance condition $G = FA$ it coincides with (17.15) and does not depend on disturbances.

17.5 Sliding mode equations

Distributed plants are particular cases of infinite-dimensional systems whose general mathematical description is given by differential equations in Banach space. The development of mathematical tools for such equations with a discontinuous right-hand part, the design of control algorithms which employ sliding modes and their practical implementation is a promising field of research. The initial problem of this field is the mathematical description of motion in sliding

$\dagger |a^T| = \sum_{i=1}^{K} |a_i| \; (a^T \; (a_1, \ldots a_k))$

modes in systems with discontinuous controls which may be applied to infinite-dimensional dynamic plants described by differential equations in Banach space.

It is important to note that such equations may be used in describing a wide class of real life control problems. Partial differential equations, integro-differential equations, and equations with delays are well known examples of differential equation in Banach space. Regardless of the abstract nature of the equations considered, this explains the applied character of the study undertaken:

$$\dot{x} = Ax + \mathscr{b}u(x,t) + f(x,t) \tag{17.20}$$

where $x(t)$ is an abstract function with the values in the reflexive Banach space B; $f(x,t)$, $u(x,t)$ are operator functions with values in B and B_1, respectively; and A is an unbounded linear operator whose domain $D(A) \subset B$ is tight (or $D(A) = B$), $\mathscr{b} \in \mathscr{L}(B_1, B)$, ($\mathscr{L}(B_1, B)$ is a space of continuous linear operators which transform B_1 into B). In particular, the B space in the heat equation is the corresponding Sobolev (reflexive) space, A is the operator of double differentiation with respect to the spatial variable, $\mathscr{b} = I$ is the identity operator, and $u(x,t)$ is the power of the distributed control. Henceforward we shall deal with discontinuous control defined as follows: a linear operator $s = cx$, $c \in \mathscr{L}(B, B_2)$ (B_2 is the Banach space) and it is assumed that no continuous operator $u^*(x,t)$ exists for which

$$\lim_{\|s\| \to 0} \|u(x,t) - u^*(x,t)\| = 0,$$

i.e. the operator $u(x,t)$ undergoes discontinuities on the surface $s = 0$. In all the above examples the controls satisfy this assumption.

Such systems obviously require certain natural assumptions to be made about the operators A, $f(x,t)$ and $u(x,t)$. By virtue of these assumptions, the solution is uniquely determined outside the surface $s = 0$; for instance, A is an abstract elliptic operator and a Lipschitz constant exists for the function $f(x,t) + \mathscr{b}u(x,t)$ in the regions outside the discontinuity surface.

The trajectory of such a system may stay in any small vicinity of the surface $s = 0$ for the time interval $[t_0, t_1]$. In addition to Sections 17.2 – 17.4 examples of such situations were given in Breger *et al.* (1980), Orlov and Utkin (1982) and Orlov (1983). The motions along the control discontinuity surface are also referred to as sliding modes for infinite-dimensional systems. Our aim is to find the differential equation of motion in the sliding mode.

Formally the equivalent control method equation is written as

$$\dot{x} = \bar{A}x + \bar{f}(x,t) \tag{17.21}$$

where $\bar{A} = gA, \bar{f} = gf, g = I - p$; I is the identity operator and $p = \mathscr{b}(c\mathscr{b})^{-1}c$ (operator $c\mathscr{b}$ is supposed to be continuously invertible). Eq. (17.21) is the result of substitution of $u_{eq} = -(c\mathscr{b})^{-1}(cAx + cf)$-the solution of $\dot{S} = cAx + cf + c\mathscr{b}u_{eq} = 0$ into (17.20) for discontinuous control u.

Since the control is a discontinuous function of the state, conventional theorems of the existence and uniqueness of the solution are inapplicable and the description of the behaviour of the system at discontinuity boundaries requires the use of *ad hoc* techniques based on the regularisation principle. According to this principle, the original system is replaced by a system resembling it, whose solution does exist in the conventional sense. Sliding mode equations are obtained by making the characteristics of the new system approach those of the original one. Eq. (17.21) should be substantiated in accordance with this principle.

Regularisation for a discontinuous dynamic system usually implies that the state trajectories do not lie in the intersection of discontinuity surfaces but in their boundary layer, and the solution exists in the conventional sense. Then by letting the boundary layer tend to zero we obtain the solution of the original system. For ordinary differential equations the validity of the equivalent control method has been substantiated under the usual assumption that the right-hand side satisfies the Lipschitz condition outside the discontinuity manifolds (Utkin, 1971).

Theoretical generalisation to infinite-dimensional plants is hindered by major difficulties, since the assumption on Lipschitzian right-hand sides of plant equations, which is quite usual for finite-dimensional systems, is not generally valid for infinite-dimensional plants. For example, in a heat equation the double differentiation operator with respect to the spatial variable is not Lipschitzian and, moreover, it is unbounded. To substantiate the equivalent control method of eq. (17.21) using the regularisation approach, let us replace the discontinuous function $u(x, t)$ in the δ-vicinity of manifold $s = 0$ by control $\tilde{u}(x, t)$ so that the solution of (17.20) corresponding to this control exists for the time interval $[t_0, t_1]$ and belongs to the boundary layer $\|s\| \leqslant \delta$.

Theorem:: Let the following conditions be satisfied:

(i) Operators $-A$ and $-\tilde{A}$ are strongly positive†
(ii) A^{-1} and $\tilde{A}^{-1}$ are compact operators
(iii Operators A and p are commutable:

$$pAx = Apx \text{ for all } x \in D(A)$$

(iv) A Lipschitz constant with respect to x exists for the nonlinear operator $f(x, t)$ and this operator is measurable with respect to t
(v) The control $\tilde{u}(x, t)$ is bounded in any bounded region, eq. (17.20) has a unique solution $\tilde{x}(t)$ under $u(x, t) = \tilde{u}(x, t)$ and this solution is bounded.

Then for the class (17.20) of systems considered, $\lim_{\delta \to 0} \|\tilde{x}(t) - x^*(t)\| = 0$ uni-

†Operator A is strongly positive, if for some $M > 0$ and any λ with a non-negative real part $\|(A - \lambda I)^{-1}\| \leqslant M/(1 + |\lambda|)$ (Krasnoselskii *et al.*, 1976)

formly for all $t \in [t_0, t_1]$ where $x^*(t)$ is the solution to eqn. (17.21) under $s(x^*(t_0)) = 0$, $x^*(t_0) \in D(\bar{A})$, $\|x^*(t_0) - \tilde{x}(t_0)\| \leqslant \delta, \tilde{x}(t_0) \in D(\bar{A})$.

The proof of the theorem is given in Orlov and Utkin (1987).

Condition (i) of the theorem can be easily generalised. Instead of strong positiveness of $-A$ and $-\tilde{A}$ it suffices that operators A and $\tilde{A}$ be abstract elliptic. Indeed, there exists a scalar λ_0 for any elliptic operator such that the operator $\lambda_0 I - A$ is strongly positive (Krasnoselskii *et al.*, 1976). Therefore eqn. (17.20) with elliptic operator A can be rewritten in the form $\dot{x} = A_0 x + \mathscr{B}u(x, t) + f_0(x, t)$ with strongly positive operator $-A_0 = \lambda_0 I - A$ and the Lipschitzian operator $f_0(x, t) = f(x, t) + \lambda_0 I$ (eq. (17.21) can be presented in a similar form).

It should be noted that the operator $-\tilde{A}$ in (17.21) does not satisfy condition (i) in the theorem. Indeed, in the sliding mode, $s = 0$, which means the existence of zero eigenvalues of operator A. Therefore it is not strongly positive. Nevertheless, according to the above generalisation, the sliding mode eq. (17.21) is valid if A is an abstract elliptic operator, which is typical for many physical processes. Let us check whether the conditions of the theorem are satisfied for the initial systems (17.21) and (17.19) and the sliding mode equation (17.15). The inverse operator for $A = I\,\partial^2/\partial x^2$ is integral and hence compact (Krasnoselskii *et al.*, 1976). Taking into account that the spectrum of the operator $I\,\partial^2/\partial x^2$ is the sequence $\{-(\pi n)^2\}$ a strongly positive operator $-I\,\partial^2/\partial x^2$ is indicated (Krasnoselskii *et al.*, 1976). This means that conditions (i) and (ii) of the theorem are true. Operators $I\,\partial^2/\partial x^2$ and $F(cF)^{-1}c$ are commutable, which results in condition (iii). The existence of the Lipschitz constant for $f(\Theta, t) = D\Theta$ [condition (iv)] is evident since D is a bounded linear operator. Condition (v) describes the class of admissible real controls 'acting' in the boundary layer. Representation of the solution of the heat equation (17.12) in an integral form using Green's function directly yields, for $\tilde{u}(\Theta(x, t))$ a measurable function of time and spacial variable, a unique and bounded solution. Thus the sliding mode in the manifold $s = 0$ is described by eq. (17.15).

Sliding mode in the oscillatary system (17.8) is described formally by the ordinary differential equation (17.11) with initial conditions depending on the spatial variable x. The regularisation approach developed for ordinary discontinuous differential equations (Utkin, 1971) may be applied directly to substantiate the validity of the sliding mode equation (17.11).

In the examples in this Chapter we have touched upon infinite-dimensional systems mainly with distributed control. Design procedures for pointwise and mobile controls based on both partial differential equations and their modal presentation may be found in Breger *et al.* (1980) and Orlov and Utkin (1982, 1986).

17.6 References

BREGER, A. M. *et al.* (1980): 'Sliding modes for control of distributed-parameter entities subjected to a mobile multi-cycle signal' *Automat. Remote Control*, **41**, pp. 346 – 355

BUTKOVSKY, A. G. (1979): 'Theory of mobile control', *Automat. Remote Control,* **40,** pp. 804 –813

DRAŽENOVIĆ, B. (1969): 'The invariance conditions in variable structure systems' *Automatica,* **5,** 287 – 295

KRASNOSELSKII, M. A. *et al.* (1976): *'Integral operators in spaces with summable functions'* (Noordholf, Leyden)

KWAKERNAAK, H., and SIVAN, R. (1972): 'Linear optimal control systems' (Wiley Interscience, NY)

ORLOV, YU. V. (1983): 'Application of Lyapunov method in distributed systems' *Automat. Remote Control,* **44,** pp. 426 – 430

ORLOV, YU. V., and UTKIN, V. I. (1982): 'Use of sliding modes in distributed systems control problems' *Automat. Remote Control,* **43**, pp. 1127–1135

ORLOV, YU V., and UTKIN, V. I. (1986): 'Existence conditions of sliding modes in control systems for distributed heat processes' *in* 'Control of plants with time-varying characteristics' (NETI, Novosibirsk) pp. 137 – 141

ORLOV, YU. V., and UTKIN V. I. (1987): 'Sliding mode control in infinite-dimensional systems', *Automatica,* **23,** pp. 753 – 757

UTKIN, V. I. (1971): 'Equations of slipping regime in discontinuous systems: I' *Automat. Remote Control,* **32**, 1897 – 1907

UTKIN, V. I. (1978): 'Sliding modes and their applications in variable structure systems' (Mir, Moscow)

UTKIN, V. I. (1981): 'Sliding modes in optimization and control' (Nauka, Moscow)

Index

Actuator, 252, 335
Approximability, 293
Assignable subspace, 60
Attractor, 256

Banach space, 358
Beam model, 161
Boundary layer, 277, 360

Canonical form, 53, 115
Caratheodory function, 222, 246, 255
Chatter, 3, 4, 6, 68, 311
Chebishev radius, 92
Composite control, 270, 280
Continuation of solutions, 84, 256
Continuous nonlinear control, 202
Controllable, 82, 84
Crane, 328

DC drive, 36, 174
Decoupling, 214
Degrees of freedom, 56
Differential inclusion, 9, 21, 80
Discontinuous control, 2, 80, 170, 352
Distributed system, 305
Duality, 341

Eigenstructure assignment, 53
Eigenvalue assignment, 100
Eigenvalues, 15
Eigenvector assignment, 59, 100
Eigenvectors, 15
Elliptic operator, 361
Equivalent control, 7, 13, 30, 41, 54, 99, 201, 291, 356
Equivalent system, 2, 53
Existence of solutions, 84, 220, 256
Existence problem, 3, 54
Exponential convergence rate, 202
Exponential observer, 335
Extension of solutions, 220

Fast controller, 277
Feedforward, 48
Finite-time attractivity, 84, 87
Flexible structure control, 160
Flight control, 96, 156, 192

Generalised feedback, 83
Global uniform attractivity, 257
Global uniform attractor, 222
Globally uniformly asymptotically stable, 222
Globally uniformly bounded, 222
Globally uniformly ultimately bounded, 232

Hausdorff metric, 265
Heat equation, 351
High gain feedback, 1, 311
Hyperstability, 170, 171

Infinite-dimensional system, 351
Invariance, 1, 17, 358
Invariant subspace, 81
Linearisation, 297
Lyapunov control, 20, 84, 220
Lyapunov equation, 278, 337
Lyapunov function, 223, 270, 336
Lyapunov functional, 352, 357

Matched uncertainty, 255
Matching conditions, 98, 202, 337
Matching uncertainty, 270
Mechanical system, 133, 237

Min-max observer, 339
Mobile control, 353
Model following, 81, 91, 180, 1
Multifunction, 80

Neglected dynamics, 252
Nonlinear system, 289

Observer, 176
Operator, 360
Oscillation process, 354
Output feedback, 144, 257

Parameter variations, 1
Pitchpointing, 96
Pointwise convergence, 352, 355
Pole assignment, 31, 42
Positive invariance, 87
Practical stabilisation, 232
Practical stability, 271, 280
Projector, 15

Quadratic minimisation, 53, 100

Range space, 145
Reachability problem, 3, 56
Reduced-order system, 253, 254
Regularisation, 360
Regulator, 29
Relay control, 7, 65
Riccati equation, 239, 255, 275
Robot, 133, 188, 203, 237
Robust eigenvalue assignment, 61, 100
Robustness, 46, 269

Self-adaptive control, 309
Sensor, 252, 335
Set-valued map, 80
Singular perturbation, 253
Singularly perturbed system, 269
Sliding, 1
Sliding hyperplane, 13, 115
Sliding mode, 1, 27, 30, 53, 56, 68, 96, 170, 340, 348, 353
Sliding motion, 2, 53, 179
Sliding subspace, 2, 53
Slow controller, 273
Smoothed continuous control, 3
Smoothed control, 311
Smoothing control, 68, 101
Spectral condition number, 63
Stabilisation, 226
State observer, 333, 347
Switching, 28
Switching manifold, 53
Switching surface, 2

Thau observer, 335
Time-varying switching hyperplane, 309
Tracking, 1, 81, 89, 201
Two-time scale system, 253, 270

Ultimate boundedness, 270
Ultimate boundedness, 273
Uniform attractivity, 84
Uniform boundedness, 84
Uniform stability, 84, 257
Uniform ultimate boundedness, 20
Uniformly stable, 222
Unit vector control, 65
Unmodelled dynamics, 345
Upper semi-continuity, 81

Variable structure control, 1, 84, 115, 133, 144, 170, 201, 289, 311, 336, 346
VASSYD, 53, 100

Zeros, 19